Radar Cross Section Analysis and Control

For a complete listing of the *Artech House Radar Library*,
turn to the back of this book . . .

Radar Cross Section Analysis and Control

Asoke K. Bhattacharyya
D.L. Sengupta

Artech House
Boston • London

Library of Congress Cataloging-in-Publication Data

Bhattacharyya, Asoke K.
Radar cross section analysis and control / Asoke K. Bhattacharyya and D.L. Sengupta.
p. cm.
Includes bibliographical references and index.
ISBN 0-89006-371-0
1. Radar targets. 2. Radar--Antennas. 3. Electromagnetic waves--Scattering. I. Sengupta, Dipak L. II. Title.
TK6580.B43 1991 91-8244
621.3848--dc20 CIP

British Library Cataloguing in Publication Data

Bhattacharyya, Asoke K.
Radar cross section analysis and control.
1. Radar
I. Title. II. Sengupta, D.L.
621.3484

ISBN 0-89006-371-0

685 Canton Street
Norwood, MA 02062

International Standard Book Number: 0-89006-371-0
Library of Congress Catalog Card Number: 91-8244

10 9 8 7 6 5 4 3 2 1

This book is dedicated to
Shibani and Anasuya Bhattacharyya
and
Sujata, Sumit and Mita Sengupta
for their love, patience, and understanding

Contents

Preface

The *radar cross section* (RCS) of a target or body plays a predominant role in its detection by a radar. Deliberate control in the form of enhancement or reduction of the RCS of a body is of no less importance in many radar applications. A thorough understanding of the electromagnetic scattering characteristics of a target is necessary for successful implementation of any desired control of its RCS. The present book is meant to provide the reader with a broad perspective on the subject of RCS and its control. To this end, it summarizes the currently available approximate techniques for high frequency analysis of scattering by bodies. Sufficient analytical, numerical, and experimental results are provided to enable the reader to judge the applicability, validity, and accuracy of a specific method applied to determine the high frequency scattering characteristics of a body. Many examples of practical significance are included so that the reader may appreciate the difficulties and limitations associated with the control of the RCS of a given object. It should be mentioned that exact analytical techniques for scattering by bodies are easily found in the existing literature and are therefore not discussed in the present book. We have also omitted the important topic of Rayleigh (or low frequency) scattering because any significant control of the RCS of an object is generally effective only in the high frequency region.

Although the material discussed in this book has appeared from time to time in several texts, many journals, reprints of journal articles, and numerous technical reports, we believe that the subject of RCS and its control has not yet been adequately treated in a single place. We hope that the present book will fill this gap.

This book is intended to provide researchers and engineers practicing in the area of electromagnetic scattering with guidelines for the design of targets with desired scattering characteristics. The reader is assumed to have had a secondary course in electromagnetic field theory. The book starts at a fairly basic level and then proceeds to a significantly advanced level; discussions given for certain topics will expose the reader to the current frontiers of scattering theory. Since the book discusses a large number of topics, in-depth discussions of specific topics are not possible within the limited space available. However, the theoretical guidelines and fairly detailed references provided should enable the interested reader to pursue a topic in great depth. It is believed that with proper choices made from among the topics discussed and some added discussions, the present book may serve as the basis for graduate level courses in electromagnetic scattering.

The references and credits given in the book are at best representative and by no means exhaustive. We offer our apologies to those whose work is not properly credited. We would appreciate having misprints and errors called to our attention so that future printings may be corrected.

We have benefited greatly from the help provided by several individuals and institutions. The first author wishes to thank Professor G. S. Sanyal, formerly of the Indian Institute of Technology, Kharagpur, India, for introducing him to the subject of electromagnetic scattering. He also wishes to express his indebtedness to his graduate students, S. Sanyal, S. K. Tandon, and A. K. Tandon, for useful discussions. The second author wishes to acknowledge that during his many years of association with the University of Michigan Radiation Laboratory he was privileged to participate actively in the development of many aspects of electromagnetic scattering. Both authors wish to express their thanks to Dr. V. V. Liepa for many discussions and for supplying them much needed information. They acknowledge the support received from the Physical Science Laboratory and the Department of Electrical Engineering of New Mexico State University, and from the Department of Electrical Engineering and Physics of the University of Detroit Mercy. The authors are indebted to the anonymous manuscript reviewers who helped clarify many points, and to Ms. Julie E. Burnham, Associate Acquisitions Editor at Artech House, for her invaluable help in the preparation and editing of the manuscript. Our grateful thanks are also extended to Ms. Juanita Longmire, Ms. Veronica Foster, and Ms. Anita Foster for their expert typing of the difficult manuscript.

Finally, the authors wish to express their deep sense of gratitude to their mothers and late fathers for their constant encouragement and appreciation.

A. K. Bhattacharyya
D. L. Sengupta
August, 1990

Chapter 1
General Considerations

1.1 INTRODUCTION

The importance of scattering of electromagnetic waves in the explanation of certain natural phenomena was appreciated almost a century ago when Lord Rayleigh investigated the problem of scattering by a perfectly conducting sphere. During the past few decades, research in electromagnetic scattering by a variety of objects or targets assumed added importance because of its direct relevance to all radar applications and, in particular, to radars used for military and space exploration where the main objectives are to locate, identify, and classify various objects. A typical radar accomplishes these objectives by processing the received signals, which generally consist of the desired signals scattered by targets and the undesired noise or clutter signals, the former containing the information characteristic of the targets. The received scattered signal power is directly proportional to the scattering or radar cross section (RCS) of the targets, and hence the importance of the RCS of objects.

During the early years of radar any acceptable methodology for calculation and measurement of RCS of objects was considered adequate for radar engineers. Literature on the analysis, computation, and measurement of RCS of variety of objects is extensive and still growing [1–7]. With better understanding of the electromagnetic scattering phenomena it is natural that consideration is now being given to deliberate control of the RCS of objects for various applications. For example, with rapid advancement in missile and space technology attention increasingly is being directed toward disguising the presence of a flying object to unfriendly radar by reducing its RCS. Also, a demand seems to be appearing for the enhancement of the RCS of communication and broadcast satellites so that they may be tracked better. The users of such satellites and other space objects may even request a complicated differential radar cross section as a function of the angular rotation of the object or the swept frequency of the tracking radar. In short, the demand for control of the RCS of various objects is increasing continuously due to its many military and civilian applications.

Apart from the estimation and measurement of RCS and its control required in the preceding applications, the RCS study finds important use in remote sensing and inverse scattering or imaging. Although in the latter two areas some inversion algorithms generally are required, the input data are the complex scattered fields derived either from experiment or a suitable analytical model.

Because modern theoretical research on RCS depends largely on the available computer power, it is appropriate to comment here on the role of the development of computer storage and speed on the capability of solving electromagnetic scattering problems in general. Although the mainframe computer power has grown at an approximately exponential rate, the prospect is not all that good for solving problems concerning electromagnetic scattering by large bodies. We cannot anticipate that the maximum target size that can be handled would increase even in proportion with the present rate of increase in computer size and speed. However, innovative methods of fast processing of large matrices, parallel processing in computers, and concepts of neural networks may help the situation.

The measurement of RCS is no less important than the theoretical modeling. A modern computerized measurement facility can always provide deeper insight into the scattering mechanisms involved in simple as well as complex targets. Recent advances in technology and measurement methods would complement the theoretical investigations to a great extent. Experimental facilities for time-domain (wideband) measurements with windowing capability to filter out unwanted background reflections and other sources of noise and accurate near-field sampling, and hence far-field prediction, have been integrated into some modern measurement systems. These should greatly enhance the understanding of the subject of RCS and its control.

This book is concerned primarily with the ways to control the RCS of human-made radar targets. A major difficulty in dealing with such a topic in the open literature is that much of the pertinent work has yet to be published. The book therefore can discuss the current state of the art in this area only within the confines of openly published information.

There are four chapters in the book. This first chapter is introductory and describes the basics of electromagnetic field theory, the concept of RCS and its dependence on various parameters, some available information on the RCS of artificial and natural objects. The second chapter briefly discusses some currently available analytical methods for estimating RCS of targets. The third chapter deals with the enhancement of RCS of objects by using passive and active means. The fourth chapter discusses the problem of RCS reduction; it describes models leading to the design of low-RCS targets and antennas. Case studies are presented to help the reader design low-RCS objects.

We should mention here that a harmonic time dependence of the form $\exp(+i\omega t)$ with $i = \sqrt{-1}$ and ω as the radian frequency, the right-handed coordinate system, and the SI system of units are used throughout the book. All of the targets, media, and relevant radar parameters considered in the book are assumed to be independent of the incident and scattered fields.

1.2 ELECTROMAGNETIC THEORY FUNDAMENTALS

Study of any electromagnetic scattering problem requires a sound knowledge of basic electromagnetic theory, which may be found in many standard textbooks [8–11]. For the benefit of the reader, a brief discussion of the fundamental laws and equations of electromagnetics including boundary conditions are discussed in this section.

1.2.1 Maxwell's Equations

Maxwell's electromagnetic field equations are mathematical descriptions of the four basic laws of electricity and magnetism. The word statements for these laws along with their mathematical representations follow.

(1) The *electromagnetic force* (emf) around a closed path L is equal to the negative of the magnetic displacement current through any stationary surface S enclosed by that path L (see Figure 1.1):

$$\text{Integral form:} \qquad \oint_L \mathbf{E}\, dl = -\int_S \frac{\partial \mathbf{B}}{\partial t} \cdot ds \tag{1.1a}$$

$$\text{Differential form:} \qquad \nabla \times \mathbf{E} = -\frac{\partial \mathbf{B}}{\partial t} \tag{1.1b}$$

where **E** and **B** are the time and space dependent electric field intensity and magnetic flux density vectors, respectively. This is Faraday's law, which states that the induced emf in a closed circuit is equal to the negative of the time rate of increase of magnetic flux linked with that circuit.

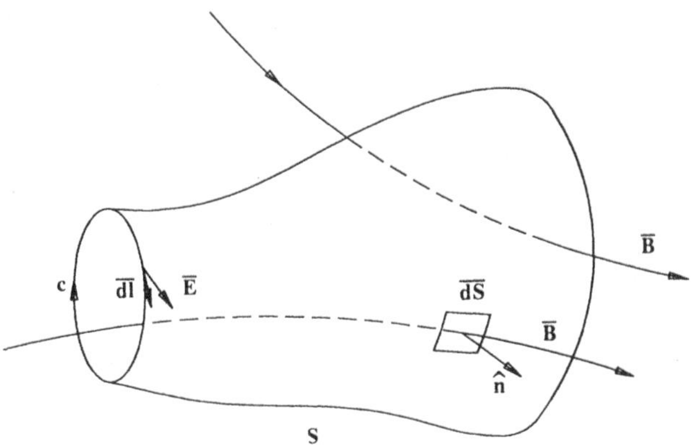

Figure 1.1 Surface S enclosed by a path L.

(2) The *magnetomotive force* (mmf) around a closed path L is equal to the total current through any stationary surface enclosed by that path L (see Figure 1.1):

$$\text{Integral form:} \qquad \oint_L \mathbf{H} \cdot d\mathbf{l} = \int_S \left(\mathbf{J} + \frac{\partial \mathbf{D}}{\partial t}\right) \cdot d\mathbf{s} \tag{1.2a}$$

$$\text{Differential form:} \qquad \nabla \times \mathbf{H} = \mathbf{J} + \frac{\partial \mathbf{D}}{\partial t} \tag{1.2b}$$

where **J** is the electric volume current density that may include the applied current density when appropriate, and D is the electric flux density (or displacement) vector. This is Ampere's law as amended by Maxwell for the time varying case by introducing the displacement current density $\partial \mathbf{D}/\partial t$ to maintain the continuity of current.

(3) The total outward electric flux (or displacement) through a closed surface S enclosing a volume V is equal to the total electric charge contained within V:

$$\text{Integral form:} \qquad \oint_S \mathbf{D} \cdot d\mathbf{s} = \int_V \rho_V \, dv \tag{1.3a}$$

$$\text{Differential form:} \qquad \nabla \cdot \mathbf{D} = \rho_V \tag{1.3b}$$

where the subscript V is the volume density of electric charge. This is a generalization of Gauss's law.

(4) The total outward magnetic flux through a closed surface S is always equal to zero:

$$\text{Integral form:} \qquad \oint_S \mathbf{B} \cdot d\mathbf{s} = 0 \tag{1.4a}$$

$$\text{Differential form:} \qquad \nabla \cdot \mathbf{B} = 0 \tag{1.4b}$$

This indicates that there is no isolated magnetic pole or charge and that the magnetic flux lines are always closed.

Another law of considerable importance is the law of conservation of electric charge, which states that the total electric current flowing through a closed surface S is equal to the time rate of decrease of electric charge contained within the stationary volume V enclosed by S:

$$\text{Integral form:} \qquad \oint_S \mathbf{J} \cdot d\mathbf{s} = -\int_V \frac{\partial \rho_V}{\partial t} \tag{1.5a}$$

$$\text{Differential form:} \qquad \nabla \cdot \mathbf{J} = -\frac{\partial \rho_V}{\partial t} \tag{1.5b}$$

Equation (1.5b) frequently is referred to as the *equation of continuity of current*.

Equations (1.1)–(1.5) together form the fundamental basis of electromagnetic theory. It is important to note that among these equations only (1.1), (1.2), and (1.5) are independent equations. All quantities in the preceding equations in general are real functions of space and time. In the time harmonic case, using phasor notations where it is assumed that all quantities are complex and space dependent and the time dependence is of the form $\exp(+i\omega t)$, we can show [8] that the differential forms of the equations given earlier reduce to the following time-independent forms:

$$\nabla \times \mathbf{E} = -i\omega\mathbf{B} \tag{1.6a}$$

$$\nabla \times \mathbf{H} = \mathbf{J} + i\omega\mathbf{D} \tag{1.6b}$$

$$\nabla \cdot \mathbf{D} = \rho_V \tag{1.6c}$$

$$\nabla \cdot \mathbf{B} = 0 \tag{1.6d}$$

$$\nabla \cdot \mathbf{J} + i\omega\rho_V = 0 \tag{1.6e}$$

In the presence of an isotropic material medium, the various field quantities are related to each other through the following constitutive relations:

$$\mathbf{D} = \epsilon\mathbf{E} \tag{1.7a}$$

$$\mathbf{B} = \mu\mathbf{H} \tag{1.7b}$$

$$\mathbf{J} = \sigma\mathbf{E} \tag{1.7c}$$

Ordinarily the parameters ϵ, μ, and σ characterizing a medium are real quantities. For a lossy dielectric medium we frequently find it convenient to use the notion of complex permittivity or (dielectric constant) represented (for $e^{i\omega t}$ time dependence) by

$$\epsilon = \epsilon_0\epsilon_r = \epsilon_0(\epsilon_r' - i\epsilon_r'') \tag{1.8}$$

where ϵ_r' is the dielectric constant and ϵ_r'' accounts for the loss associated with the dielectric. The term ϵ_r'' is related to ϵ_r' through the loss tangent of the dielectric material defined as

$$\tan\delta = \epsilon''/\epsilon' \tag{1.9}$$

For a lossy dielectric having dielectric constant ϵ_r' and conductivity σ the following relationship holds:

$$\sigma = \omega\epsilon_0\epsilon_r'' \tag{1.10}$$

Similarly, the notion of complex permeability $\mu = \mu_0\mu_r = \mu_0(\mu_r' - i\mu_r'')$ is also used for a lossy magnetic medium.

Equations (1.6) and (1.7) are the basic sets of equations that provide general solutions to any electromagnetic problem in the time-harmonic case. Solutions to arbitrary time dependence can be obtained from the harmonic solutions by the use of Fourier transform techniques [9–11].

1.2.2 Wave Equations

In any electromagnetic problem the field quantities **E** and **H** must satisfy wave equations obtained from Maxwell's equations. For simplicity let us assume a source-free case involving an isotropic and homogeneous medium characterized by $\mu = \mu_0\mu_r$, $\epsilon = \epsilon_0\epsilon_r$, and σ. After taking the curl (i.e., $\nabla\times$) of (1.6a) and using the vector identity $\nabla\times\nabla\times\mathbf{E} = \nabla\cdot\nabla\mathbf{E} - \nabla^2\mathbf{E}$, and making use of (1.6b), (1.7a), and (1.7b) we obtain the time-independent homogeneous vector wave equation:

$$\nabla^2\mathbf{E} - \gamma^2\mathbf{E} = 0 \tag{1.11}$$

where γ is the complex propagation constant in the medium given by

$$\begin{aligned}\gamma &= \left[i\omega\mu(\sigma + i\omega\epsilon)\right]^{1/2} \\ &= ik\left(1 - i\frac{\sigma}{\omega\epsilon_0\epsilon_r}\right)^{1/2}\end{aligned} \tag{1.12}$$

with

$$k = k_0\sqrt{\mu_r\epsilon_r} \tag{1.13}$$

$k_0 = \omega\sqrt{\mu_0\epsilon_0}$ being the free space propagation constant. If this lossy dielectric medium is characterized by $\mu = \mu_0$ and $\epsilon = \epsilon_0(\epsilon_r' - i\epsilon_r'')$ then γ is given by

$$\gamma = ik\left(1 - i\frac{\epsilon_r''}{\epsilon_r}\right) \tag{1.14}$$

with

$$k = k_0\sqrt{\epsilon_r'} \tag{1.15}$$

Similarly, we can show that H satisfies the wave equation

$$\nabla^2\mathbf{H} - \gamma^2\mathbf{H} = 0 \tag{1.16}$$

with γ as given earlier.

The solution of (1.11) and (1.16) provides mathematical representation of an electromagnetic wave. Application of boundary conditions appropriate for a given problem then makes the solution unique [10]. Thus, the boundary conditions play an important role in the solution of all electromagnetic problems. These are discussed next.

1.2.3 Exact Boundary Conditions

When an electromagnetic wave passes through an interface (or boundary) between two dissimilar media, the electric and magnetic field quantities in the immediate vicinity of the interface must satisfy certain identities or boundary conditions involving the fields, the material properties of the media, and the charge or current densities on the interface. These conditions can be obtained from the basic electromagnetic laws [8–10]. Referring to Figure 1.2, we give without proof the boundary conditions on the tangential components of **E** and **H**, and the normal components of **D** and **B** at the interface between two media, where it is assumed that a surface charge density ρ_S and a surface current density $\mathbf{J}_S$ exist on the interface:

$$\hat{n} \times (\mathbf{E}_2 - \mathbf{E}_1) = 0$$

$$E_{2t} = \mathbf{E}_{1t} \tag{1.17a}$$

$$\hat{n} \cdot (\mathbf{D}_2 - \mathbf{D}_1) = \rho_S$$

$$D_{2n} - D_{1n} = \rho_S \tag{1.17b}$$

$$\hat{n} \times (\mathbf{H}_2 - \mathbf{H}_1) = \mathbf{J}_S$$

$$H_{2t} - H_{1t} = J_S \tag{1.17c}$$

$$\hat{n} \cdot (\mathbf{B}_2 - \mathbf{B}_1) = 0$$

$$B_{2n} = B_{1n} \tag{1.17d}$$

where t and n in the subscripts indicate the tangential and normal components, respectively.

The conditions on the tangential components of **D** and **B** and the normal components of **E** and **H** can be obtained, if desired, from (1.10) by using the

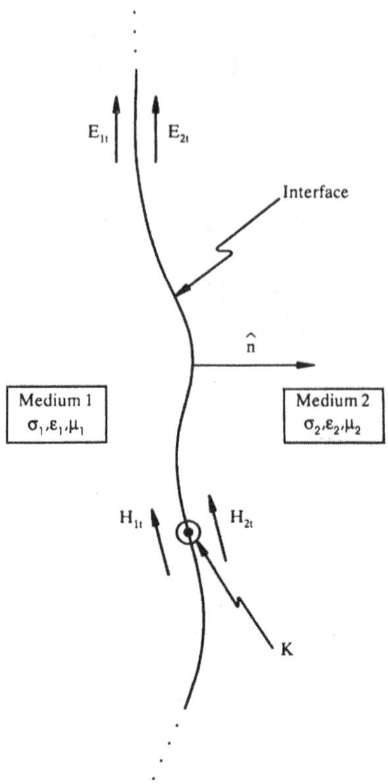

Figure 1.2 The interface between two dissimilar media.

constitutive relations in (1.7). It is important to note that in (1.10) ρ_S and $\mathbf{J}_S$ can be different from zero only when one of the media is conducting infinitely (i.e., a perfect conductor). Thus, at the interface of two media, neither of which is conducting infinitely, the tangential components of **E** and **H** and the normal components of **D** and **B** are continuous at the interface. The case of one of the media being perfectly conducting is of considerable practical importance and hence merits special treatment. Assuming $\sigma_1 = 0$, all field quantities in medium 1 are then equal to zero. Under these conditions, the boundary conditions at the interface between a perfectly conducting medium (medium 1) and a nonconducting medium (medium 2) reduce to

$$\begin{aligned} \hat{n} \times \mathbf{E} &= 0, \quad \hat{n} \cdot \mathbf{D} = \rho_S \\ \hat{n} \times \mathbf{H} &= \mathbf{J}_S, \quad \hat{n} \cdot \mathbf{B} = 0 \end{aligned} \tag{1.18}$$

where we have removed the subscript 2 from the field quantities in medium 2.

1.2.3 Impedance Boundary Conditions

Problems of electromagnetic scattering from bodies that are imperfectly conducting, highly absorbing, conducting but coated with absorbing material, or conducting but with a rough surface are of considerable interest. In solving such problems we find it convenient to use an approximate boundary condition, named the *impedance boundary condition* (IBC) [12–15], also known as the *Leontovitch boundary condition* [12]. At the interface between such a body and the surrounding medium, the IBC is expressed as

$$\hat{n} \times (\hat{n} \times \mathbf{E}) = -Z_S(\hat{n} \times \mathbf{H}) \tag{1.19}$$

where

$\hat{n}$ = the unit normal directed into the surrounding medium,
$\mathbf{E}, \mathbf{H}$ = the fields on the surrounding medium side,
Z_S = the surface impedance of the body defined as the ratio of tangential components of E and H on the surface.

As an example, for a plane wave traveling in the negative z-direction and incident on an imperfectly conducting material half-space defined by $z \leq 0$, (1.19) yields

$$E_x = -Z_s H_y \quad \text{and} \quad E_y = Z_s H_x \tag{1.20}$$

and in this case

$$Z_s = \frac{1 + i}{\sigma \delta_s} \quad \Omega \tag{1.21}$$

where $\delta_s = (2/\omega\mu_0\sigma)^{1/2}$ is the skin depth of the material.

The advantage of using the IBC is that it enables us to evaluate the scattered fields without knowing the fields inside the target. However, the IBC has its own limitations and region of validity that are discussed in [14, 15]. The possible errors involved in field calculations using the IBC depend on the assumptions made regarding the surface impedance. Figure 1.3 shows the region of validity of the IBC. It turns out that for a smooth surface, the exact boundary conditions can be replaced [14] by IBC if

$$\begin{aligned} &|N| \gg 1 \\ &\mathrm{Im}(N) \geq \frac{2.3}{k_0 a} = \frac{2.3}{2\pi}\frac{\lambda}{a}, \quad \lambda \ll a \end{aligned} \tag{1.22}$$

where $N = \sqrt{\mu_r \epsilon_r}$ is the complex refractive index of the material of the scatterer, a is the minimum radius of curvature of its surface, and k_0 is the free space propagation constant.

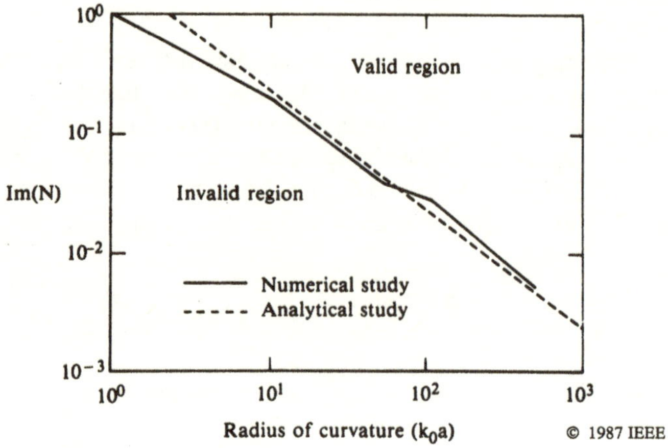

Figure 1.3 The region of validity of the IBC (from [14]).

1.3 TYPICAL ELECTROMAGNETIC SCATTERING PROBLEM

The geometry of a typical electromagnetic scattering problem is depicted in Figure 1.4, where the transmitter (or source), usually located at a large distance from the target, illuminates the target, which now acts as a secondary radiator to produce scattered fields at the receiver or observation point P, also usually located at a long distance from the target. Monostatic or backscattering occurs when the transmitter and receiver are collocated, and bistatic scattering occurs when they are separated in space by an angle

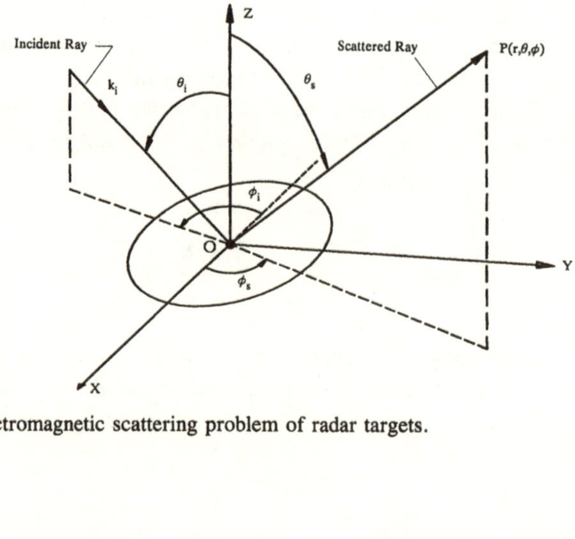

Figure 1.4 The electromagnetic scattering problem of radar targets.

called the *bistatic angle*. Although the scattering results reported in the literature are predominantly monostatic in nature, bistatic scattering also is important to many applications.

Depending on the size, shape, and the material constituent of the target, a wide variety of analytical techniques, summarized in Chapter 2, is available for analysis of the scattering characteristics of a target. Generally, we try to determine the current distribution on the target induced by the incident field and then determine the scattered field as the field due to the induced current. For example, a smooth conducting body in free space illuminated by an electromagnetic wave will sustain an induced surface current distribution $\mathbf{J}_S(\mathbf{r}')$ where $\mathbf{r}'$ indicates the position vector of a point on the surface of the body. The magnetic vector potential $\mathbf{A}$ due to this current distribution at an observation point P whose position vector is denoted by r is [8, 9, 10]

$$\mathbf{A} = \frac{\mu_0}{4\pi} \int_S \frac{\mathbf{J}_s(\mathbf{r}') e^{-ik_0|\mathbf{r}-\mathbf{r}'|}}{|\mathbf{r} - \mathbf{r}'|} \, ds' \tag{1.23}$$

where S is the total surface of the scattering body. With an incident plane wave $\mathbf{E}^i, \mathbf{H}^i$, the high-frequency or physical optics approximation to current distribution is

$$\mathbf{J}_S(\mathbf{r}') = 2\hat{n} \times \mathbf{H}^i|_S \tag{1.24}$$

with $\hat{n}$ as the unit positive normal at the surface. With knowledge of $\mathbf{A}$, the scattered fields at P now can be obtained by using the following [8–10]:

$$\begin{aligned} \mathbf{E}^S &= -\frac{i\omega}{k_0^2}\left[k_0^2 \mathbf{A} + \nabla\nabla \cdot \mathbf{A}\right] \\ \mathbf{H}^S &= \frac{1}{\mu_0}[\nabla \times \mathbf{A}] \end{aligned} \tag{1.25}$$

In the far zone (i.e., $r \to \infty$), the scattered field expressions given in (1.25) simplify as follows:

$$\begin{aligned} \mathbf{E}^S &= -i\omega \mathbf{A} \\ \mathbf{H}^S &= E^S/\eta_0 \end{aligned} \tag{1.25a}$$

As we can see, the fundamental task in all scattering problems is the determination of the induced current, which often is most difficult and cannot be obtained exactly. Frequently we formulate the problem in terms of an integral equation for the unknown current and then try to solve the equation either by approximate analytical methods or by numerical techniques.

As discussed in Chapter 2, for a limited number of scattering problems involving targets with edges or curvature or material discontinuities it is possible to estimate the scattered field by using the concepts of diffraction coefficients and scattering center theory [16].

In general, with unit amplitude incident field the scattered electric field at an observation point located in the far zone (i.e., $r \to \infty$) of the target can be written as

$$\mathbf{E}^S = \mathbf{S}\frac{e^{-ikr}}{kr} \tag{1.26}$$

where

k = the propagation constant in the medium $= 2\pi/\lambda$,

$\mathbf{S}$ = a dimensionless complex vector function depending on the aspect angles, target characteristics, and incident polarization.

The function $\mathbf{S}$ that is independent of r but orthogonal to $\mathbf{r}$ is referred to as the *scattering function*, and $|\mathbf{S}|$ gives the scattering pattern for the target.

Exact analytical determination of $\mathbf{S}$ often is very difficult, if not impossible, except for some simple targets. For most targets of practical interest approximate analytical or numerical techniques are used to determine $\mathbf{S}$.

1.3.1 Near-Zone and Far-Zone Considerations

Near zone and far zone concepts play an important role in scattering by objects and in the design of measurement system for scattering. In the far zone of the target the scattered fields are orthogonal to the direction of energy propagation and their amplitudes decrease inversely with the distance from the target, whereas in the near zone there are nonzero components of the scattered fields in the direction of propagation and their amplitudes fall off faster than the inverse distance. There is no clear-cut definition for near-zone and far-zone boundaries. Customarily, an observation point is said to be in the near or far zone of the scatterer (target) if its distance from the scatterer is less or greater, respectively, than $2L^2/\lambda$, where L is the maximum dimension of the target; this criterion is based on the maximum allowable phase error of $\lambda/8$ (or $\pi/4$) at the receiving aperture.

During measurement, the near-zone and far-zone question is decided by the requirements of the user and the minimum range to the target need not necessarily be $2L^2/\lambda$. Generally, the target or its model should be placed in the far zone if we want detailed information about the target. However, this may give rise to interference due to reflections from the floor, ground or multiple reflections from the surroundings. Using a range less than $2L^2/\lambda$ is not of great concern if we decide to use "average" values from the measurement. An organized discussion on these key observations is available in [17]. Significant research on various aspects of range requirement are reported in

[17–21]. In the area of antenna measurement, antennas with low sidelobes have been found to require ranges longer than $2L^2/\lambda$ [21–22] which is equally applicable to targets with low scattering sidelobes. For many low- and ultralow-sidelobe antennas or scatterers a practical range distance, such as $2L^2/\lambda$, still can be used to accurately measure wide angle sidelobes, but at the expense of the first one or two sidelobes. The range requirements for large antenna or scatterer are discussed in [23].

1.4 RADAR CROSS SECTION

The IEEE definition [24] of radar cross section is as follows: "For a given scattering object, upon which a plane wave is incident, that portion of the scattering cross section corresponding to a specified polarization component of the scattered wave." Clearly if the receiver is polarization matched to the scattered wave then the radar cross section and scattering cross section become identical, and we define the radar or scattering cross section in the direction (θ, φ), that is $\hat{r}$ (Figure 1.4) as

$$\begin{aligned}\sigma &= 4\pi \frac{\text{scattered power per unit solid angle at target}}{\text{incident power per unit area at target}} \\ &= \lim_{r\to\infty} 4\pi r^2 \frac{\text{scattered power per unit area at target}}{\text{incident power per unit area at target}}\end{aligned} \tag{1.27}$$

where r is the distance from the target to the receiver and note that σ has the dimension of area, that is, in square meters, m^2; frequently, σ is expressed as dBm^2. With receiver in the far zone of the target

$$\begin{aligned}\sigma &= \lim_{r\to\infty} 4\pi r^2 \left|\frac{\mathbf{E}^S}{\mathbf{E}^i}\right|^2 \\ &= \lim_{r\to\infty} 4\pi r^2 \left|\frac{\mathbf{H}^S}{\mathbf{H}^i}\right|^2\end{aligned} \tag{1.28}$$

where $\mathbf{E}^i, \mathbf{H}^i$ are the incident (plane wave) electric and magnetic fields at the target, respectively.

$\mathbf{E}^S, \mathbf{H}^S$ are the scattered electric and magnetic fields at the receiver, respectively. We shall use (1.28) as the definition of RCS and it indicates that σ is independent of r. But it depends on the following: frequency; target size; shape and composition; target aspect (i.e., direction of illumination); viewing aspect (i.e., direction of scattering); and the incident and observed polarizations. Monostatic (or back) scattering cross section occurs when the transmitter and receiver are collocated, and a bistatic cross section occurs when they are separated in angle.

With the far-zone scattered field given by (1.26), and assuming $\mathbf{E}^i$ has unit strength at target, we obtain

$$\sigma = \frac{\lambda^2}{\pi} |\mathbf{S}|^2 \tag{1.29}$$

where $\mathbf{S}$ is the scattering function defined earlier, and λ is the wavelength in the medium.

In general, the scattered field may contain a component whose polarization is parallel to that of the incident field and another with cross polarization; that is, orthogonal to the former. Thus, we can define the parallel and cross-polarized radar cross sections of $\sigma_{\parallel}$ and $\sigma_{\perp}$, respectively, as:

$$\sigma_{\parallel} = \frac{\lambda^2}{\pi} |\mathbf{S} \cdot \hat{E}|^2 \tag{1.30}$$

$$\sigma_{\perp} = \frac{\lambda^2}{\pi} |\mathbf{S} \cdot \hat{r} \times \hat{E}^i|^2 \tag{1.31}$$

where

$\hat{r}$ = the unit vector pointed from the target to the receiver,
$\hat{E}^i$ = the unit vector defining the polarization of the incident wave at the target.

It follows from (1.29)–(1.31) that

$$\sigma = \sigma_{\parallel} + \sigma_{\perp} \tag{1.32}$$

If one prefers to distinguish between scattering and radar cross sections of a target, as defined by IEEE [24], then (1.32) may be interpreted as stating that the scattering cross section of a target is the sum of its parallel and cross-polarized radar cross sections.

Assuming that σ is the scattering or radar cross section defined earlier (i.e., Eq. (1.26)), we can show that σ plays a key role in the well-known radar equation giving the power received by a radar receiver [4, 5]:

$$P_R = \left[\left(\frac{P_T G_T}{4\pi R_T^2}\right)\frac{1}{L_{MT} L_T}\right]\left[\left(\frac{\sigma}{4\pi R^2}\right)\frac{1}{L_{MR} L_R}\right]\left(\frac{G_R \lambda^2}{4\pi}\right)\frac{1}{L_P} \tag{1.33}$$

where

P_R = received power,
P_T = transmitted power,
G_T = gain of the transmitting antenna,
G_R = gain of the receiving antenna,
R_T = distance between the transmitting antenna and the target,
R = distance between the receiving antenna and the target,
L_T = loss factor in transmitter channel,
L_R = loss factor in receiver channel,
L_{MT} = material loss factor in the propagating medium due to dissipation between the target and transmitter,
L_{MR} = material loss factor in propagating medium due to dissipation between the target and receiver,
L_P = polarization loss factor due to improper polarization match between transmitting and receiving antennas (if the radar is not monostatic), scatter depolarization, and effect of medium Faraday rotation, if any.

1.4.1 Absorption and Extinction Cross Sections

Part of the energy intercepted by a target may be absorbed or dissipated as heat, and the remainder of the energy is scattered in all directions. Various scattering cross sections are used to account for this distribution of energy.

Let us introduce $\sigma(\theta, \varphi)$ as the bistatic scattering cross section with $\sigma(\theta_F, \varphi_F)$ defined as such scattering in the forward direction. The total scattering cross section σ_T as a measure of total scattered power is defined as

$$\sigma_T = \frac{1}{4\pi} \int_0^{\pi} \int_0^{2\pi} \sigma(\theta, \varphi) \sin\theta \, d\theta \, d\varphi \tag{1.34}$$

which indicates that if $\sigma(\theta, \varphi) = \sigma$, that is, it is the same in all directions, then $\sigma_T = \sigma$.

As mentioned earlier, targets may absorb energy as well as scatter. A measure of total energy extracted from the incident wave is the extinction cross section σ_e. From the forward scattering theorem that relates the extinction cross section to the forward scattered field we obtain

$$\sigma_e = -\frac{4\pi}{k^2} \operatorname{Im} \mathbf{S}(\theta_F, \varphi_F) \cdot \mathbf{E}^i \tag{1.35}$$

A measure of the energy absorbed is the absorption cross section σ_a, given by

$$\sigma_a = \sigma_e - \sigma_T \tag{1.36}$$

Alternatively, σ_a may be calculated by using the fields inside the target relative to the incident field as follows:

$$\sigma_a = \eta \int_V \sigma |\mathbf{E}(r')|^2 \, dv' \tag{1.37}$$

where

η = the intrinsic impedance of the outside medium,
V = the volume of the nonconducting target,
$\sigma = \omega\epsilon_0\epsilon_r''$ = the conductivity of the target material.

Because (1.37) requires a knowledge of the internal fields, it usually is difficult to compute σ_a following this method.

1.5 SCATTERING MATRIX AND POLARIZATION

We have seen earlier that the scattering function **S** relates $\mathbf{E}^S$ to $\mathbf{E}^i$, and it contains information about the target. For a given incident polarization a target may or may not change the polarization of the scattered wave. Knowledge of the polarization characteristics of radar targets can lead to their significant identification information about the targets. Use of optimum polarization in the radar system may considerably enhance the target signal and suppress the clutter signal in the most likely case of detection of targets in the presence of rain, ground, or sea clutter.

A recent text [25] discusses various aspects of polarization very well. Here we mention only some of the key issues. It is appropriate to begin with a discussion of the transformation between linear and circular polarization. A linearly polarized plane electromagnetic wave of amplitude E_0 propagating in the $-\hat{r}$ direction can be expressed as superposition of two linearly polarized waves as follows:

$$E = \hat{h}E_h \exp[i(kr + \omega t)] + \hat{v}E_V \exp[i(kr + \omega t)] \tag{1.38}$$

where

$E_0 = [E_h^2 + E_v^2]^{1/2}$,
$\hat{h}, \hat{v}$ = the two mutually orthogonal unit vectors, named *horizontal* and *vertical polarization*, respectively, and both orthogonal to the direction of propagation.

The same wave can be expressed also as the superposition of a left and a right circularly polarized wave propagating in the same direction as

$$E = \hat{l}_c E_{l_c} \exp[\mathrm{i}(\mathbf{kr} + \omega t)] + \hat{r}_c E_{r_c} \exp[\mathrm{i}(kr + \omega t)] \tag{1.39}$$

$$\begin{bmatrix} \hat{l}_c \\ \hat{r}_c \end{bmatrix} = \frac{1}{\sqrt{2}} \begin{bmatrix} 1 & \mathrm{i} \\ \mathrm{i} & 1 \end{bmatrix} \begin{bmatrix} \hat{h} \\ \hat{v} \end{bmatrix} \tag{1.40}$$

The amplitudes of the various wave components in (1.38) and (1.39) are related by

$$\begin{bmatrix} \mathbf{E}_h \\ \mathbf{E}_v \end{bmatrix} = \frac{1}{\sqrt{2}} \begin{bmatrix} 1 & \mathrm{i} \\ \mathrm{i} & 1 \end{bmatrix} \begin{bmatrix} \mathbf{E}_{l_c} \\ \mathbf{E}_{r_c} \end{bmatrix} \tag{1.41}$$

A complete description of the interaction of the incident wave and the target is given by the scattering matrix that relates the incident and scattered fields. For linear polarization the two fields are related by

$$\begin{bmatrix} \mathbf{E}_h^S \\ \mathbf{E}_v^S \end{bmatrix} = \frac{\mathrm{e}^{-ikr}}{kr} \begin{bmatrix} S_{hh} & S_{hv} \\ S_{vh} & S_{vv} \end{bmatrix} \begin{bmatrix} \mathbf{E}_h^i \\ \mathbf{E}_v^i \end{bmatrix} \tag{1.42}$$

where $\mathbf{E}_h^i, \mathbf{E}_v^i$ and $\mathbf{E}_h^S, \mathbf{E}_v^S$ are the horizontally and vertically polarized components of the incident and scattered fields, respectively, and the matrix $[S]$ defined as

$$[S] = \begin{bmatrix} S_{hh} & S_{hv} \\ S_{vh} & S_{vv} \end{bmatrix} \tag{1.43}$$

is called the *scattering matrix*, whose elements are in general complex quantities. If desired, the magnitude of each element of the scattering matrix can be related to a corresponding radar cross section by using (1.29)–(1.31). For example, $\sigma_{hh} = \lambda^2/\pi \, | S_{hh} |^2$ and similarly for the other elements.

For circular polarization, the scattered and incident fields can also be obtained from (1.41) and (1.42), provided the subscripts h and v are replaced by l_c and r_c, respectively.

The scattering matrix contains all measurable information about the scattering properties of the target. In the bistatic case there are seven measurable quantities: four amplitudes and three relative phases. In the monostatic case using a feed that is capable of receiving only the same polarization as that of the transmitter, there are five measurable quantities: three amplitudes and two relative phases. Note that the four complex elements of the scattering matrix depend on the choice of orthogonal axes, the aspect angle of the target, the scattering properties of the target, and of course on the frequency of operation of the radar. In recent years, an urgent need arose for cross-polarized information in a variety of applications, like terrain mapping, and target identification and classification. Therefore, any modern scattering model should be capable of predicting the depolarization characteristics of radar targets.

1.6 DISCUSSION OF RCS

Available information about the measured RCS of a variety of targets is described briefly in this section. The purpose of the discussion is to provide a perspective on the need for control of RCS of targets in various practical situations. The literature on observed RCS of targets is vast [1–7], hence only those results are mentioned here that are considered pertinent to bringing out the dependence of RCS on various key parameters. Generally, the RCS of a target depends on its electrical size, shape, and constituent material and on the frequency, aspect angle, and the transmitting and receiving polarizations associated with the radar. The RCS of a target will depend on the range if the observations are made in the near-zone of the target, and on time if there is a relative motion between the target and the radar.

Radar targets can be divided broadly into two classes: artificial or human-made, like aircraft, and natural, like terrain. Depending on their structural complexity, targets are also classified as simple, like a conducting sphere, or complex, like an aircraft. In the following sections the relevant information regarding the observed or measured RCS of a variety of objects is discussed briefly in the light of the preceding facts. For more detailed information on scattering from a wide variety of artificial and natural targets the reader is referred to [26] and references cited earlier [e.g., 3–5].

1.6.1 Target Size and Shape

Depending on the maximum dimension of a three-dimensional body relative to the wavelength, the scattering phenomena are frequently described as occurring in the low frequency (or Rayleigh), resonance, and high-frequency (or optical) regions. This is illustrated by Figure 1.5, which shows the RCS (backscattering) of a perfectly conducting sphere of radius as a function of $k_0 a$, where $k_0(= 2\pi/\lambda_0)$ is the free space propagation constant. The low-frequency, resonance, and high-frequency regions, indicated in the figure, may be defined as $k_0 a < 1$, $1 \leq k_0 a < 10$ and $k_0 a \geq 0$, respectively. Notice that in the optical region (i.e., $\lambda_0 \rightarrow \infty$), $\sigma = \pi a^2$ = the geometrical optical cross section of the sphere. Although these three regions of scattering may not be as clear-cut for all three-dimensional targets, the analyses still may be classified as such. For the conducting sphere, σ increases with frequency in the low-frequency region ($\sigma \propto f^4$), is an oscillating function of frequency in the resonance region and assumes a limiting value in the high frequency region. The RCS of a variety of three dimensional smooth bodies of revolution and other shapes are discussed extensively in the literature [1–7].

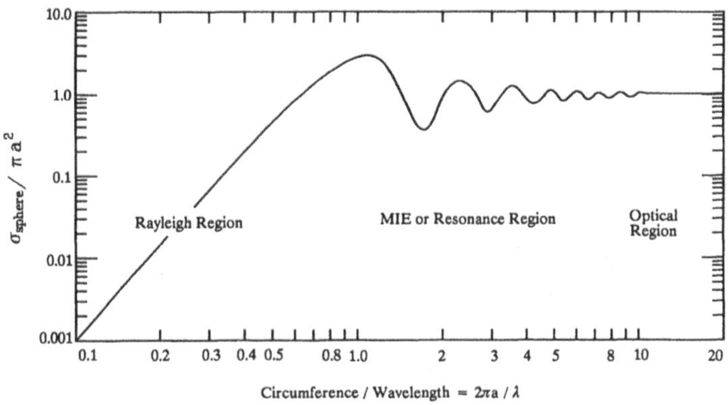

Figure 1.5 Scattering by a sphere: low-frequency, resonance, and optical regions.

1.6.2 Artificial Targets

Table 1.1 gives head-on RCS (backscattering) for some common targets at microwave frequencies. One of the earliest measurements of RCS of a ship was conducted by the US Naval Research Laboratory [27]. X-band measurements of RCS were carried for a loaded tanker YAGR picket ship and the sea at elevation angles from 0° to 90° and various aspect angles. The system calibration was done by using an aluminum sphere dropped from an aircraft. Both co- and cross-polarization measurements were performed. The radar used was an airborne radar installed in a Navy aircraft with pulsewidth = 0.3 μs, repetition rate (pulse repetition frequency, PRF) = 1000 pulses/s and half-power beamwidth = 5°. Tracking and fixed elevation runs were performed. In the tracking run, the antenna was pointed at the ship during the time the aircraft was flying; and for the fixed elevation run, the antenna was set at a known value and the pilot flew the aircraft directly off the ship approaching at a predetermined aspect angle.

Table 1.1

Head-on RCS for Some Common Targets at Microwave Frequencies, 3–10 GHz (from Skolnik [4])

Target	RCS in m^2
Pickup Truck	200
Automobile	100
Jumbo Jet	100
Large Bomber or Commercial Jet	40
Cabin Cruiser Boat	10
Large Fighter Aircraft	6
Adult Man	1

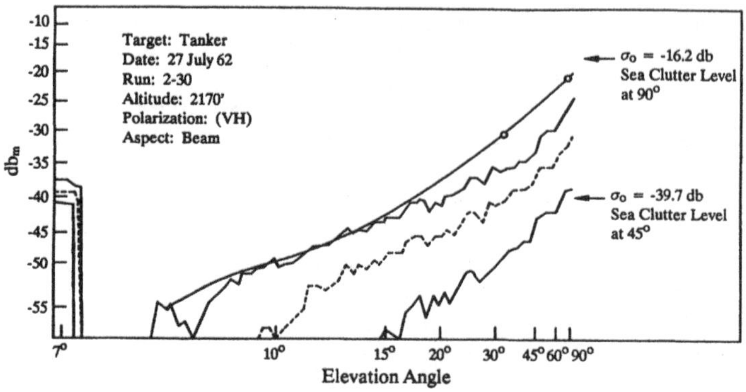

Figure 1.6 Receiver power in dB/mW with elevation angle for a ship (from [27]).

Figure 1.6 shows the received relative signal power for a tanker at beam aspect for VH or CODSS polarization from an altitude of 2170′. Figures 1.7 and 1.8 show plots obtained from typical runs with VV polarization and for different elevation angles but at the same height.

RCS data for full-scale aircraft, particularly for the low observables, are scarce; some data obtained from scale model measurements are available. Results obtained from polygonal shape modeling of a rocket shape, the Concorde aircraft, and the Boeing 747 aircraft are described in [28, 29]. Figures 1.9 to 1.11 show the horizontally polarized RCS for the rocket, Concorde and Boeing 747, respectively. Figure 1.12 gives the phase of the horizontally polarized received scattered field in the azimuthal

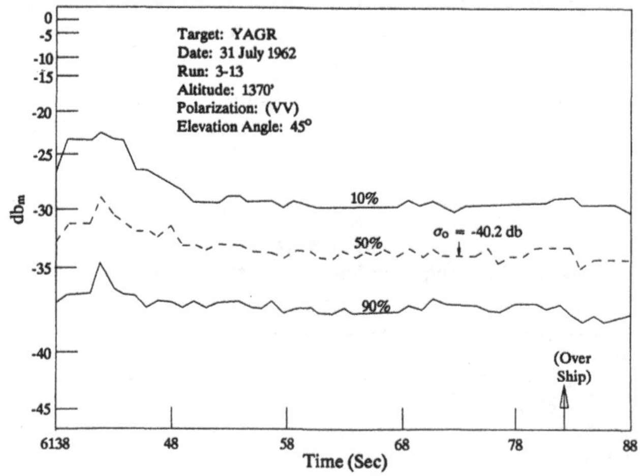

Figure 1.7 Receiver power in dB/mW with time for elevation angle 45° for VV polarization at an altitude of 1370 ft. (from [27]).

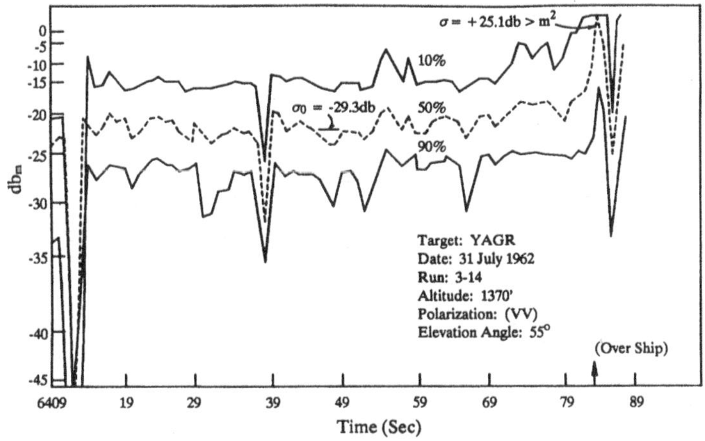

Figure 1.8 Receiver power in dB/mW with time for elevation angle 55° for VV polarization at an altitude of 1370 ft. (from [27]).

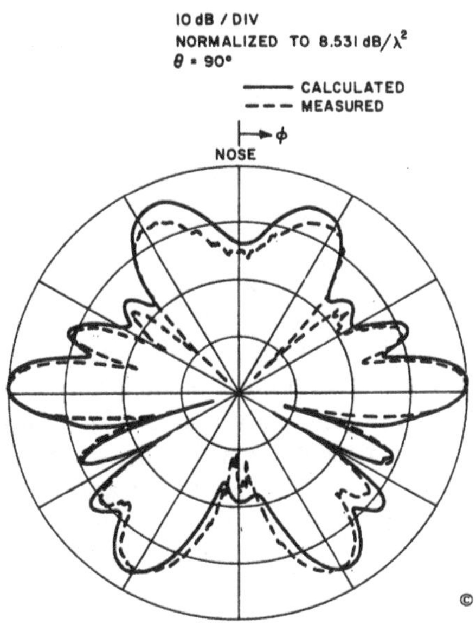

Figure 1.9 $\hat{\phi}$-Polarized RCS of a rocket in the azimuth plane (from [28]).

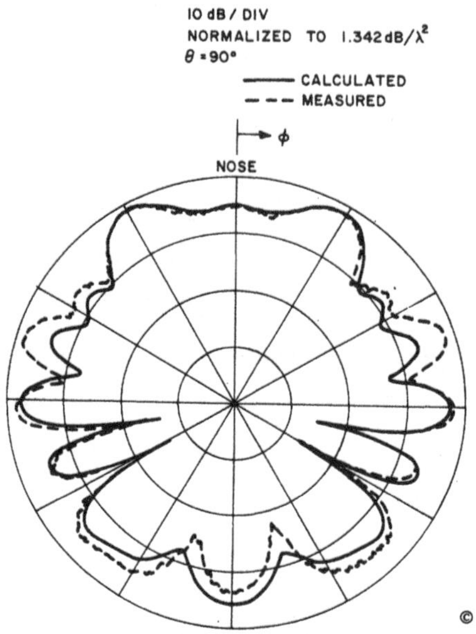

Figure 1.10 $\hat{\phi}$-Polarized RCS of the Concorde in the azimuthal plane (from [28]).

plane for the Concorde. A comparison of the computed (solid curves) and experimental RCS of the B-47 jet aircraft as a function of aspect angle and frequency is shown in Figure 1.13.

A radar engineer often is interested in the scattering from urban structures such as buildings or even a city. Table 1.2 shows the average RCS per m^2 for small houses and cities measured at selected frequencies and grazing angles. The parameter presented is the average backscattering cross section per unit surface area and is designated as γ in dB. The RCS of cities with parallel and perpendicular roads are found to be much larger than those of cities with irregular roads when the plane of incidence coincides with one of the two road locations. This may be because corner reflectors are formed between the ground and the outer walls of the buildings.

1.6.3 Natural Targets

The earth's surface behaves as a source of clutter for ground-based as well as space-based radar. Hence, we often must know, at least roughly, the order of magnitude of radar returns from desert, cultivated land, open woods, wooded hills, *et cetera*. Table 1.3 shows the measured values of the scattering cross section in dB with reference to 1 m^2 for various natural targets.

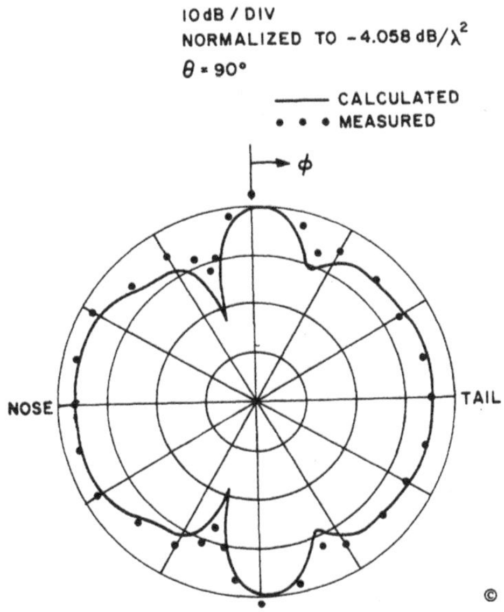

Figure 1.11 $\hat{\phi}$-Polarized RCS of a Boeing 747 in the azimuthal plane (from [28]).

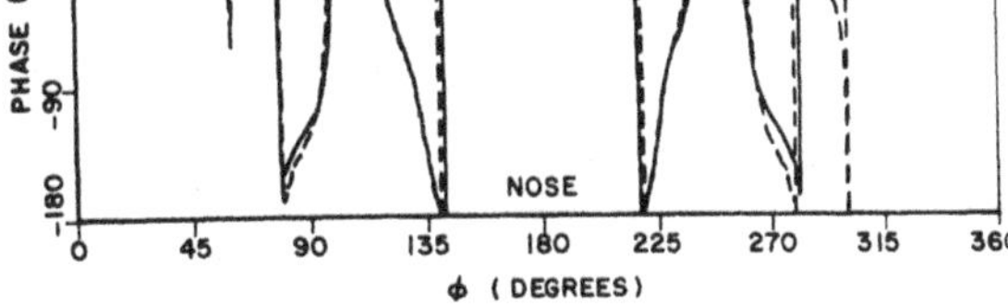

Figure 1.12 Polarized RCS phase of the Concorde in the azimuthal plane (from [28]).

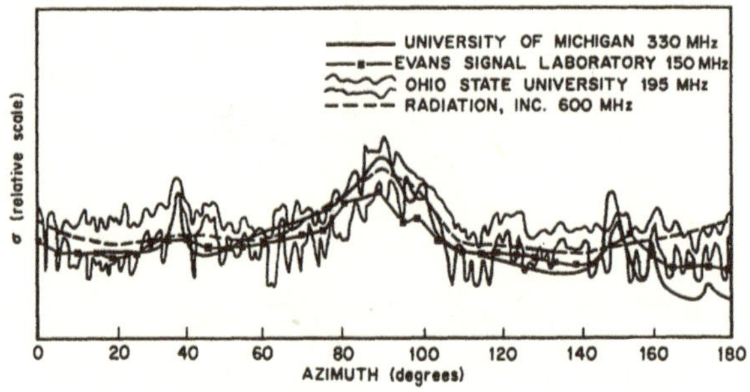

Figure 1.13 Comparison of the computer generated (solid curves) and experimental RCS of the B-47 jet aircraft as a function of aspect angle and frequency (from [4]).

Table 1.2
Measured Average RCS per m^2 for Small Buildings and Cities, with λ in dB for various terrain and grazing angles (from Ruck *et al.* [3], Vol. 1, p. 867)

Frequency (Grazing Angles)	Small Buildings		Cities	
	0–1°	10°	0–1°	10°
HF (3.28 MHz)				−8
UHF (0.5 GHz)		−23	−30 to −22	−9
S-band (3.0 GHz)	−35	−23		−15
X-band (9.3 GHz)	−36 to −30	−23	−24	−12
K_a-band (35 GHz)				

Table 1.3
Measured Average RCS per m^2 γ for Various Terrain Types (from Ruck *et al.* [3])

Frequency (Grazing Angle)	Desert		Cultivated Land		Open Woods		Wooded Hills	
	0.1°	10°	0–1°	10°	0–1°	10°	0–1°	10°
HF (32.8 MHz)	−6 to 55			−75 to −65	−60	−65 to −55	−50 to −30	−60 to −45
UHF (0.5 GHz)	−40		−38 to −30	−36		−23	−34 to −24	−22
S-band (3.0 GHz)		−25	−25 to −35	−21	−33	−23	−47 to −32	
X-band (9.3 GHz)	−38	−26	−36	−20	−30	−23	−36 to −30	−25
K_a-band (35 GHz)		−22	−20	−20	−25	−19	−17	−19

Another natural target with an RCS of considerable importance is the sea surface. Information about scattering from the sea's surface is needed for detection, identification, and discrimination of radar targets against the background of the sea. Sea surface conditions may differ significantly throughout the day and the year or due to wind or sea ice. The literature on sea-surface scattering is vast, and we shall just barely touch on the topic. A balanced discussion on the early analytical models for sea surface scattering are available in [4, Chapter 26] and a rigorous analytical model is discussed in [30–32]. As an illustration, Table 1.4 shows the average copolarized RCS per square meter for

Table 1.4

Measurements of Average Radar Cross Section per Unit Area for Sea Surface, with γ, or σ, in dB for Various Grazing, or Depression, Angles (from Ruck *et al.* [3])

Frequency	Polarization States	0.3°	1.0°	3.0°	10°	30°	60°
Sea State 1 (5 knot wind)							
HF: see middle section							
UHF	VV		-70	-60		-38	-23
0.5 GHz	HH		-84	-70			-22
S-band	VV	-62	-56	-52		-40	-24
3.0 GHz	HH	-74	-65	-59			-25
X-band	VV	-58	-50	-45	-42	-39	-28
9.36 GHz	HH	-66	-51	-48	-51		-26
K_a-band	VV			-41	-38	-37	-26
35 GHz	HH		-40	-43			
Sea State 3 (14 knot wind)							
HF*	VV		-40	-41	-42	-40	-40
32.8 MHz	HH				-47		
UHF	VV		-58	-43	-34	-28	-18
0.5 GHz	HH		-76	-61	-50	-40	-21
S-band	VV	-55	-48	-43	-34	-29	-19
3.0 GHz	HH	-58	-48	-46	-46	-38	
X-Band	VV	-45	-39	-38	-32	-28	-17
9.36 GHz	HH	-46	-49	-39	-37	-34	-21
K_a-band	VV		-34	-34	-31	-23	-14
35 GHz	HH		-36	-37	-31		
Sea State 5 (22 knot wind)							
HF: see middle section							
UHF	VV	-75			-25		
0.5 GHz	HH		-65	-53	-46		
S-band	VV	-50	-38	-35	-28		
3.0 GHz	HH	-44	-42	-37	-38		
X-band	VV	-39	-33	-31	-26	-20	-10
9.36 GHz	HH	-39	-33	-32	-31	-24	-12
K_a-band	VV		-31	-30	-26	-20	-4
35 GHz	HH				-27	-20	

*HF sea return has not been distinguished as to the sea state. The numbers given here are believed to represent an average state, somewhere between 1 and 3.

the sea surface measured at various frequencies and grazing angles and for variable wind speeds. Figures 1.14 and 1.15 show the normalized RCS of the ocean for both polarizations and at frequencies 428 MHz and 4.455 GHz, respectively, the results shown are in dB where 0 dB is 1 m^2. The RCS data shown are derived from Daley *et al.* [30] and the theoretical results are from composite surface scattering model [31].

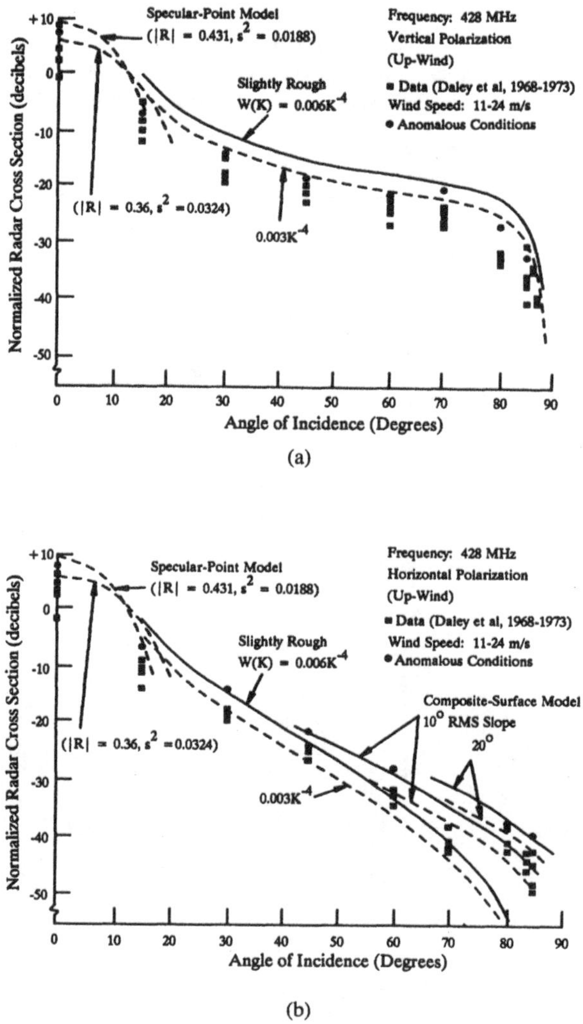

Figure 1.14 Comparison between theoretical and experimental RCS of sea surface for radar frequency 428 MHz for vertical polarization (from [31]).

The conditions indicated in Figures 1.14 and 1.15 refer to a snowy day when the air was colder than water.

Large numbers of insects and birds, particularly during migration, produce radar clutter, and in fact the radar may even get saturated if the number is large enough. Problems associated with birds and insects are discussed in [33, 34]. Figure 1.16 shows a high temporal resolution time-height display as the targets (birds and insects) fly through the antenna beam, obtained with a 100 MHz waveform digitizer. Figure 1.17 presents the RCS of insects and birds as functions of the weight of the different species. Some experimental results on the RCS of ducks and chickens are available in [35].

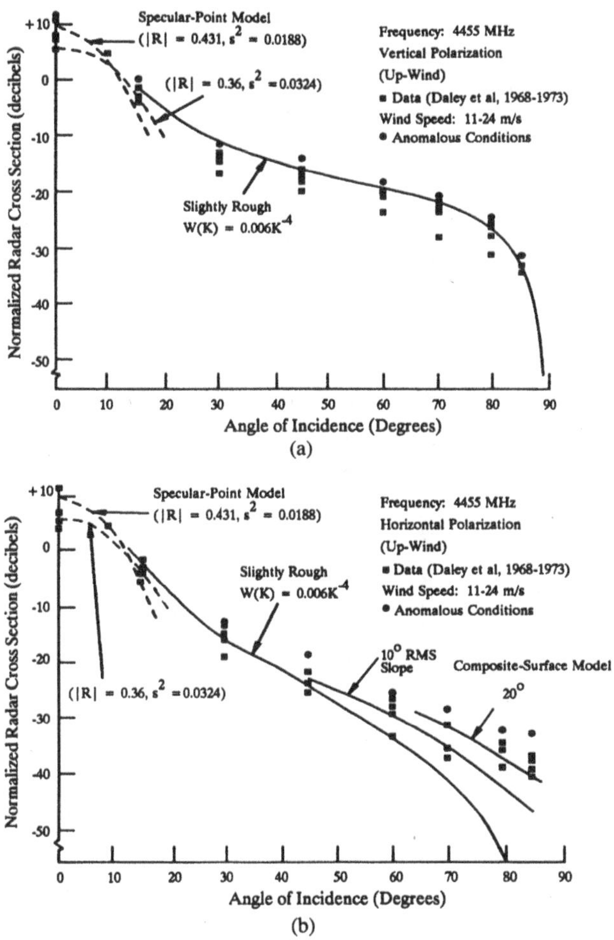

Figure 1.15 Comparison between theoretical and experimental RCS of sea surface for radar frequency 4455 MHz for vertical polarization (from [31]).

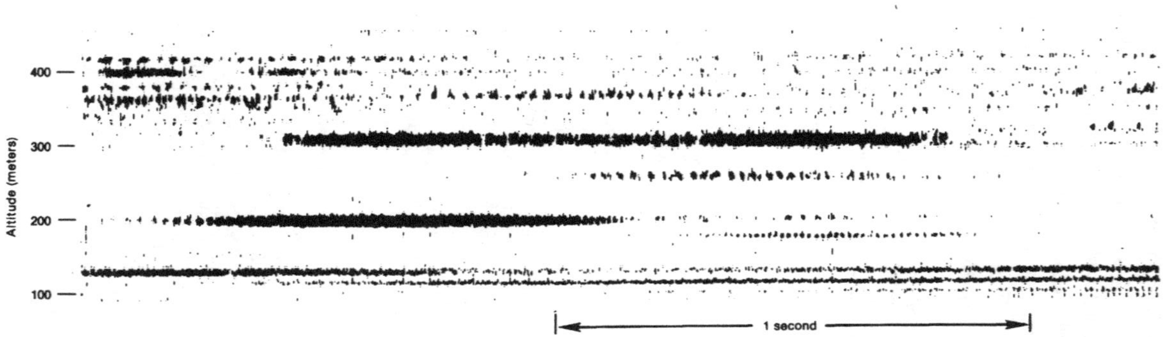

Figure 1.16 Time-height display of migrating birds and insects (from [33]).

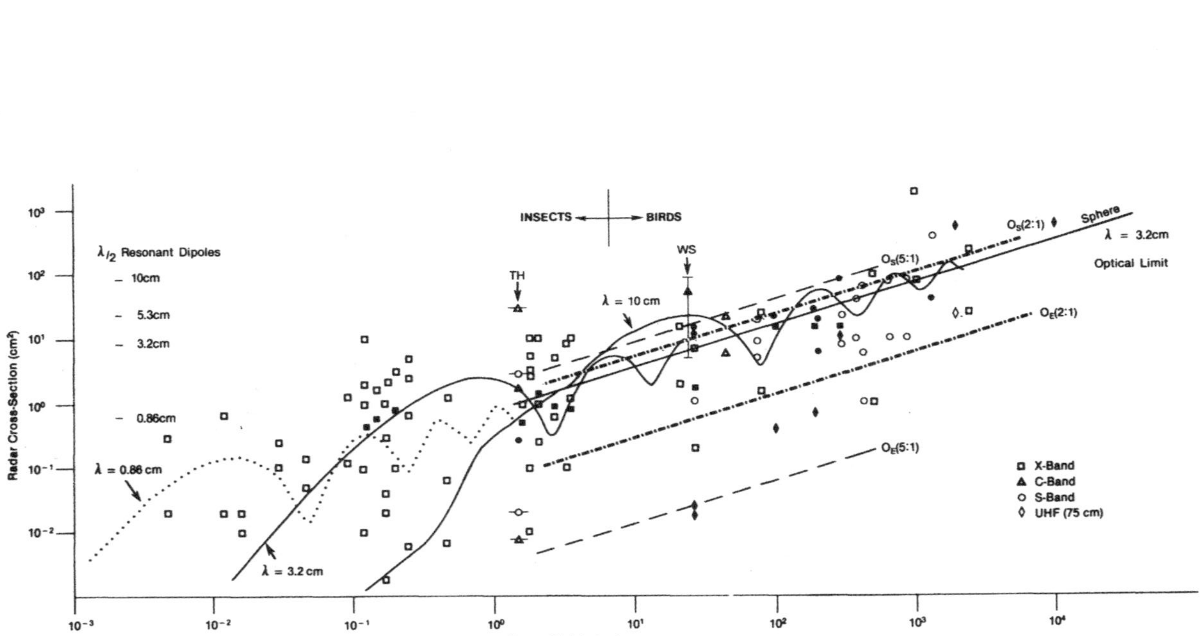

Figure 1.17 RCS *versus* target weight for different types of birds and insects (from [33]).

1.7 DYNAMIC RCS

In many practical situations there may be relative motion between the target and the radar. In such cases, the resultant doppler shift and the rapid change of aspect angle and the angle of polarization make the received radar signal more complicated and the associated signal processing significantly different. Under these conditions the target is said to have a dynamic RCS. Figure 1.18 obtained from [36; 3, Vol. 1, p. 3] shows a plot of pulse to pulse RCS in m^2 *versus* time for the satellite 1957 Beta measured at 440 MHz.

The variation of dynamic RCS with time for moving targets is best described by a suitable statistical distribution represented by a *probability density function* (PDF). The most complete reference on this topic is by Swerling [37] who extended Marcum's treatment [38,39] and studied different types of statistical distributions describing variations of RCS for moving targets. Formulas for detection probabilities are made available and curves of detection probability *versus* range are given for different target fluctuation models. A brief account of this is also available in [5].

REFERENCES

1. P. C. Fritsch, guest ed., Special Issue on Radar Reflectivity, *Proc. IEEE*, Vol. 53, No. 8, August 1965.
2. J. W. Crispin, Jr., and K. M. Siegel, eds., *Methods of Radar Cross-Section Analysis*, Academic Press, New York, 1968.
3. G. T. Ruck, ed., *Radar Cross Section Handbook*, Vols. 1 and 2, Plenum Press, New York, 1970.
4. M. L. Skolnik, ed., *Radar Handbook*, McGraw-Hill, New York, 1970.
5. E. F. Knott, J. F. Shaeffer, and M. T. Tuley, *Radar Cross Section*, Artech House, Dedham, MA, 1985.
6. W. R. Stone, guest ed., Special Issue on Radar Cross Sections of Complex Objects, *IEEE Trans. on Antennas and Propagation*, Vol. AP-37, No. 5, May 1989.
7. W. R. Stone, guest ed., Special Issue on Radar Cross Sections of Complex Objects, *Proc. IEEE*, Vol. 77, No. 5, May 1989.
8. J. D. Kraus, *Electromagnetics*, 3d Ed., McGraw-Hill, New York, 1984.
9. R. F. Harrington, *Time-Harmonic Electromagnetic Fields*, McGraw-Hill, New York, 1961.
10. J. A. Stratton, *Electromagnetic Theory*, McGraw-Hill, New York, 1941.
11. J. Van Bladel, *Electromagnetic Fields*, McGraw-Hill, New York, 1964.
12. T. B. A. Senior, "Impedance Boundary Conditions for Imperfectly Conducting Surfaces," *Applied Science Research*, Sec. B., Vol. 8, 1960, pp. 418–436.
13. T. B. A. Senior, "Approximate Boundary Conditions," *IEEE Trans. on Antenna and Propagation*, Vol. AP-29, No. 5, September 1981, pp. 826–829.
14. D. S. Wang, "Limits and Validity of the Impedance Boundary Condition on Penetrable Surfaces," *IEEE Trans. on Antennas and Propagation*, Vol. AP-35, No. 4, April 1987, pp. 453–457.
15. S. W. Lee, "How Good is the Impedance Boundary Conditions?," *IEEE Trans. on Antennas and Propagation*, Vol. AP-35, No. 11, November 1987, pp 1313–1315.
16. J. L. James, *Geometrical Theory of Diffraction*, Peter Pereguinus, Hertz, England, 1976.
17. R. G. Kouyoumjian and L. Peters, Jr., "Range Requirements in Radar Cross Section Measurements," *Proc. IEEE*, Special Issue on Radar Reflectivity, Vol. 53, No. 8, August 1965, pp. 920–928.

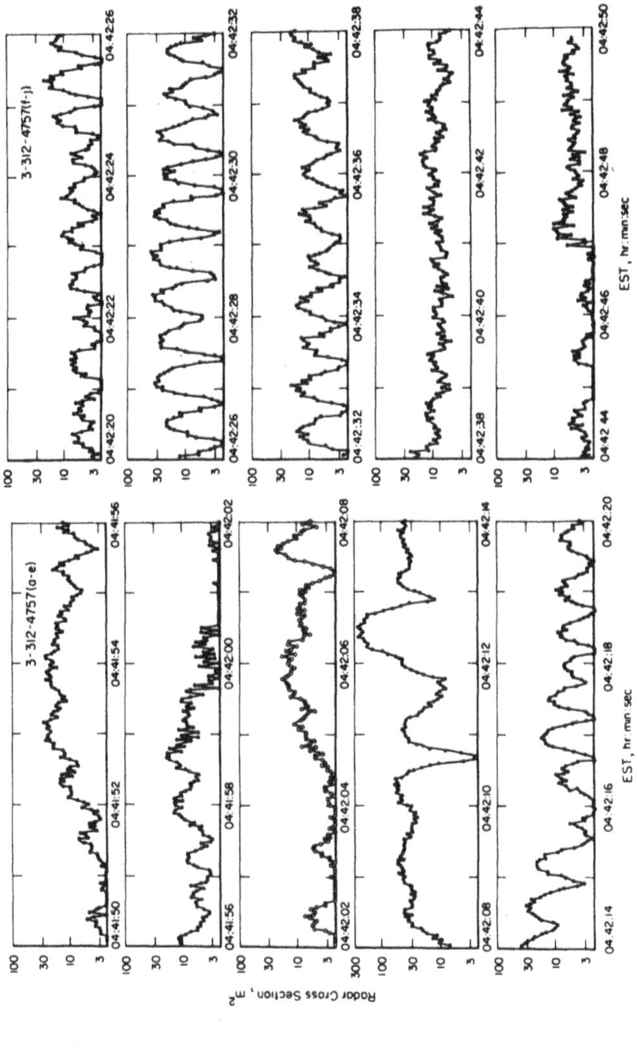

Figure 1.18 Plot of RCS in m^2 *versus* time for the satellite, 1957 β measured at 440 MHz.

18. J. R. Mentzer, *Scattering and Diffraction of Waves*, Pergamon Press, New York.
19. D. R. Rhodes, "On Minimum Range for Radiation Pattern," *Proc. IRE*, Vol. 42, September 1954, pp. 1348–1410.
20. E. V. Jull, "An Investigation Near-Field Radiation Patterns Measured with Large Antennas," *IRE Trans. Antennas and Propagation*, Vol. AP-10, 1962, pp. 363–369.
21. E. F. Knott and T. B. A. Senior, "How Far Is Far?" *IEEE Trans. on Antennas and Propagation*, Vol. AP-22, No. 5, September 1974, pp. 732–734.
22. P. S. Hacker and H. E. Schrank, "Range Distance Requirement for Measuring Low and Ultralow Sidelobe Antenna Patterns," *IEEE Trans. on Antennas and Propagation*, Vol. AP-30, No. 5, September 1982, pp. 956–966.
23. T. Uno and S. Adachi, "Range Distance Requirements for Large Antenna Measurements," *IEEE Trans. on Antennas and Propagation*, Vol. AP-37, June 1989, pp. 707–720.
24. "IEEE Standard Definitions of Terms for Antennas," *IEEE Trans. on Antennas and Propagation*, Vol. AP-31, No. 6, November 1983.
25. H. Mott, *Polarization in Antennas and Radar*, John Wiley and Sons, New York, 1986.
26. H. R. Raemer and A. K. Bhattacharyya, *Analytical Modeling and Computer Simulation of Radar Clutter and Multipath Signal*, Vol. 1. *Background Analysis and Description of Software Excluding Scattering Details* (1989); Vol. 2. *Details of Scattering and Clutter Models* (forthcoming), Center for Electromagnetic Research, Northeastern University, Boston.
27. J. T. Ransome, Jr., W. T. Davis, and J. C. Daley, "Radar Echoes from Ships," NRL Memo. Rep. 1368, U.S. Naval Research Laboratory, Washington, DC, October 30, 1962.
28. E. H. Newman, P. Alexandropoulous, and E. K. Walton, "Polygonal Plate Modeling of Realistic Structures," *IEEE Trans. on Antennas and Propagation*, Vol. AP-32, No. 7, July 1984, pp. 742–747.
29. P. Alexandropoulous, "Computer Modelling of Arbitrary Surfaces by Polygonal Plates and Wires," Masters Thesis, Ohio State University, Department of Electrical Engineering, 1983.
30. J. C. Daley *et al.*, "Sea Clutter Measurement on Four Frequencies," Rep. 6806, Naval Research Laboratory, Washington, DC, 1968.
31. G. R. Valenzuela, "Theories for the Interaction of Electromagnetic and Oceanic Waves—A Review," *Boundary-Layer Meteorology*, Vol. 13, 1978, pp. 61–85.
32. G. R. Valenzuela, "Scattering of Electromagnetic Waves from a Tilted Slightly Rough Surface," *Boundary-Layer Meteorology*, Vol. 3, No. 11, November 1968, pp. 1057–1066.
33. C. R. Vaughan, "Birds and Insects as Radar Targets—A Review," *Proc. IEEE*, Vol. 73, No. 2, February 1985, pp. 205–227.
34. J. R. Riley, "Radar Cross Section of Insects," *Proc. IEEE*, Vol. 73, No. 2, February 1985, pp. 228–232.
35. P. Blacksmith, Jr., and R. B. Mack, "On Measuring Radar Cross Sections of Ducks and Chickens," *Proc. IEEE*, Special Issue on Radar Reflectivity, Vol. 53, No. 8, August 1955, p. 1125.
36. G. H. Pettengill and L. G. Kraft, "Earth Satellite Observations Made with the Millstone Hill Radar," in *Avionics Research: Satellite and Problems of Long Range Detection and Tracking*, Pergamon Press, New York, 1960, p. 125.
37. P. Swerling, "Probability of Detection for Fluctuating Targets," *IRE Trans. on Information Theory*, Vol. IT-6, No. 2, April 1960, pp. 269–308.
38. J. I. Marcum, "A Statistical Theory of Target Detection by Pulsed Radar," the RAND Corporation, Research Memo. RM-754, December 1, 1947.
39. J. I. Marcum, "Mathematical Appendix," the RAND Corporation, Research Memo. RM-753, July 1, 1948.

Chapter 2
Methods For RCS Estimation

2.1 INTRODUCTION

In radar applications, we frequently must compute the RCS of a wide variety of targets, ranging from simple objects (perfectly conducting spheres, infinite cylinders, *et cetera*), to objects having intermediate complexity (a finite cylinder, a finite curved plate, *et cetera*), to complicated objects (aircraft, ship, *et cetera*). The targets may consist of dielectric or magnetic materials, partially dielectric and magnetic materials, or a number of layers of such materials. Although such combinations of shapes and materials offer more degrees of freedom in controlling the RCS of targets, they also increase the complexity of the analytical and numerical computations.

A large variety of analytical techniques have been developed to determine the RCS of targets having wide variations in shape, size, and constituent material. These techniques can be broadly divided into two classes: exact methods and approximate methods. An exact method formulates a boundary value problem, wherein a solution is obtained by using the pertinent exact solution of the wave equation (Section 1.2.2) subject to the appropriate exact boundary conditions (Section 1.2.3). Because exact methods have been adequately discussed in the standard literature, for example in [1, 2], we will not dwell on these here. However even with exact methods, the final numerical results are obtained by making some form of approximation, like truncating an infinite series. To this extent, the results of an otherwise exact method are approximate, although they may be made sufficiently accurate for the particular problem.

An approximate method, as its name implies, is developed for situations where an exact solution is either impossible to obtain or less effective. In electromagnetics, very few problems of practical interest can be solved by an exact method. This is because exact solutions are possible only for problems having geometry conforming to coordinate systems in which the wave equation is separable. Hence, a number of approximate but accurate methods have been developed, particularly during the past three decades. They generally are based on some asymptotic solution of Maxwell's equations [3] or on some simplifying assumptions related to the boundary conditions. Approximate methods are used to solve the majority of problems having practical interest. However, some of these methods have severe limitations and require revision or correction to accommodate the tasks for specific cases. One such task is to account for the finiteness of a structure and the associated edge effects or surface traveling (or creeping) waves and multiple diffraction effects.

Despite a number of approximate methods available, no single method is sufficient for all cases. For this reason, hybrid techniques [4, 5] have been proposed that combine the advantages of two or more techniques. Scattering by many complicated bodies of practical interest generally are not amenable to solution even by approximate analytical methods, and frequently they are solved by numerical methods. Such methods often start by numerically solving for the unknown induced current contained in an integral equation.

Analytical methods for estimating the RCS of targets are discussed in several books [1–3, 6–9] and review papers [10–12]. An excellent source of information and bibliography prior to 1964 is available in [13], where classified references on a large number of target types also are available. Our objective in this chapter is not to develop various approximate analytical and numerical techniques but to summarize them sufficiently for application by the reader and later use in Chapters 3 and 4. We shall also touch on the advantages and disadvantages of various available analytical techniques. A number of case studies will be presented. Some of the simple targets like conducting spheres, infinite cylinders, and cones, which have been thoroughly discussed in the literature [14], will not be presented here. The chapter is written so that given a target, the reader would be able to decide which technique to use for efficiently obtaining the desired results subject to any special requirement.

2.2 GEOMETRICAL OPTICS METHOD

The *geometrical optics* (GO) method is a high-frequency (or zero wavelength) approximation where the scattering mechanism and energy propagation are accounted for by classical tubes of rays. Therefore it also is called the *ray optics method*. The complex electric or magnetic field associated with a transverse electromagnetic (TEM) wave is written as

$$u = A\,e^{-iks} \tag{2.1}$$

where A is the complex amplitude, and s is the distance, measured along a ray, between the observation and reference planes.

When narrow beam radars are used and a target is illuminated by an electromagnetic wave, the radar return appears to come from specific points on the target if the target size is large compared to the wavelength of the radar. These points (sometimes called *flash points*) are the points at which the waves suffer specular reflection (i.e., where the angle of incidence equals the angle of reflection), and they may be considered as scattering centers. Under this approximation, we need not consider the scattering contributions from the entire body; the contributions from the approximate scattering centers when suitably combined are sufficient to account for the scattering in a given direction. Given the target, the incident and observation planes, and the bistatic angle, we can locate the scattering centers using Fermat's principle of optimum path between points a and b:

$$\delta \int_a^b n \, \mathrm{d}l = 0 \tag{2.2}$$

where

$\delta =$ the variation,
$n =$ the refractive index of the medium.

The relationship between the fields at the start and end of a tube of ray can be obtained by using the law of conservation of energy. For example, the electric field intensity $\mathbf{E}(s)$ at the observation end is related to the electric field intensity $\mathbf{E}(s)$ at the reference end ($s = 0$) by the following

$$\mathbf{E}(s) = \mathbf{E}(0)\Gamma \mathrm{e}^{-iks} \tag{2.3}$$

where the divergence factor Γ is given by

$$\Gamma = \sqrt{\frac{R_1 R_2}{(R_1 + s)(R_2 + s)}} \tag{2.4}$$

with R_1 and R_2 as the principal radii of curvature of the waveform at the end of the tube. For a plane wave, $R_1 = R_2 = \infty$, and it follows from (2.4) that $\Gamma = 1$; similarly, for a spherical wave, $R_1 = R_2 = R$ and

$$\Gamma = \frac{R}{R + s}$$

and for a cylindrical wave, $R_1 = \infty$, $R_2 = \rho$, and hence

$$\Gamma = \frac{\rho}{\sqrt{\rho + s}}$$

where ρ is the radius of curvature of the cylindrical wavefront. The divergence factor accounts for the spread of energy and consequent decrease in the field intensity as the wave propagates, which is not accounted for by the usual reflection and transmission coefficients.

The GO field is the leading term ($n = 0$) of the Luneberg-Kline Series [3] expressed as inverse powers of ω:

$$\mathbf{E}(\mathbf{r}, \omega) = e^{ik\psi(r)} \sum_{n=0}^{\infty} (i\omega)^{-n} \mathbf{E}_n(\mathbf{r}) \tag{2.5}$$

where $\psi(r)$ is the phase factor. In GO approximation the region of space is divided into distinct illuminated and shadow regions, and hence the method cannot account for any diffraction effect.

2.3 PHYSICAL OPTICS METHOD

As indicated earlier (Section 1.3), a fundamental problem in all scattering formulation is the determination of induced currents on the target from which the scattered fields and other quantities are calculated by using standard methods. The *physical optics* (PO) method is a high-frequency technique wherein the target is replaced by induced surface currents that then serve as the source of scattered fields. The basic assumptions involved in obtaining the currents are (1) the radii of curvature of the surface are large compared to wavelength, and (2) currents exist only in the area that is directly illuminated by the incident wave and currents on the illuminated surface have the same characteristics as of those on an infinite plane tangent to the surface at the point of incidence.

With a conducting body, the physical optics currents on the illuminated side of the body is given by eq. (1.24), and the scattered fields can then be determined by using (1.23), (1.25), and (1.26). For a general target, the required currents can be obtained by using Schelkunoff's equivalence principle [1] subject to the two basic physical optics assumptions just given.

Depending on the choice of current distribution on the target surface and other characteristics, twelve versions of the PO method have been delineated [15]. Three possible choices of the currents on the target surface are

$$E-H\text{-formulation: } \mathbf{J}_S = \hat{n} \times (\mathbf{H}^1 + \mathbf{H}^s); \quad \mathbf{M} = (\mathbf{E}^i + \mathbf{E}^s) \times \hat{n} \tag{2.6}$$

$$H\text{-formulation: } \mathbf{J}_S = 2(\hat{n} \times \mathbf{H}^i); \quad \mathbf{M} = 0 \tag{2.7}$$

$$E\text{-formulation: } \mathbf{J}_S = 0; \mathbf{M} = 2(\mathbf{E}^s \times \hat{n}) \tag{2.8}$$

where $\mathbf{E}^i$,$\mathbf{H}^i$ and $\mathbf{E}^s$, $\mathbf{H}^s$ are the incident and scattered fields at the target surface and are related by the appropriate boundary conditions.

The physical optics current, in general, is embedded as an unknown in an integral equation (Section 2.6) whose solution provides the required current for field calculations. From this point of view, and ignoring the effects of any surface current in the direction of observation, the Stratton-Chu equations (Section 2.6) are modified [8, Chap. 5] to give the scattered fields:

$$\mathbf{E}^s = +\mathrm{i}\frac{\mathrm{e}^{-k_0 r}}{r}\int_S \hat{s} \times (\mathbf{M} + \eta_0 \hat{s} \times \mathbf{J}_S) \exp\left[\mathrm{i}k_0 \mathbf{r}' \cdot (\hat{i} - \hat{s})\right] \mathrm{d}s' \tag{2.9}$$

$$\mathbf{H}^s = -\mathrm{i}\frac{\mathrm{e}^{-\mathrm{i}k_0 r}}{r}\int_S \hat{s} \times \left(\mathbf{J}_S - \frac{1}{\eta_0}\hat{s} \times \mathbf{M}\right) \exp\left[-\mathrm{i}k_0 \mathbf{r}' \cdot (\hat{i} - \hat{s})\right] \mathrm{d}s' \tag{2.10}$$

where $\hat{i}$ and $\hat{s}$ are unit vectors in the directions of incidence and observation, S represents the illuminated portion of the target surface and the other parameters as explained in Figure 2.1. Note that in (2.9) and (2.10) J_S, $\mathbf{M}$ represent the vector amplitudes of currents on S given by (2.6); their phases have been separated out and included in the exponential terms. Equations (2.9) and (2.10) may be considered the generalized physical optics formulation and are used to obtain the scattered field with a knowledge of the currents $\mathbf{J}_S$ and $\mathbf{M}$ obtained either by solving appropriate integral equations or by suitable approximations. For a conducting body, a popular choice is to use $\mathbf{J}_S = 2(\hat{n} \times \mathbf{H}_i)$, $\mathbf{M} = 0$ in (2.9) and (2.10).

Physical optics is a popular method for solving high-frequency RCS problems because it requires no special consideration of caustics or shadow boundaries. Increased computer power in recent times has absorbed the difficulties associated with its extensive numerical integration. The accuracy of the PO method can be improved by improving the assumed current distributions. It cannot account for the presence of

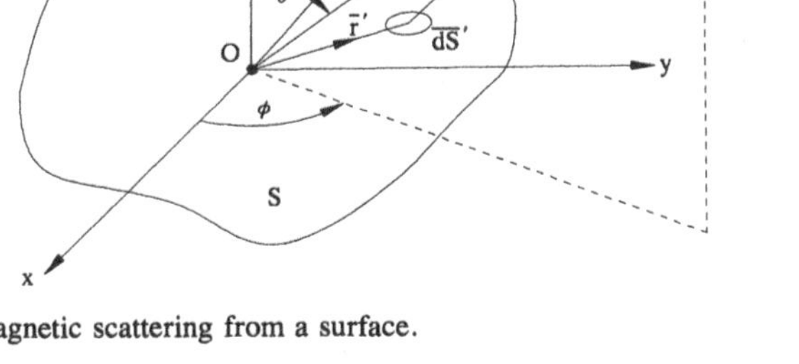

Figure 2.1 Electromagnetic scattering from a surface.

discontinuities on a target and cannot predict depolarization effects for monostatic single scatterers; however, for monostatic multiple scattering, it can predict the depolarization effects. In general, the fields obtained by the PO method do not obey reciprocity. But the average of the two solutions, one with a given position of the transmitter and receiver and another with their positions interchanged, satisfies reciprocity. Hence, monostatic or backscattering solutions automatically satisfy reciprocity. A good discussion on the different and special computation of various quantities associated with backscattering by complicated objects is available in [16].

2.4 GEOMETRICAL THEORY OF DIFFRACTION

The first major contribution to surmount the shortcomings of the GO approximation was provided by J. B. Keller, who from considerations of scattering of electromagnetic waves by a straight wedge formulated the *geometrical theory of diffraction* (GTD) to account for the diffraction and polarization effect [17]. By analogy from reflection and transmission, he introduced the terms *diffraction coefficients* and *diffracted rays*. By comparing the approximate fields with those obtained after asymptotic evaluation of the diffraction integrals in the exact Sommerfeld solution for straight wedge, Keller obtained explicit expressions for the diffraction coefficients.

Keller's diffraction coefficients for a straight wedge (Figure 2.2) is asymptotically given by [17]

$$D^{s,h} \sim \frac{(2/n)\sin(\pi/n)}{\sqrt{(8/\pi k)}\,\sin\beta_0}\left\{\left[\cos\left(\frac{\pi}{n}\right) - \cos\left(\frac{\varphi - \varphi_0}{n}\right)\right]^{-1} \mp \left[\cos\left(\frac{\pi}{n}\right) - \cos\left(\frac{\varphi + \varphi_0}{n}\right)\right]^{-1}\right\} \tag{2.11}$$

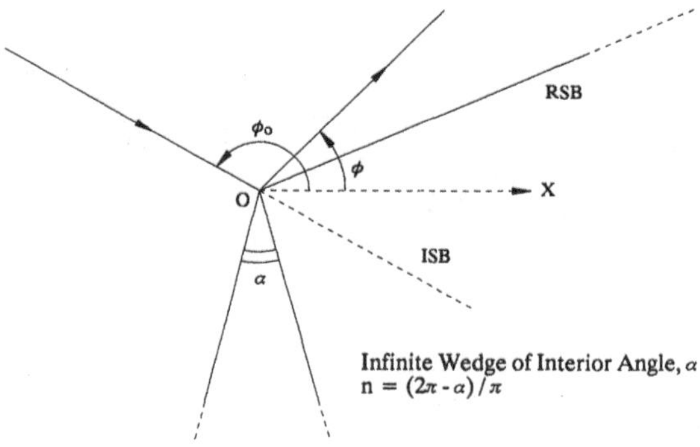

Figure 2.2 Diffraction of a plane wave by an infinite plane wedge.

where

$n = (2\pi - \alpha)/\pi$ with α as the interior wedge angle (Figure 2.2),

β_0 = the smaller angle between the direction of incidence and the tangent at the point of incidence,

φ_0, φ = the angles of incidence and scattering, respectively, and superscripts s and h stand for soft (i.e., tangential $\mathbf{E} = 0$) and hard (i.e., $\partial \mathbf{E}/\partial n = 0$) wedges, s and h correspond to plus and minus, respectively.

For a general incident wave, the diffracted field can be related to the incident field by the following matrix relationship:

$$\begin{bmatrix} \mathbf{E}_{\parallel}^{d} \\ E_{\perp}^{d} \end{bmatrix} = \begin{bmatrix} D^{s} & 0 \\ 0 & D^{h} \end{bmatrix} \begin{bmatrix} E_{\parallel}^{i} \\ E_{\perp}^{i} \end{bmatrix} \frac{e^{-iks}}{\sqrt{s}} \tag{2.12}$$

where s is the distance between the edge and the observer, and $\parallel$ and $\perp$ refer to the electric fields parallel and perpendicular to the plane of incidence, respectively.

Keller's theory yields good results for backscattering by right circular cone and cylinder [18, 19]. In an experiment, the agreement of the GTD results was much better than that provided by the PO calculations. Although the GTD is being utilized with success in many scattering problems of practical interest, it suffers from a number of limitations:

1. The diffracted field becomes infinite at both incident and reflected shadow boundaries.
2. The diffraction coefficient becomes infinite at the edge, thereby violating the edge condition.
3. The expression for diffraction coefficient has been obtained only by comparison with available exact solution of a canonical problem and, therefore, does not follow as an integral part of the theory.
4. The higher-order terms in the expression for the diffracted wave cannot be determined.
5. The solution becomes singular at a caustic.

In spite of this, Keller's GTD has had a profound impact on later development of a variety of asymptotic theories of diffraction in electromagnetics.

2.5 UNIFORM THEORIES OF DIFFRACTION

A great amount of research was directed at improving upon the limitations of the GTD described in the previous section. The goal was to obtain uniformity of diffraction solutions at or near the shadow boundaries. Basically, two successful approaches have emerged. The first approach developed by Ahluwalia, Lewis, and Boersma [20] has been named the *uniform asymptotic theory* (UAT) of *diffraction*. This was developed based on the ansatz (assumption) that the field solution in an edge diffraction problem can be expanded in a particular asymptotic series involving a Fresnel integral. A systematic procedure to determine the coefficients of the asymptotic series for all orders of the wave number k is described in [20]. The UAT solution is uniformly valid close to and away from the shadow boundaries. UAT also has been applied to diffraction by curved wedge [21, 22] and recently to the three-dimensional half-plane diffraction [23].

The second, and alternative, popular approach, known as the *uniform theory of diffraction* (UTD), has been developed by Kouyoumjian and Pathak [24] and also is discussed by McNamara, Pistorius, and Malherbe [25]. In the UTD Keller's diffraction coefficient is modified by a multiplication factor involving a Fresnel integral, and the results are extended to the general case of a curved wedge, where at least one surface is curved. The multiplication factor has the property that it approaches zero when the field point approaches the shadow boundary, thereby compensating for the singularity in Keller's diffraction coefficient and leading to a finite field at the shadow boundary. Both UAT and UTD along with their applications are well documented in the literature. The following two subsections present the key analytical features of UAT and UTD.

2.5.1 Uniform Asymptotic Theory

The uniform asymptotic theory of diffraction provides uniformity of the solution at and around shadow boundaries by modifying the geometrical optics term and expressing it as a product of the original Keller's GO term and one consisting of the difference between a Fresnel integral F and a series function $\hat{F}$. The total field satisfies Maxwell's equations asymptotically, and it remains finite at and around the shadow boundaries.

Consider the conducting half-plane diffraction problem shown in Figure 2.3. Assuming a plane wave with $\mathbf{E}^i$ polarized along the edge and incident on the half-plane,

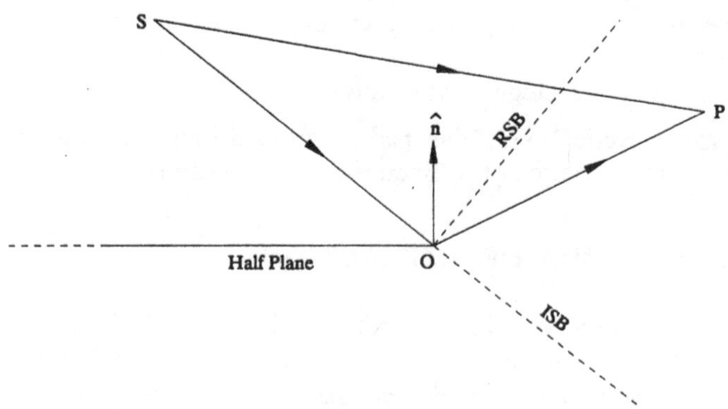

Figure 2.3 Half-plane illuminated by an incident plane wave.

the UAT asymptotically represents the total field $\mathbf{E}^t(r)$ at P as

$$\mathbf{E}^t(\mathbf{r}) = \mathbf{E}^G(\mathbf{r}) + \mathbf{E}^d(\mathbf{r}) + 0(k^{-1}) \tag{2.13}$$

where
$\mathbf{E}^G(\mathbf{r})$ = the new GO field,
$\mathbf{E}^d(\mathbf{r})$ = the diffracted field.

According to the UAT ansatz [20], the dominant term of $\mathbf{E}^G(r)$ has the form

$$\mathbf{E}^G(\mathbf{r}) = \left[F(\xi^i) - \hat{F}(\xi^i)\right]\mathbf{E}^i(\mathbf{r}) + (i \rightarrow r) + 0(k^{-1}) \tag{2.14}$$

with the second term being same as the first with i changed to r, where $(\xi^i)^2$, called the *phase detour parameter*, is defined as the difference between the phase of the diffracted ray via the edge and the direct ray from the transmitter to receiver. The detour parameter ξ^i is the square root of the phase detour and is given by

$$\xi^{i,r}(\mathbf{r}) = \epsilon^{i,r}(\mathbf{r})\left|\sqrt{k_0\left[s^d(\mathbf{r}) - s^{i,r}(\mathbf{r})\right]}\right| \tag{2.15}$$

where $s^d(r)$ and $s^{i,r}(r)$ are the phase paths associated with the incident and diffracted or reflected (i.e., i, r) rays, and $\epsilon^{i,r}(\mathrm{r}) = 1$ for the lit region and 0 for the shadow region.

The Fresnel function F in (2.14) is

$$F(x) = (\pi)^{-1/2}\,\mathrm{e}^{\pm i\pi/4}\int_x^{\infty} \mathrm{e}^{-i\tau^2}\,\mathrm{d}\tau \tag{2.16}$$

for real x. For large x, $F(x)$ can be expanded asymptotically as

$$F(x) \sim H(-x) + \hat{F}(x) + 0(x^{-3}) \tag{2.17}$$

where, $H(x)$ is a unit step function such that

$$H(x) = \begin{cases} 1 & x > 0 \\ 0 & x < 0 \end{cases} \tag{2.18}$$

and $\hat{F}(x)$ is the dominant term of the expansion, given by

$$\hat{F}(x) = \frac{1}{2x\sqrt{\pi}}\,\mathrm{e}^{-(x^2+\pi/4)} \tag{2.19}$$

Note that the key features of UAT are (1) the total field as a function of r is given by a single expression, by (2.13), and it is continuous across both the incident and reflected shadow boundaries; (2) when r is far away from the incident and reflected shadow boundaries and the edge, that is, $|\xi^{i,r}| \to \infty$, the total field given by (2.13) reduces to Keller's form.

On or near the incident of reflection shadow boundaries where ξ^i is small or zero and both $\mathbf{E}^G$ and $\mathbf{E}^d$ become infinite in such a way that their singular parts cancel out each other in (2.17) and $\mathbf{E}^t$ remains finite. This is the basic improvement provided by UAT on Keller's GTD.

2.5.2 Uniform Theory of Diffraction

Figure 2.4 shows the geometry of diffraction by a curved conducting wedge. The diffraction coefficients uniform at and around the shadow boundaries are given by [24]

$$\begin{aligned} D^{s,h}(\varphi, \varphi', \beta_0) = & \frac{-\exp(i\pi/4)}{2n\sqrt{2\pi k}\sin\beta_0} \\ & \times\Bigg(\Bigg\{F\big[kLa^+(\varphi-\varphi')\big]\cot\frac{\pi+(\varphi-\varphi')}{2n} \\ & +F\big[kLa^-(\varphi-\varphi')\big]\cot\frac{\pi-(\varphi-\varphi')}{2n}\Bigg\} \\ & \mp\Bigg\{F\big[kLa^+(\varphi+\varphi')\big]\cot\frac{\pi+(\varphi+\varphi')}{2n} \\ & +F\big[kLa^-(\varphi+\varphi')\big]\cot\frac{\pi-(\varphi-\varphi')}{2n}\Bigg\}\Bigg) \end{aligned} \tag{2.20}$$

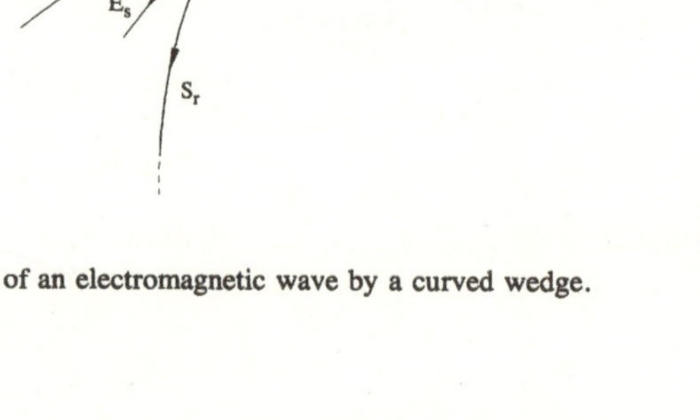

Figure 2.4 Diffraction of an electromagnetic wave by a curved wedge.

where

$$F(x) = 2\mathrm{i}\sqrt{x}\,\exp^{\mathrm{i}x}\int_{\sqrt{x}}^{\infty}\exp^{-\mathrm{i}\tau^2}\,d\tau$$

$$a^{\pm}(\beta) = 2\cos^2\frac{2n\pi N^{\pm} - (\beta)}{2} \tag{2.21}$$

where $N^{\pm}$ are the integers that most nearly satisfy these equations:

$$2\pi n N^{+} - (\beta) = \pi \quad \text{with } \beta = \varphi + \varphi'$$

$$2\pi n N^{-} - (\beta) = \pi \quad \text{with } \beta = \varphi - \varphi'$$

L is a distance parameter defined as

$L \quad = S^2 \sin\beta_0$ for plane wave incidence,

$L \quad = \dfrac{rr'}{r + r'}$ for cylindrical wave incidence,

$L \quad = \dfrac{ss'}{s + s'}\sin^2\beta_0$ for conical and spherical wave incidence,

$r' \quad$ = radius of curvature of the cylindrical wave normally incident on the edge,
$r \quad$ = perpendicular distance of the field point from the edge,
$S \quad$ = distance along the ray from the edge to the observation point,
$N^{\pm}$ = functions of β and n (see [21] for details),
$\beta_0 \quad$ = the smaller angle between the direction of incidence,

and the tangent is drawn at the point of incidence on the edge.

$F(x)$ given by (2.21) is a modified Fresnel integral that may be approximated as follows:

$$F(x) \simeq \left[\sqrt{\pi x} - 2x\exp(\mathrm{i}\pi/4) - \frac{2}{3}x^2\exp(-\mathrm{i}\pi/4)\right]\exp\left[\mathrm{i}\left(\frac{\pi}{4} + x\right)\right] \quad \text{for } x \ll 1 \tag{2.22}$$

$$F(x) \simeq \left(1 + \frac{\mathrm{i}}{2x} - \frac{3}{4}\frac{1}{x^2} - \mathrm{i}\frac{15}{8}\frac{1}{x^2} + \frac{75}{16}\frac{1}{x^4}\right) \quad \text{for } x \gg 1 \tag{2.23}$$

Note that if $x = 10$, (2.23) yields $F(x) \simeq 1 + \mathrm{i}0.05$.

Special cases of a half-plane ($n = 2$) diffraction and the associated scaler diffraction coefficients easily follow from (2.20) and are given by eqs. (52) or (55) in [24, 25].

When the source is very near the edge, the coefficient in (2.20) needs a correction due to the variation of incident field intensity along the edge of the wedge. This is known as ***uniform slope diffraction correction*** [26]. The uniform slope diffraction coefficient is given by [26]

$$
\begin{aligned}
&\frac{\partial D_{s,h}(\phi, \phi'; \beta_0)}{\partial \phi'} \\
&\quad = -\frac{e^{-\pi/4}}{4n^2\sqrt{2\pi k}\,\sin\beta_0} \\
&\quad \times \left(\left\{ \csc^2\left[\pi + \frac{(\phi - \phi')}{2n}\right] F_s[kLa^+(\phi - \phi')] \right.\right. \\
&\quad \left. - \csc^2\left[\frac{\pi - (\phi - \phi')}{2n}\right] F_s[kLa^-(\phi - \phi')] \right\} \\
&\quad \pm \left\{ \csc^2\left[\frac{\pi + (\phi + \phi')}{2n}\right] F_s[kLa^+(\phi + \phi')] \right. \\
&\quad \left.\left. - \csc^2\left[\frac{(\pi - \phi + \phi')}{2n}\right] F_s[kLa^-(\phi + \phi')] \right\} \right)
\end{aligned} \tag{2.24}
$$

where

$$F_s(x) = 2\mathrm{i}x + 4x^{3/2}\,e^{\mathrm{i}x} \int_{\sqrt{x}}^{\infty} e^{-\mathrm{i}\tau^2}\, d\tau$$

or

$$F_s(x) = 2\mathrm{i}x[1 - F(x)]$$

At a shadow boundary, the following limit is used:

$$\csc^2\left(\frac{\pi \pm \beta}{2n}\right) F_s[kLa^{\pm}(\beta)] = 4\mathrm{i}kLn^2\left(1 - \epsilon\sqrt{\frac{\pi kL}{2}}\, e^{\mathrm{i}\pi/4}\right)$$

where

$$\beta = 2\pi n N^{\pm} \mp (\pi - \epsilon)$$

and ϵ is small.

UTD has been successfully applied to many scattering and antenna problems and, in fact, is one of the most popular high-frequency asymptotic methods having applications for practical problems.

2.6 INTEGRAL EQUATION TECHNIQUE

Frequently, electromagnetic scattering problems are formulated as integral equations for the unknown field quantities. In fact, many powerful approximate analytical and popular numerical techniques, like the *method of moments* (MOM), currently used to solve scattering problems are based on integral equations of one kind or another.

Consider an arbitrary scattering body of closed surface S illuminated by some known fields $\mathbf{E}^i$, $\mathbf{H}^i$ as shown in Figure 2.5. The illuminating or incident field may be due to some confined source outside S or an incoming plane electromagnetic wave. In general, a portion of the incident field may reach the field point P, and another also may reach P after being scattered by S. Thus, the total fields at P can be formally written as

$$\begin{aligned} \mathbf{E}(\mathbf{r}) &= \mathbf{E}^i(\mathbf{r}) + \mathbf{E}^s(\mathbf{r}) \\ \mathbf{H}(\mathbf{r}) &= \mathbf{H}^i(\mathbf{r}) + \mathbf{H}^s(\mathbf{r}) \end{aligned} \tag{2.25}$$

where $\mathbf{E}^i(r)$, $\mathbf{H}^i(r)$ are the incident electric and magnetic fields at P in the absence of the scatterer, and $\mathbf{E}^s(r)$, $\mathbf{H}^s(r)$ are the scattered electric and magnetic fields at P.

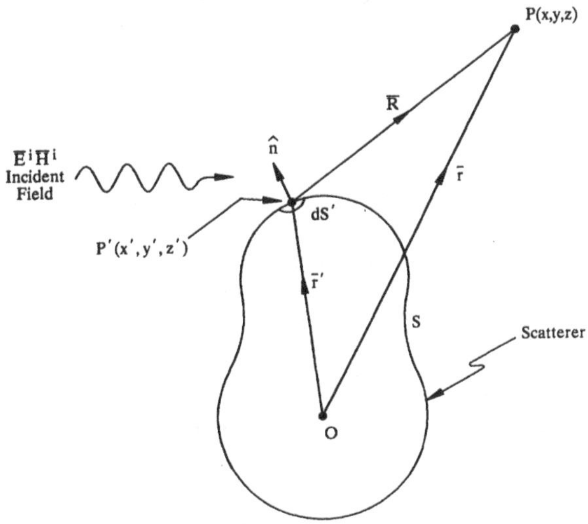

Figure 2.5 Geometry of scattering.

We can show that the scattered fields at P are given by [27, 28]

$$\mathbf{E}^s(r) = \mathbf{E}(r) - \mathbf{E}^i(r)$$
$$= \frac{1}{4\pi} \oint_S \{-i\omega\mu[\hat{n} \times \mathbf{H}(\mathbf{r}')]G + [\hat{n} \times \mathbf{E}(\mathbf{r}')]\nabla' G$$
$$+ [\hat{\mathbf{n}} \cdot \mathbf{E}(\mathbf{r}')]\nabla' G\} \, ds' \tag{2.26}$$

$$\mathbf{H}^s(\mathbf{r}) = \mathbf{H}(\mathbf{r}) - \mathbf{H}^i(\mathbf{r})$$
$$= \frac{1}{4\pi} \oint_S \{i\omega\epsilon[\hat{n} \times \mathbf{E}(\mathbf{r}')]G + [\hat{n} \times \mathbf{H}(\mathbf{r}')] \times \nabla' G$$
$$+ [\hat{n} \cdot \mathbf{H}(\mathbf{r}')]\nabla' G\} \, ds' \tag{2.27}$$

with

$$G = \frac{e^{-ikR}}{R} \tag{2.28}$$

$$R = \left[(x - x')^2 + (y - y')^2 + (z - z')^2\right]^{1/2} \tag{2.29}$$

The term ∇' represents operation with respect to the primed coordinates, and $P(x, y, z)$, $P'(x', y', z')$ are the field and source points, and all other notations are as explained in Figure 2.5. Equations (2.26) and (2.27) known as *Stratton-Chu equations* indicate that the scattered fields can be completely determined from a knowledge of the total tangential and normal components of the fields at the surface of the scatterer. In the important case of a conducting scatterer, where $\hat{n} \times \mathbf{E} = 0$ on S, (2.27) indicates that only the tangential $\mathbf{H}$ on S is sufficient for the determination of $\mathbf{H}^s$, but (2.26) indicates that it still requires knowledge of the normal $\mathbf{E}$ on S. However, $\mathbf{E}^s$ can be obtained from $\mathbf{H}^s$ through use of Maxwell's equation, thereby making the knowledge of $\hat{n} \cdot \mathbf{E}$ on S unnecessary.

An alternative integral formulation by Franz [29] gives the scattered field as

$$\mathbf{E}^S(\mathbf{r}) = \frac{1}{i\omega\epsilon} \nabla X \nabla X \frac{1}{4\pi} \oint_S [n \times \mathbf{H}(\mathbf{r})] G \, ds'$$
$$+ \nabla X \oint_S [\hat{n} \times \mathbf{E}(\mathbf{r}')] G \, ds' \tag{2.30}$$

$$\mathbf{H}^S(\mathbf{r}) = \nabla X \oint_S [\hat{n} \times \mathbf{H}(\mathbf{r})] G \, ds'$$
$$- \frac{1}{i\omega\mu} \nabla X \nabla X \oint_S [\hat{n} \times \mathbf{E}(\mathbf{r}')] G \, ds' \tag{2.31}$$

Franz formulation has the advantage that only the tangential components of E and H on S suffice to determine the scattered fields at any point.

In (2.26), (2.27), (2.30), and (2.31), it is customary to interpret the tangential magnetic and electric fields on S as the surface electric and magnetic currents, respectively. The physical optics approximation assumes that the surface fields (currents) exist only on the illuminated portion of S and are zero otherwise, and then determines the scattered fields by using the preceding equations with the required surface fields obtained by tangent approximation. For example, in a conducting body the surface current at a point P located on the illuminated portion of S is assumed to be

$$\begin{aligned} \mathbf{J}_S(\mathbf{r}) &= \hat{\mathbf{n}} \times \left[\mathbf{H}^i(\mathbf{r}') + \mathbf{H}^S(\mathbf{r}')\right] \\ &= 2\hat{n} \times \mathbf{H}^i(\mathbf{r}') \end{aligned} \tag{2.32}$$

The scattered field in the far zone ($r \to \infty$ in Figure 2.5) obtained under this approximation is as given earlier (see (2.9) and (2.10)).

2.6.1 Integral Equations

For a given field $\mathbf{E}^i, \mathbf{H}^i$ incident on a conducting body having a smooth surface, the induced surface current density $\mathbf{J}_S$ can be determined by solving one of two integral equations, commonly known as the *electric field integral equation* (EFIE) and the *magnetic field integral equation* (MFIE) [30, 31]. The two equations can be derived from Stratton-Chu or Franz equations in conjunction with the boundary conditions.

$$\hat{n} \times (\mathbf{E}^i + \mathbf{E}^S) = 0$$

$$\hat{n} \times (\mathbf{H}^i + \mathbf{H}^S) = \mathbf{J}_S$$

Using the boundary condition on the electric field and (2.30) we obtain the EFIE [30]:

$$\hat{n} \times \mathbf{E}^i(\mathbf{r}) = \frac{\hat{n}}{2\pi i\omega\epsilon} \times \oint_S \left[-\omega^2\mu\epsilon\mathbf{J}_S(\mathbf{r}')G + \left[(\nabla' \cdot \mathbf{J}_s(r')G\right] ds' \tag{2.33}$$

and by using the boundary condition on the magnetic field and (2.31) we obtain the MFIE [30, 31]:

$$\mathbf{J}_S(\mathbf{r}) = 2\hat{n} \times \mathbf{H}^i(\mathbf{r}) + \hat{n} \times \frac{1}{2\pi} \oint_S \left[\mathbf{J}_S(\mathbf{r}') \times \nabla' G\right] ds' \tag{2.34}$$

where the integrals on the right-hand side of (2.33) and (2.34) are to be interpreted in the principal value sense. The MFIE is also referred to as *Maue's integral equation*

[31]. The EFIE is an integral equation of the first kind, where the unknown current appears only inside the integral; whereas the MFIE is an integral equation of the second kind, where the unknown current appears both inside and outside the integral.

Except for frequencies corresponding to the internal body resonances either EFIE or MFIE can be used to determine the surface current, although the MFIE has been found to be popular and more convenient for smooth bodies. However, for thin wires and thin cylinders the EFIE must be used.

When the radius of curvature of S is large compared to the operating wavelength, we can show that the dominant contribution to $\mathbf{J}_S(r)$ in (2.34) comes from the first term on the right-hand side, and the contribution from the second term generally is small. Thus, neglecting the second term in (2.34), we obtain the tangent approximation used in the physical optics approximation:

$$\mathbf{J}_S(\mathbf{r}) \simeq \mathbf{J}_S^0 = 2\hat{n} \times \mathbf{H}^i(\mathbf{r}) \tag{2.35}$$

where $\mathbf{J}_S^0$ is referred to as the *uniform part* of the current induced by the incident field. The basis of the physical theory of diffraction, discussed later, is (2.34), where attention is now devoted to analytical determination of the nonuniform part of the current due to the second term on the right-hand side.

2.6.2 Method of Moments

For many practical problems the desired current distributions are obtained by numerically solving the integral equations discussed in the previous section. The method of moments is one of the most popular techniques used for such a purpose. The MOM essentially converts the integral equation to a set of algebraic equations; that is to a matrix equation that can be solved by standard matrix inversion algorithms. Because the method of moments as applied to electromagnetic scattering and antenna problems is adequately discussed in the classic treatise by Harrington [32], here we shall only briefly outline the basic steps involved in the analysis.

Let us rewrite the MFIE for the unknown current density $\mathbf{J}_S(r)$ as an inhomogeneous equation in operational form:

$$L(\mathbf{J}_S) = 2\hat{n} \times \mathbf{H}^i(\mathbf{r}) \tag{2.36}$$

where the right-hand side is known and $L(J_S)$ is an integrodifferential linear operator defined as

$$L(\mathbf{J}_S) = \mathbf{J}_S(\mathbf{r}) - \hat{n} \times \frac{1}{2\pi} \oint_S \mathbf{J}_S(\mathbf{r}') X \nabla' G \, ds' \tag{2.37}$$

To simplify the discussion we shall describe the solution of the inhomogeneous scalar equation:

$$L(f) = g \tag{2.38}$$

where $f = f(x)$ is the unknown function to be determined, $g(x)$ is a known function, and L is a linear operator. Although (2.38) is much simpler than (2.36), the basic steps involved in its solution will be applicable as well for the solution of (2.36). The steps (1 to 7) involved in obtaining the solution of (2.38) are as follows:

1. Expand f as

$$f = \sum_{n=1}^{N} \alpha_n f_n \tag{2.39}$$

where the values of α_n are unknown complex coefficients, and those of f_n are known functions of x, called *expansion* or *basis functions*.

2. Introduce (2.39) into (2.38) to obtain

$$\sum_{n=1}^{N} \alpha_n L(f_n) = g \tag{2.40}$$

3. Define a suitable inner product $\langle f, g \rangle$ defined in the range L of x:

$$\langle f, g \rangle = \int_L f(x) g(x)\, \mathrm{d}x \tag{2.41}$$

4. Define a set of weighting or testing functions w_m, $m = 1, 2, \ldots, N$ in the range L. Take the inner product of (2.40) with each w_m and obtain

$$\sum_{n=1}^{N} \alpha_n \langle w_m, L(f_n) \rangle = \langle w_m, g \rangle \tag{2.42}$$

with $m = 1, 2, \ldots, N$.

5. Express the set of algebraic equations given by (2.42) in the matrix form:

$$[l_{mn}][\alpha_n] = [g_m] \tag{2.43}$$

where $[l_{mn}]$ is the matrix

$$[l_{mn}] = \begin{bmatrix} \langle W_1, L(f_1)\rangle & \langle W_1, L(f_2)\rangle & \cdots & \langle W_1, L(f_N)\rangle \\ \langle W_2, L(f_1)\rangle & \langle W_2, L(f_2)\rangle & \cdots & \langle W_2, L(f_n)\rangle \\ \vdots & \vdots & & \vdots \\ \langle W_N, L(f_1)\rangle & \langle W_N, L(f_2)\rangle & \cdots & \langle W_N, L(f_N)\rangle \end{bmatrix}$$

Here α_n and g_n are the column matrices given by

$$[\alpha_n] = \begin{bmatrix} \alpha_1 \\ \alpha_2 \\ \vdots \\ \alpha_N \end{bmatrix}$$

$$[g_m] = \begin{bmatrix} \langle W_1, g\rangle \\ \langle W_2, g\rangle \\ \vdots \\ \langle W_N, g\rangle \end{bmatrix}$$

6. Assuming $[l_{mn}]$ is not singular, use the standard matrix inversion algorithm to determine $[\alpha_n]$ from (2.43):

$$[\alpha_n] = [l_{mn}]^{-1}[g_m] \tag{2.44}$$

7. With knowledge of α_n, determine the unknown function f by using (2.39).

In the preceding we have assumed that the basis functions and the weighting functions are different. Where the two functions are assumed to be the same, the solution is referred to as *Galerkin's solution*.

Although the numerical method just outlined is straightforward, due to the required matrix inversion involved the available computer capacity severely restricts the size of the problem that can be handled. Generally, we know that the MOM method can provide solutions for bodies having a maximum dimension of a few wavelengths.

2.7 PHYSICAL THEORY OF DIFFRACTION

An alternative way of solving the edge discontinuity problems associated with scattering calculations without using an edge diffraction coefficient or a uniform asymptotic expansion is the *physical theory of diffraction* (PTD) discovered by Ufimtsev. A summary of PTD is available in [10, 33]. Ufimtsev observed that the total surface current on a conducting surface is equal to the sum of two currents given by

$$\mathbf{J}_S^t = \mathbf{J}_S^0 + \mathbf{J}_S^1 \tag{2.45}$$

where $\mathbf{J}_S^0$ is the physical optics current discussed earlier, in (2.35), and called the *uniform part* of the total current; $\mathbf{J}_S^1$ is the fringe current or nonuniform part of the current due to a discontinuity in some form.

Once $\mathbf{J}_S^1$ is estimated, $\mathbf{J}_S^t$ then becomes a better approximation to the exact total surface current than $\mathbf{J}_S^0$. This leads to more accurate results than provided by the PO approximation alone. The advantage of the PTD approximation is that, for a given finite total current, the fields would be finite everywhere including the incident and reflection shadow boundaries and also on the caustic. The disadvantages of the method are the need to estimate $\mathbf{J}'_S$ for wedge diffraction, and the additional task of finding the fields from the current. The fringe field $\mathbf{E}^1$ is formally given by

$$\mathbf{E}^1(\mathbf{r}) = \int_{\Sigma_{12}} \overline{\overline{G}} \cdot \mathbf{J}'_S \, ds' \tag{2.46}$$

where Σ_{12} is the surface involved to produce J'_S, and $\overline{\overline{G}}$ is the dyadic Green's function associated with the geometry.

In field calculations, Ufimtsev uses only the dominant term of the order $k^{-1/2}$ [34]. Generally, determination of $\mathbf{E}'(\mathbf{r})$ is quite involved and will not be discussed here.

2.8 FIELD ESTIMATION AT A CAUSTIC

Due to their basic nature of formulation, the ray techniques (GTD, UAT, and UTD) encounter a singularity at the focal plane of a structure and consequently fail to provide correct and finite fields at the caustics. Examples of caustics are the axial direction of a circular aperture and a disk or a ring slope discontinuity on a body of revolution. The fields at the caustic have to be estimated separately; and there are different ways of achieving this.

First, by comparing the asymptotic expansion of the scalar wave equation with the divergent result obtained after using the diffraction coefficient and ray optics, Keller developed [35] a correction factor. The GTD solution when multiplied by this correction factor yields a reasonable estimate of the fields at the caustic.

The second method uses the PO method described in Section 2.3.

The third method uses the edge current method [36]. This method directly incorporates the equivalent electric and magnetic currents to remove the singularity from the GTD solution. We need not write a new program for each shape. For example, consider a plane wave having its E-field parallel to the xy plane incident on a circular metallic disk oriented in the xy plane, as shown in Figure 2.6. We can show that the equivalent electric and magnetic edge currents I^e and I^m, respectively, are given by [36]

$$I^e = \mathrm{i}\frac{2}{\eta_0 k}\left[T(n,\psi^-) - T(n,\psi^+)\right]\left(-E_x^i \sin\varphi' + E_y^i \cos\varphi'\right) \mathrm{e}^{\mathrm{i}ka\sin\theta\cos\varphi} \tag{2.47}$$

$$I^m = \mathrm{i}\frac{2\eta_0}{k}\left[T(n,\psi^-) + T(n,\psi^+)\right]\left(H_y^i \cos\varphi' - H_z^i \sin\varphi'\right) \mathrm{e}^{\mathrm{i}ka\sin\theta\cos\varphi} \tag{2.48}$$

where

$$T(n,\psi^{\pm}) = \frac{1}{n}\sin\frac{\pi}{n}\left(\frac{1}{\cos\pi/n - \cos\psi^{\pm}/n}\right) \tag{2.49}$$

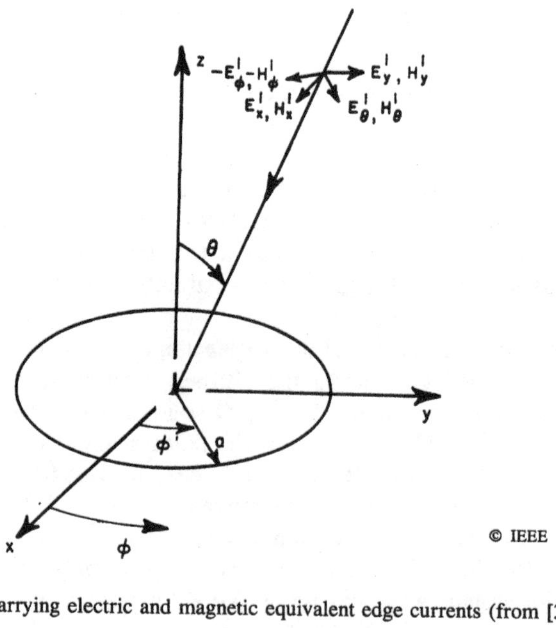

Figure 2.6 Loop of wire carrying electric and magnetic equivalent edge currents (from [36]).

where n is the wedge angle defined earlier, and E_x^i, E_y^i, H_y^i, H_z^i are the incident field component values at the point of incidence on the edge. These currents do not include GO terms. For practical situations, the currents are used along the rim of the aperture or disk to determine the desired caustic fields. For example, the fields backscattered by a circular loop illuminated by a plane wave, shown in Figure 2.6, may be obtained by using an electric current I^e along the loop. The fields are then given by

$$E_\theta^r = \frac{-\mathrm{i}\omega\mu a \cos\theta \sin\varphi}{4\pi r} \mathrm{e}^{-\mathrm{i}kr} \int_0^{2\pi} I^e(\varphi') \sin(\varphi - \varphi')\, \mathrm{e}^{\mathrm{i}ka\cos(\varphi-\varphi')}\, \mathrm{d}\varphi' \quad (2.50)$$

$$E_\varphi^r = -\frac{\mathrm{i}\omega\mu a \sin\theta}{4\pi r} \mathrm{e}^{-\mathrm{i}kr} \int_0^{2\pi} I^e(\varphi') \cos(\varphi - \varphi')\, \mathrm{e}^{\mathrm{i}ka\cos(\varphi-\varphi')}\, \mathrm{d}\varphi' \quad (2.51)$$

The fields of the loop carrying magnetic current can be found by the principle of duality [32].

The *E*-plane RCS pattern of a conically capped metallic cylinder, calculated by using the equivalent edge currents, is shown in Figure 2.7. The results do not consider higher-order diffraction. The axial RCS results for finite cones obtained by using equivalent edge currents and higher-order diffraction terms are available in [37].

2.9 HYBRID TECHNIQUES

So far, we have discussed a number of high-frequency approximate analytical methods for solving electromagnetic scattering problems and described a popular numerical method, the MOM, that is commonly used to obtain numerical solution of a variety of problems. Each of these methods has its own range of application along with certain advantages and disadvantages. Hybrid techniques generally combine the advantage of one method with that of another to determine the solution of a problem efficiently. A number of such methods [38–41] have been proposed. We shall briefly describe two of them here.

The MOM is limited to bodies having a maximum dimension of usually less than few wavelengths, whereas the GTD generally yields good results for bodies much larger than those of one wavelength. Hence, the two methods can be combined to efficiently solve radiation and scattering problems involving bodies of intermediate size, and sizes in the resonance region and beyond. The method proposed in [39] combines MOM and GTD to determine the impedance matrix associated with a radiation or scattering problem. The moment method generally obtains an impedance matrix for a smooth surface. When the smooth surface contains some discontinuity, the usual MOM matrix is modified by using GO or GTD. Details of the method are given in [39], where the proposed method was found to work well for antennas like a monopole at the center of a conducting disk, a monopole near a conducting wedge, and a monopole at the

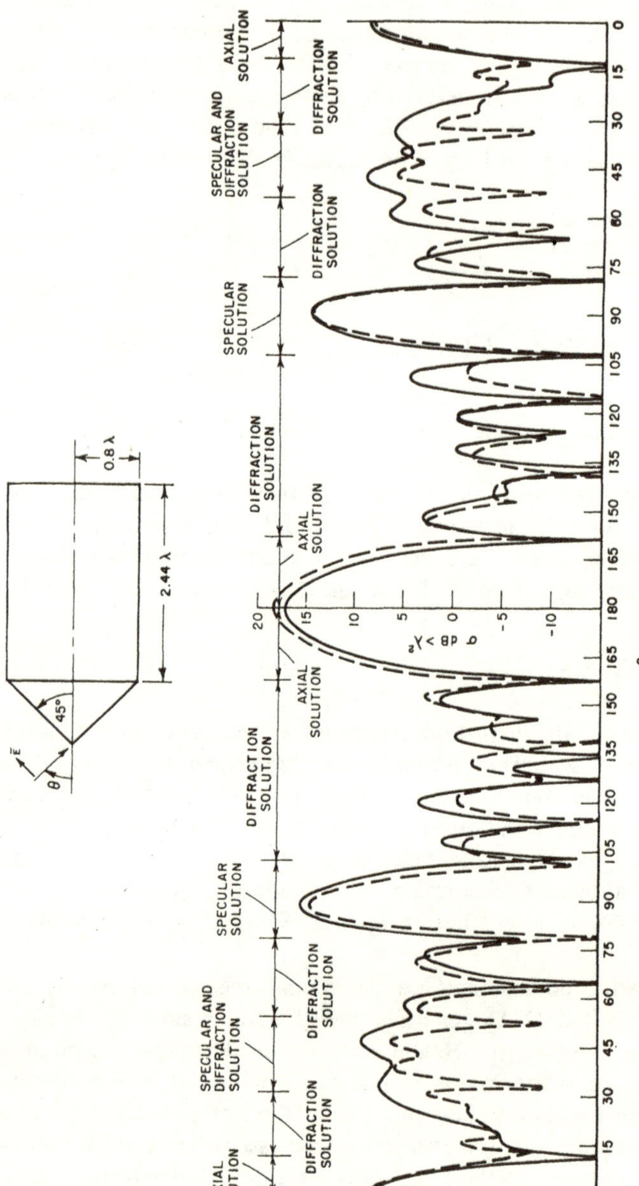

Figure 2.7 E-plane RCS of a conically capped cylinder (from [36]).

center of a four-sided or an eight-sided conducting plate. Apparently, no specific results are available for RCS problems.

Numerical solution of electromagnetic scattering problems has been obtained by combining GTD and MOM in another fashion [3]. We shall illustrate the method by considering the scattering of plane waves by a 90°-angle conducting wedge structure shown in Figure 2.8. In this method [39], the MOM currents are postulated around the edge and GTD field is matched at certain points. This substantially reduces the number of MOM current samples required for the solution. The procedure leads to a set of simultaneous equations that when solved provide the complex amplitudes of the MOM currents and the GTD numerical coefficients.

Consider a plane wave having its H-field polarized in the z-direction incident on a conducting right-angled wedge as in Figure 2.8. Under GTD approximation the total current on the wedge surfaces is given by

$$\begin{aligned} \mathbf{J}^{\mathrm{GTD}} &= (\hat{n} \times \hat{z})\left[\mathbf{H}^{i} + \mathbf{H}^{r} + \mathbf{H}_{z}^{d}\right] \\ &\simeq \mathbf{J}^{i} + \mathbf{J}^{r} + (\hat{n} \times \hat{z}) C \frac{e^{-ik\rho}}{\sqrt{\rho}}, \quad \text{for } \rho \geq \lambda/4 \end{aligned} \tag{2.52}$$

where

H^{i}, H^{r}, H^{d} = the amplitude of the incident, reflected, and diffracted H-fields on the wedge surface,

$\mathbf{J}^{i}, \mathbf{J}^{r}$ = the surface currents due to the incident and reflected parts of the H-field,

C = an unknown constant to be determined, it represents the diffraction coefficients D_x and D_y in the present case, the other parameters are as shown in Figure 2.8.

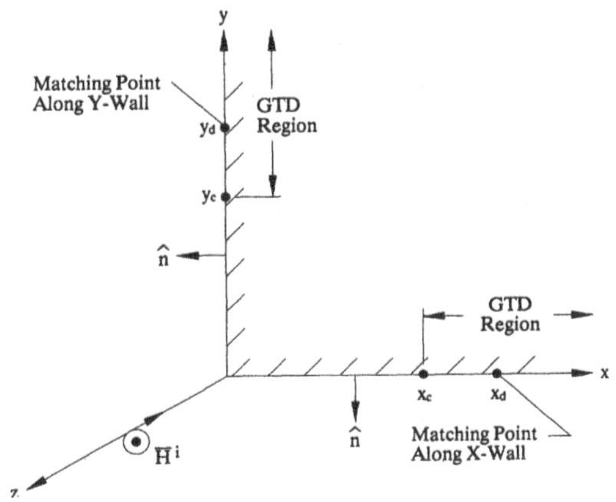

Figure 2.8 Geometry of plane wave scattering by a right-angled conducting wedge.

The appropriate MFIE for the total surface current $\mathbf{J}_S$ is [32]

$$\mathbf{J}_S = -\left[\mathbf{H}_z^i + \hat{z} \cdot \nabla X \int_S \mathbf{J}_S \frac{H_0^{(2)}(k\,|\,\rho - \rho'\,|)}{4i}\,\mathrm{d}l'\right] \tag{2.53}$$

where $H_0^{(2)}$ is the standard notation for the zeroth order Hankel function of the second kind. In component form the total surface current density is assumed to be

$$\begin{aligned} J_S &= J_y^{\mathrm{MOM}} && 0 \le y \le y_c,\ x = 0 \\ &= J_y^{\mathrm{GTD}} && y_c \le y \le \infty,\ x = 0 \\ &= J_x^{\mathrm{MOM}} && 0 \le x \le x_c,\ y = 0 \\ &= J_x^{\mathrm{GTD}} && x_c \le x \le \infty,\ y = 0 \end{aligned} \tag{2.54}$$

where it can be seen that the currents are obtained by the MOM in the ranges $0 \le x \le x_c$ and $0 \le y \le y_c$, and the GTD currents given by (2.52) are valid away from the diffracting edge, that is, beyond these two ranges.

Now, MOM components of currents around the edge can be defined by simple basis functions such as

$$J^{\mathrm{MOM}} = \sum_{m=1}^{N} \alpha_m P_m(l - l_m) \tag{2.55}$$

where $P_m(l - l_m)$ are orthogonal pulse functions with weight α_m. By point matching, we get a set of simultaneous equations:

$$L_n(J_S) = \sum_{m=1}^{N} l_{mn}\alpha_m + L_n\left(J_x^{\mathrm{GTD}}\right) + L_n\left(J_y^{\mathrm{GTD}}\right) = g_n, \quad \text{for } 1 \le n \le N \tag{2.56}$$

where $L_n(J^{\mathrm{GTD}})$, l_{mn}, and g_n are given in [39].

To determine the two diffraction coefficients D_x and D_y associated with J_x^{GTD} and J_y^{GTD}, respectively (represented by C in (2.52)), two additional matching points at $(X_d, 0)$ and $(0, Y_d)$ in Figure 2.8, both located in the GTD region, are chosen. This provides two more equations similar to (2.56). In this manner $(N + 2)$ equations are obtained for $(N + 2)$ unknown constants (N current pulse coefficients in (2.58) plus the two diffraction coefficients).

This technique has been applied to a number of problems [39]. Figure 2.9 (a)–(d) show the current along the x-wall of a right-angled metallic wedge illuminated by a plane wave. Figure 2.10 shows the current around the cross-sectional circumference and the far-zone scattered field in the cross-sectional plane for a $2\lambda \times 2\lambda$ square metallic cylinder of infinite length illuminated by a plane wave as shown in the inset. Figure 2.11 compares the theoretical and experimental results for the 30° bistatic RCS

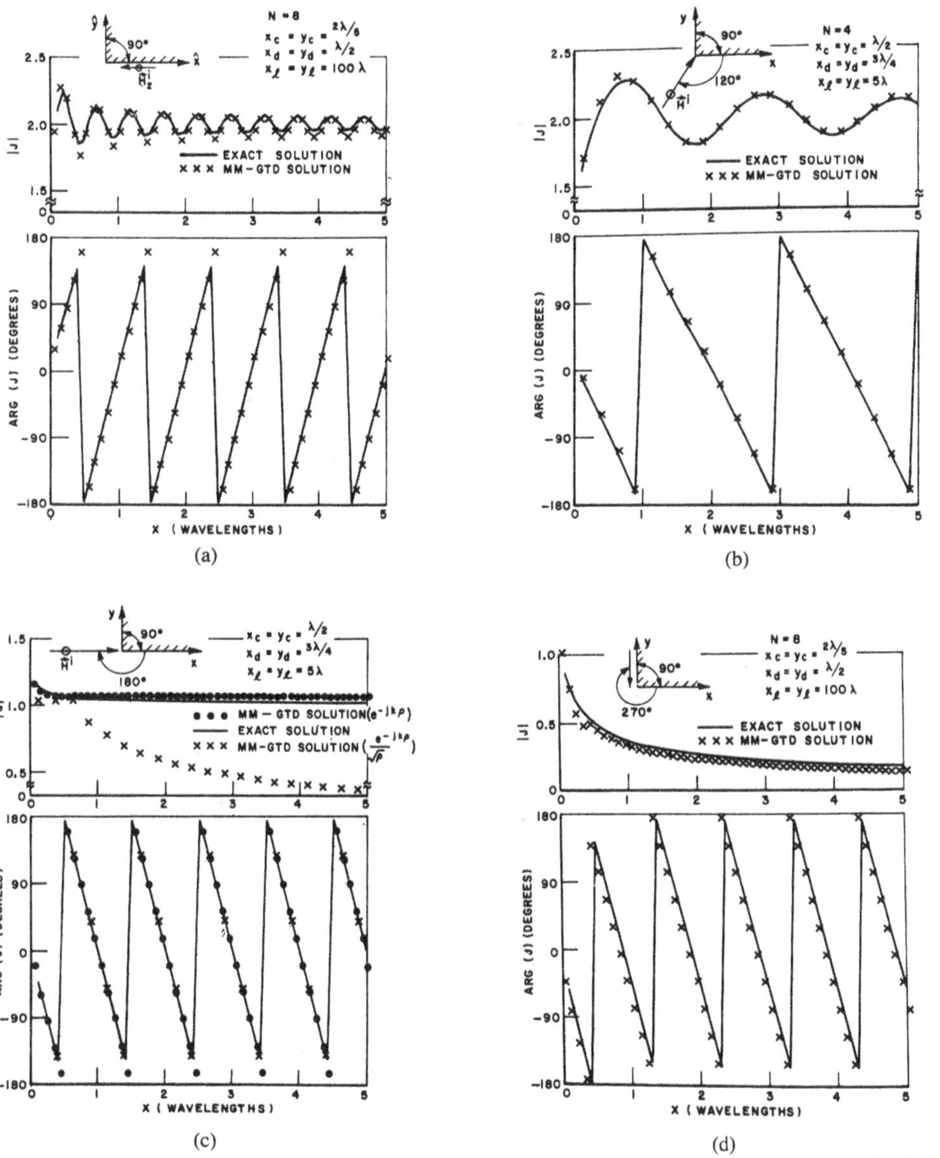

Figure 2.9 Current along x-wall of 90° wedge (from [4]): (a) $N = 8$, $x_c = y_c = 2\lambda/5$; $x_d = Y_d = \lambda/2$ $x_1 = y_1 = 100\lambda$; (b) $N = 4$, $x_c = y_c = \lambda/2$; $x_d = y_d = 3\lambda/6$; (c) $N = 8$, $x_c = y_c = \lambda/2$; $x_d = y_d = 3\lambda/4$; $x_1 = y_1 = 5\lambda$; (d) $N = 8$, $x_c = y_c = 2\lambda/5$; $x_d = y_d = \lambda/2$; $x_1 = y_1 = 100\lambda$.

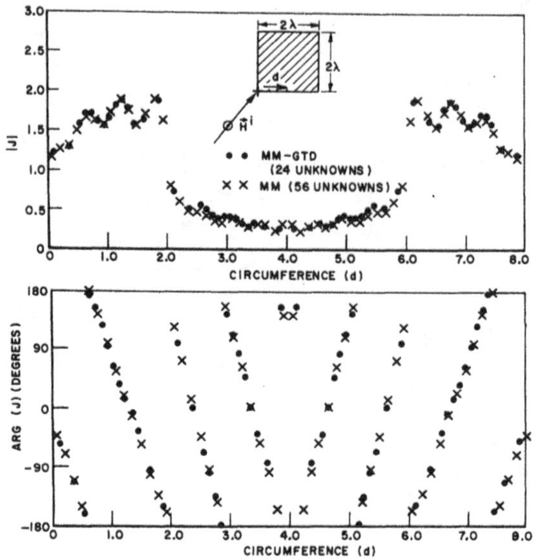

(a)

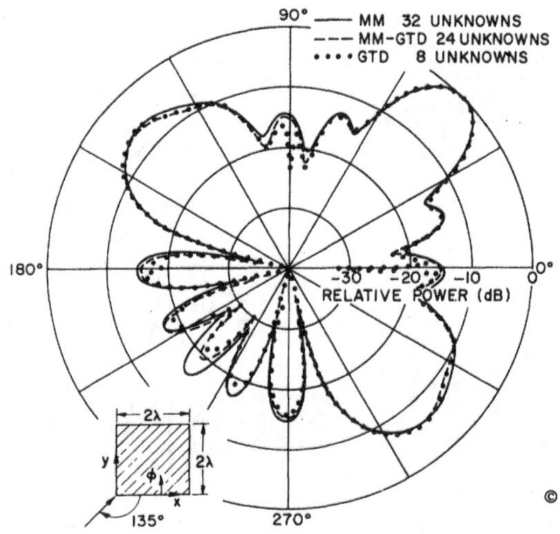

(b)

Figure 2.10 (a) Current around the circumference of a square cylinder; (b) far-zone scattered field by 2 λ square cylinder (from [4]).

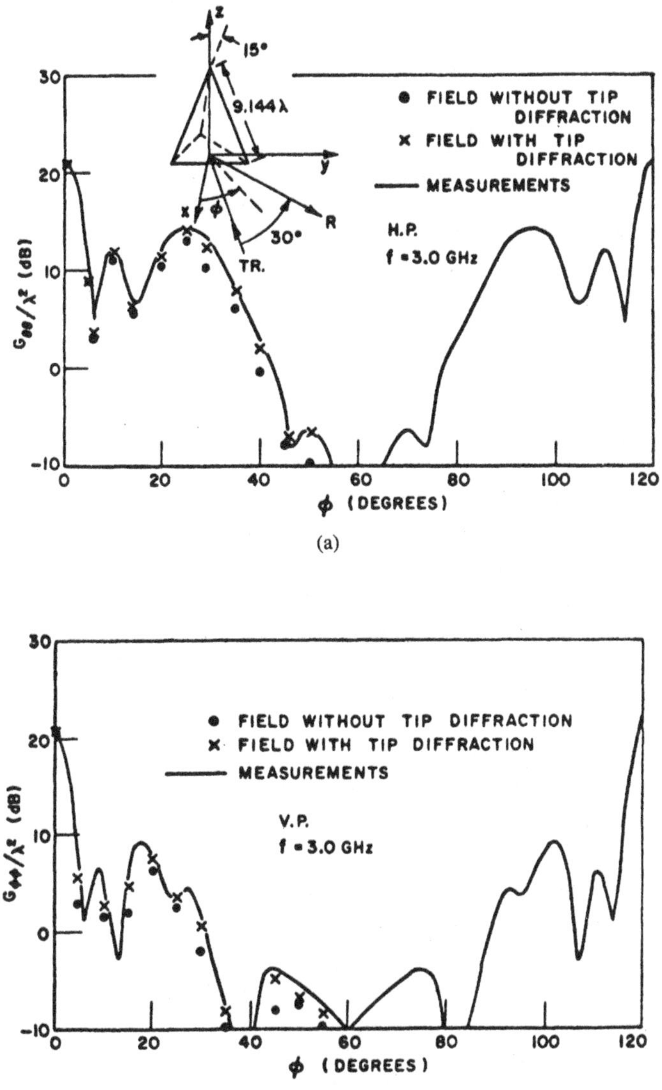

Figure 2.11 Bistatic (30°) RCS of a three-sided pyramid: (a) horizontal and (b) vertical polarization (from [5]).

of three-sided metallic pyramid. The results indicate clearly that the technique works well not only for two-dimensional but also for three-dimensional bodies.

2.10 RADAR CROSS SECTIONS OF TARGETS

This section reviews the available analytical results related to electromagnetic scattering from and RCS of simple and complex targets, including the results obtained from complex target modeling. Wherever possible, theoretical and experimental results are compared. All theoretical results have been obtained by application of the techniques described in the previous sections. For reference, theoretical expressions for the RCS of a number of targets are also given without derivation.

2.10.1 Simple Targets

2.10.1.1 Wire Scatterers

Scattering from metallic wire scatterers are of interest either as single scatterers or as a random cloud of wire scatters commonly called *chaff*, under the ideal zero-coupling condition. Information on the monostatic cross section is used to determine the ability of chaff to mask targets and on bistatic chaff either for masking or communication analysis. Chaff also can be used to support communication beyond the horizon.

Chang and Liepa [41] carried out a comprehensive experimental study of RCS of thin wires; some of their significant results are discussed in [8, p. 168] and will not be reproduced here.

The broadside and average backscattering cross sections of straight wires with length-to-diameter ratio of 100 *versus* normalized wire length are shown in Figure 2.12, and the analytical expressions for these results are available in [42]. Borison [43] developed a generalized bistatic scattering model for a randomly oriented dipole. The ensemble average cross section for such a dipole is given by

$$\langle \alpha \rangle = \frac{\sigma_0(kl)}{15}\left[1 + 2(\cos \alpha_t \cos \alpha_r + \cos \beta \sin \alpha_t \sin \alpha_r)\right] \qquad (2.57)$$

where l is the length of the dipole, and α_t, α_r, and β are as explained in Figure 2.13. Palermo and Banner [44] estimated average bistatic cross sections for chaff dipoles for three distributions of dipole orientations: uniform, horizontal, and vertical. Figure 2.14 shows the average theoretical values of bistatic RCS for the average variations, and they found that maximum bistatic RCS occurs for low elevation angles but drops off at large elevation. Measured and theoretical average RCS results for horizontal distributions are shown in Figure 2.15, which indicates that the two results agree within about 3 dB.

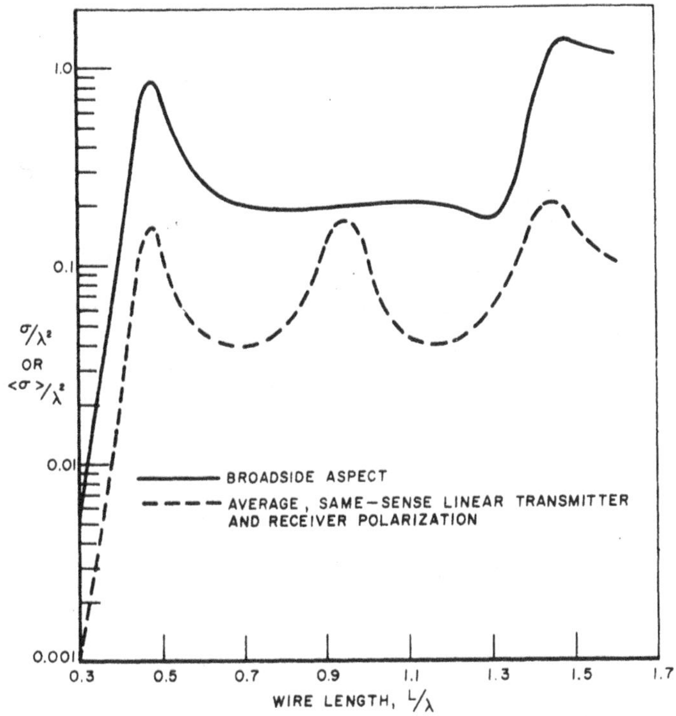

Figure 2.12 Broadside and average backscattering cross section of straight wires with length-to-diameter ratio: $1/2a = 12.1300$ (from [42]).

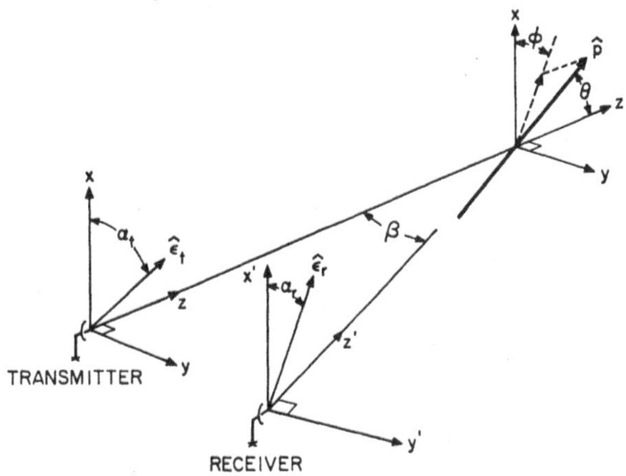

Figure 2.13 Bistatic scattering by a randomly oriented dipoles (from [43]).

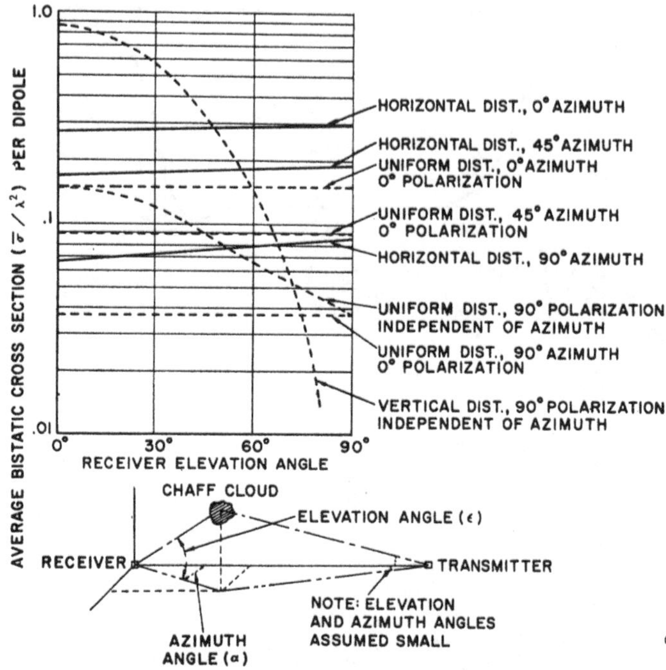

Figure 2.14 Theoretical values of average bistatic cross section (from [44]).

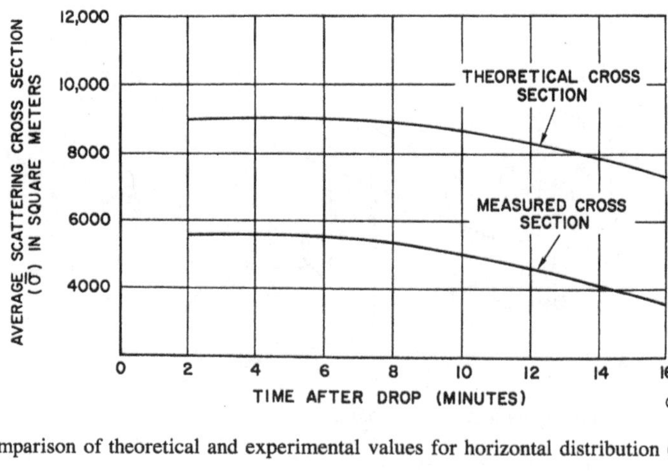

Figure 2.15 Comparison of theoretical and experimental values for horizontal distribution (from [44]).

Scattering cross sections of thin wire scatterers also have been determined [32] numerically by applying the method of moments to the standard equations appropriate for such scatterers. The broadside backscattering cross section as a function of a normalized length for a straight wire with a length-to-diameter ratio of 72.4 is shown in Figure 2.16, which also shows the results obtained by an alternate method. The bistatic scattering cross section patterns for a straight wire with $l/2a = 74.2$ and $l = 2\lambda$ are shown in Figure 2.17, where the results were obtained numerically by the method of moments [32]. In Figure 2.17 the arrow on each curve indicates the angle of incidence, and semicircular coordinate line represents a cross section of $0.5\lambda^2$.

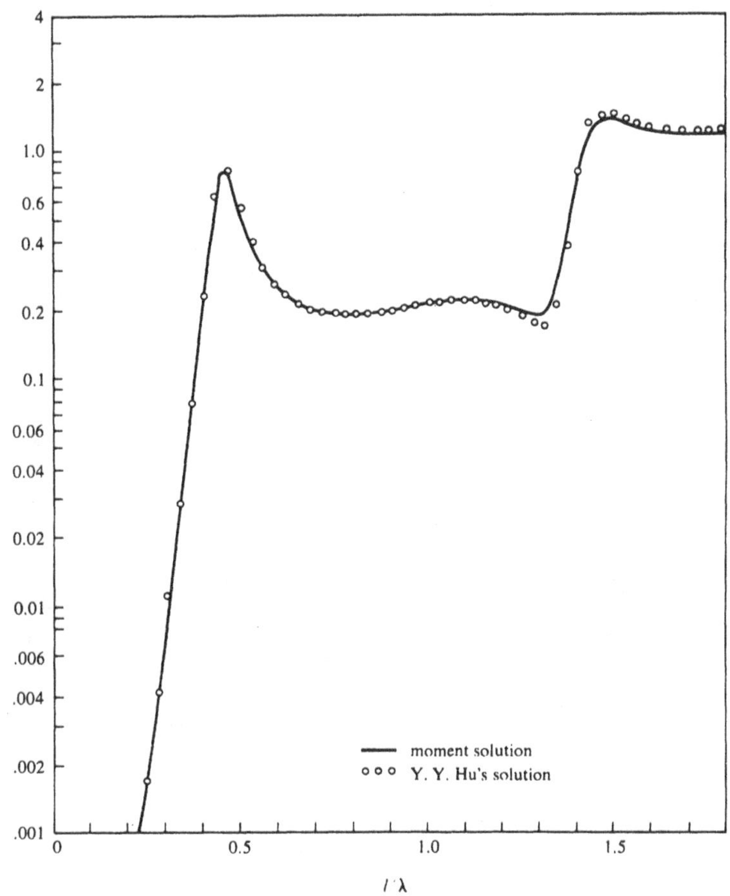

Figure 2.16 Broadside backscattering cross section for a straight wire: $1/2a = 74.2$ (from [32]).

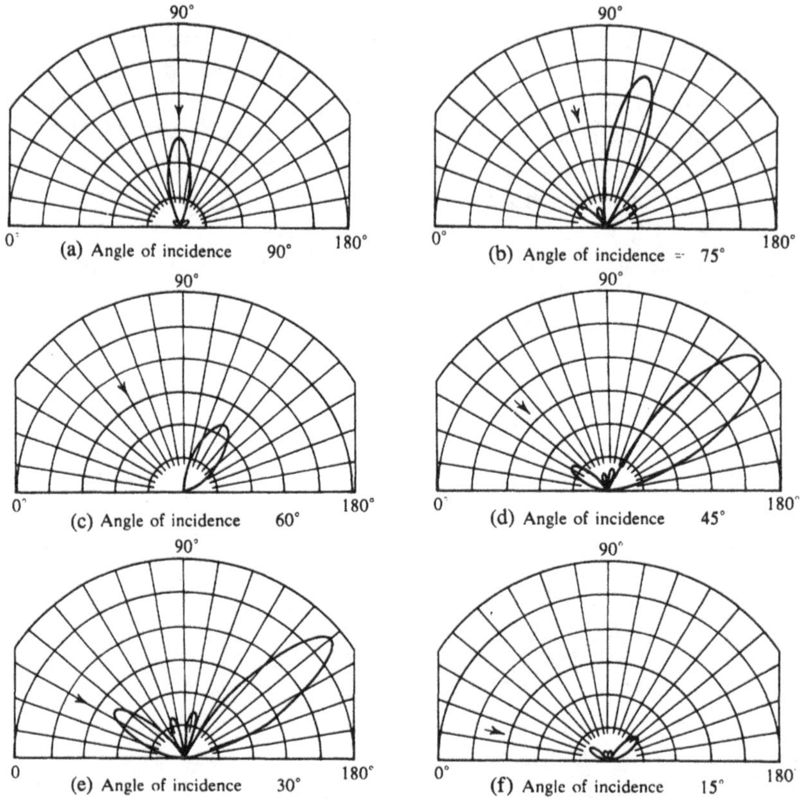

Figure 2.17 Bistatic scattering cross section patterns for a straight wire scatterers: $1/2a = 74.2$, $1 = 2\lambda$ *versus* angle of incidence (from [32]).

2.10.1.2 Wire Loops

The RCS of a thin wire loop can be found [7, Chap. 5] by integrating scattering from a straight piece of thin wire over the complete loop of radius placed in the *xy* plane shown in Figure 2.6. Without entering into the details, the RCS σ is given by [7]

$$\begin{aligned}\sigma &= \frac{1}{\pi}\left| a\int_u^{2\pi} \exp(\mathrm{i}2ka\sin\theta\cos\varphi')\left(\cos^2\theta\sin^2\varphi'\sin^2\gamma\right.\right.\\ &\qquad \left.\left.+2\cos\theta\sin\varphi'\cos\varphi'\sin\gamma\cos\gamma+\cos^2\varphi'\cos^2\gamma\right)\mathrm{d}\varphi'\right|^2 \\ &= \pi a^2\left|\left(\cos^2\theta\sin^2\gamma+\cos^2\gamma\right)J_0(2ka\sin\theta)\right.\\ &\qquad \left.+\left[\left(\cos^2\theta\sin^2\gamma-\cos^2\gamma\right)J_2(2ka\sin\theta)\right]^2\right|^2\end{aligned} \tag{2.58}$$

where

γ = the angle between the polarization vector $\mathbf{E}^i$ (incident) and the y-axis,
a = the radius of the loop,
θ = the angle of observation (or incidence),
J_0, J_2 = the standard notations for Bessel functions of the first kind and orders 0 and 2, respectively.

For normal incidence, $\theta = 0°$ and $\sigma = \pi a^2$. If $2ka \sin\theta \gg 1$ and $\theta > 0°$, we can show that the RCS is then given by [7]

$$\sigma \sim \frac{a\lambda}{\pi \sin\theta}\left[\frac{\cos^2\gamma + 8\cos^4\theta \sin^4\gamma - 8\cos^2\theta \sin^2\gamma \cos^2\gamma - \cos^4\gamma}{32(ka \sin\theta)^2}\right] \tag{2.59}$$

The preceding expressions neglect the wire radius, although the results shown are based on the assumption that the loop wire radius is $\lambda/85$.

For a loop of wire with wire radius b and loop radius a, the backscattering cross section in the normal direction is given by [45]

$$\frac{\sigma}{\pi a^2} \sim \frac{(\pi/2)^2 + [\ln(85/\gamma)]^2}{(\pi/2)^2 + [\ln(\lambda/\gamma\pi b)]^2}, \tag{2.60}$$

where $\gamma = 1.78$.

RCS results at normal incidence for loops of various size obtained by using (2.60) have been compared with those obtained by other analytical means [7, Chap. 5]. The backscattering and average cross sections for a circular loop [45] with $a/b = 100$ are shown in Figure 2.18, where all the loop resonances are not visible because the loop is oriented at a broadside aspect.

The monostatic radar cross section of an unloaded loop of wire in terms of the Fourier coefficients for induced currents is given in [32] as

$$\frac{\sigma}{\lambda^2} = 16\pi\eta_0^2 \,|\nabla y'_{r,t}|^2 \tag{2.61}$$

where

$$\nabla y_{r,t} = -\left(\frac{kb}{4\pi}\right)^2 \sum_{n=-\infty}^{\infty} \frac{\alpha_n^r \alpha_{-n}^t}{\beta_n} \tag{2.62}$$

β_n = the Fourier coefficients,
b = the mean radius of the loop,

$$\alpha_n^{t,r}(\mathrm{TE}) = 1/2\mathrm{i}^{n+1}\mathrm{e}^{-in\varphi_{t,r}}\left[J_{n+1}(kb\sin\theta_{t,r}) - J_{n-1}(kb\sin\theta_{t,r})\right] \tag{2.63}$$

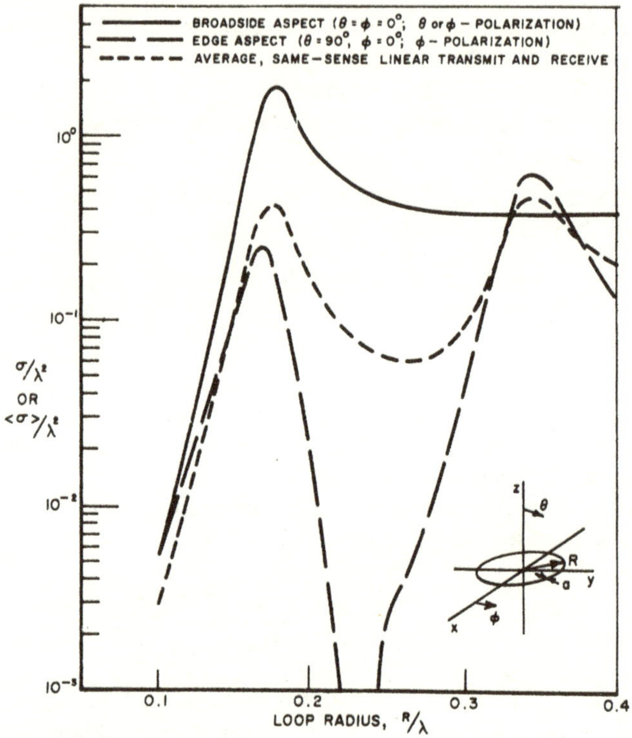

Figure 2.18 Backscattering and average cross section of circular loops with a loop-radius-to-wire-radius ratio: $a/b = 100$ (from [42]).

and J_{n+1}, J_{n-1} are the standard notations for Bessel functions of order $n + 1$ and $n - 1$, respectively, and $\alpha_n^{t,r}$(TM) are also given by (2.63).

The TE monostatic radar cross section pattern of a loop in the grazing direction is shown in Figure 2.19. The analysis [32] can take care of the case when the loop is loaded, which will be discussed in Chapter 4. Burns and Senior [45] considered corrections for circulating currents and used second-order GTD considerations to analyze the scattering by a thin wire loop under H-polarized illumination by electromagnetic plane waves. The magnitude of the far-zone scattered field as a function of angle of incidence is shown in Figure 2.20, where the corresponding results obtained with NEC code are also shown for comparison.

Narrow-band resonance phenomenon occurs in the scattering when a straight wire is bent into a nearly closed loop [46]. Such cases usually are analyzed by approximating the loop into N straight wire segments. Figures 2.21 (a) and (b) show the broadside

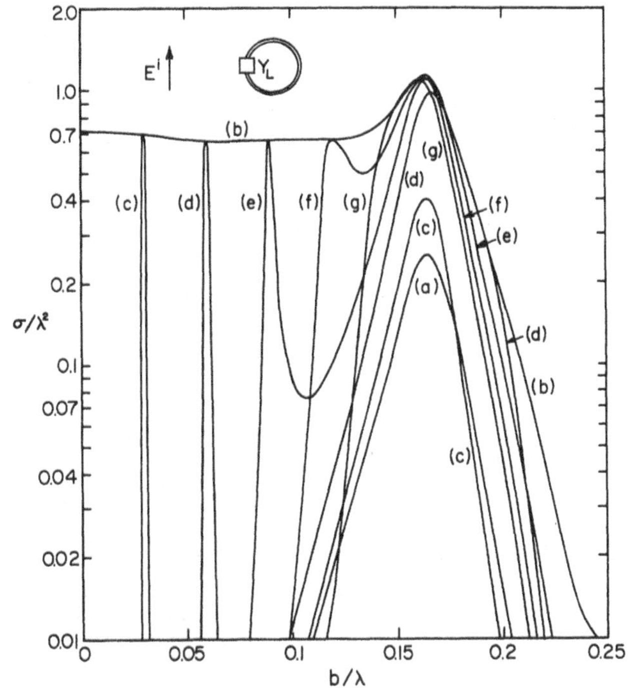

Figure 2.19 TE monostatic radar cross section of a loop in grazing direction in the (zx) plane (from [32]): $b/a = 100$; (a) complete loop; (b) continuously resonated; (c) resonated by capacitor at $b/\lambda = 0.03$; (d) resonated by capacitor at $b/\lambda = 0.06$; (e) resonated by inductor at $b/\lambda = 0.09$; (f) resonated by inductor at $b/\lambda = 0.12$; (g) resonated by inductor at $b/\lambda = 0.15$.

backscattering cross sections *versus* the radius of the loop calculated at 5.75 GHz; for comparison, corresponding results obtained from measurements at 5.70 and 5.75 GHz also are shown in the figure.

2.10.1.3 Thin Plates

Scattering by thin metallic rectangular plates can be analyzed by PO, GTD, or integral equation methods. The numerical solution of RCS obtained from the wire grid model of a rectangular plate is discussed in [47]. A comparative study of iterative techniques, moment method, and the spectral Galerkin approach for solving the problem of electromagnetic scattering by flat plates is discussed in [48]. The integral equation solutions and associated numerical computations for the RCS of rectangular plates are given in [49, 50]. A variational method also has been used to calculate the RCS of flat plates [32].

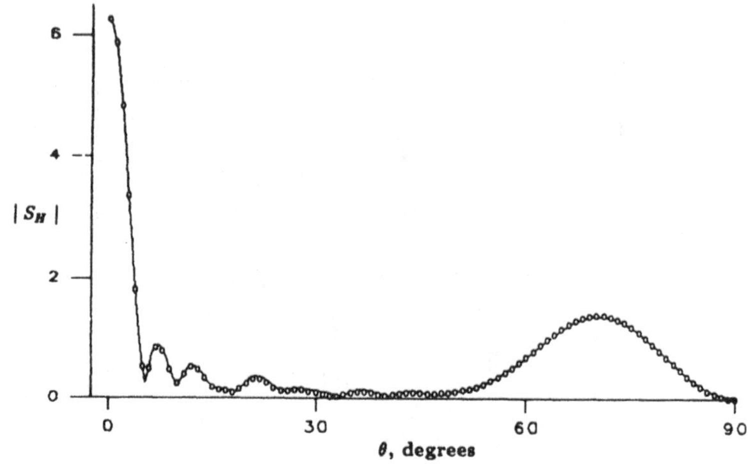

Figure 2.20 The magnitude of far field as a function of angle of incidence for $ka = 20$ of a loop present theory; 0000 NEC code (from [45]).

The geometry of electromagnetic scattering by a flat plate is shown in Figure 2.22, where we assume that the incident plane wave is in the vertical *zy* plane and is polarized in the *y*-direction. The PO backscattering cross section (independent of polarization) of the metal plate of width W and length L is given by

$$\sigma = \frac{k^2 \cos^2\theta}{\pi} \left[\int_S e^{i2ky \sin\theta'} \, ds' \right]^2$$

$$= \frac{4\pi W^2 L^2}{\lambda^2} \cos^2\theta \left[\frac{\sin(kW \sin\theta)}{kW \sin\theta} \right]^2 \tag{2.64}$$

As $\theta \to 0$, (2.64) indicates that for normal incidence the RCS is given by the familiar equation

$$\sigma = \frac{4\pi A^2}{\lambda^2} \tag{2.65}$$

where $A = WL$ is the physical area of the plate.

The RCS results *versus* its azimuth aspect angle for a rectangular plate (Figure 2.22) obtained from measurements and from PO and GTD calculations are shown in Figure 2.23. As expected, because the PO analysis assumes that the current induced on the plate is identical with that for an infinitely large plate and that the current exists only

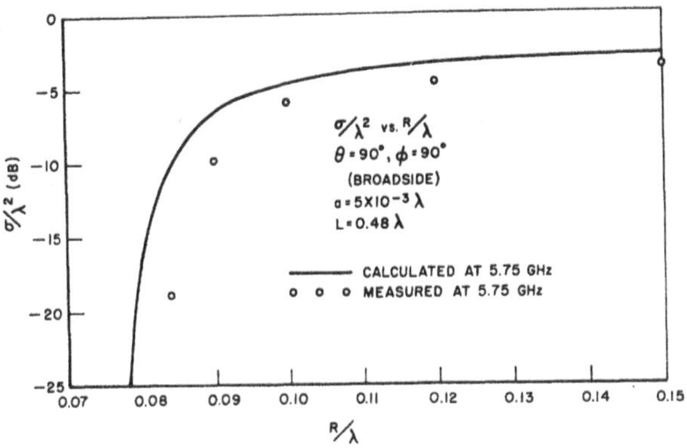

(a)

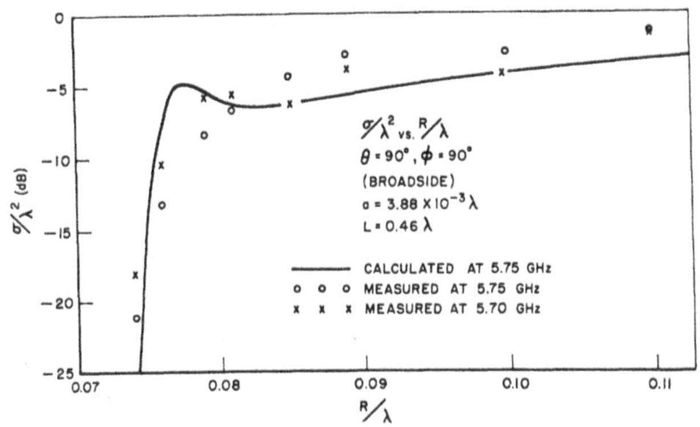

(b)

Figure 2.21 (a) Theoretical results of backscattering cross section *versus* its electrical radius compared with experiment at 5.76 GHz (from [44]); (b) theoretical results of backscattering cross section at 5.75 GHz *versus* its electrical radius compared with experiment at 5.75 and 5.70 GHz (from [46]).

on the illuminated portion of the plate, the PO results become inaccurate for incidence angles near the grazing incidence.

For accurately predicting the RCS of a flat plate over a wide range of angles, we must employ a technique that accounts for the effects of edge diffraction; that is, in effect we must use either PTD or GTD. In this case the latter method is more convenient due to the availability of appropriate diffraction coefficient expressions. Based on Keller's diffraction coefficients, Ross obtained analytical expressions [51] for the RCS of a flat plate as a function of aspect angle. With the vertically oriented zy

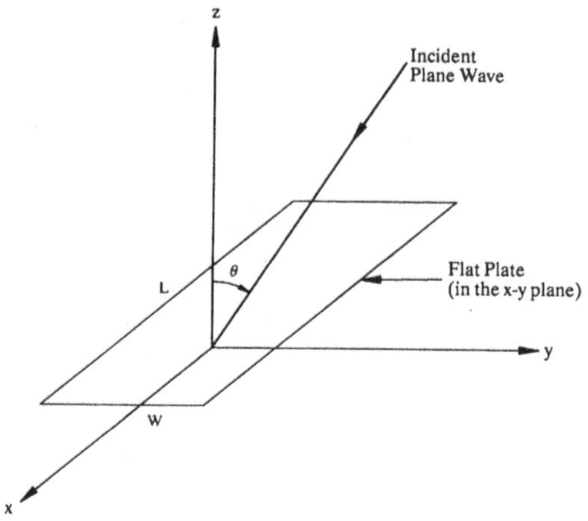

Figure 2.22 Electromagnetic scattering by a flat plate: plane of incidence $\phi = \pi/2$.

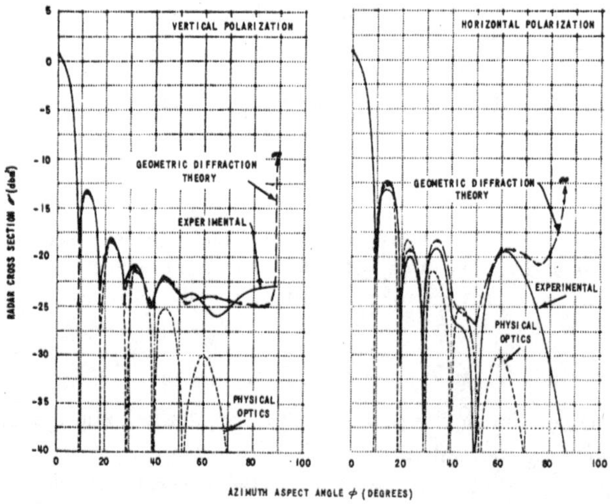

Figure 2.23 Radar cross section of a flat rectangular plate: comparison of PO, GTD formulas, and experimental results (from [51]).

plane as the plane of incidence and a wave incident on the plate having $L = 2a$, $W = 2b$ (Figure 2.22), the backscattering cross sections $\sigma_V(\theta)$ and $\sigma_H(\theta)$ for vertical and horizontal polarizations, respectively, obtained under GTD approximation are [51]:

$$\sigma_V = \frac{4b^2}{\pi}\left|\left[\cos(2ka\sin\theta) - i\frac{\sin(2ka\sin\theta)}{\sin\theta}\right] - \frac{e^{i(2ka-\pi/4)}}{\sqrt{2\pi}(2ka)^{3/2}}\right.$$

$$\times\left[\frac{1}{\cos\theta} + \frac{e^{i(2ka-\pi/4)}}{4\sqrt{2\pi}(2ka)^{3/2}}\frac{(1+\sin\theta)e^{-i2ka\sin\theta}}{(1-\sin\theta)^2}\right. \tag{2.66a}$$

$$\left.+\frac{(1-\sin\theta)e^{i(2ka\sin\theta)}}{(1+\sin\theta)^2}\right]\left[1 - \frac{e^{i(4ka-\pi/2)}}{8\pi(2ka)^3}\right]^{-1}\Bigg|^2$$

$$\sigma_H = \frac{4b^2}{\pi}\left|\left[\cos 2ka\sin\theta + i\frac{\sin(2ka\sin\theta)}{\sin\theta}\right] - \frac{4e^{i(2ka+\pi/4)}}{\sqrt{2\pi}(2ka)^{1/2}}\right.$$

$$\left[\frac{1}{\cos\theta} - \frac{e^{i(2ka+\pi/4)}}{\sqrt{2\pi}(2ka)^{1/2}}\frac{e^{-i2ka\sin\theta}}{1-\sin\theta} + \frac{e^{i2ka\sin\theta}}{1+\sin\theta}\right] \tag{2.66b}$$

$$\times\left[1 - \frac{e^{i(4ka+\pi/2)}}{2\pi(2ka)}\right]^{-1}\Bigg|^2$$

which are not valid for grazing incidence $\theta = \pi/2$.

For electrically large plates the double and triple diffraction terms can be neglected. Experimental RCS patterns for a flat plate, oriented in the horizontal *xy* plane (Figure 2.22), are compared with corresponding results obtained from the PO and GTD analysis [52], as shown in Figure 2.23. As expected, the PO results agree well with experimental results near the broadside region but disagree significantly in the region away from broadside. Over most of the off-broadside region the GTD results agree much better with the experimental results; however, it should be observed that near grazing incidence the agreement is poor for horizontal polarization. Although exactly at grazing incidence results are poor for both polarization.

RCS of flat plates has also been computed by using the integral equation technique [49]; the results are shown in Figures 2.24 and 2.25, where *this method* refers to the integral equation method. Results obtained from PO, GTD, and integral equation methods are shown in Figure 2.24. We can see that for $\theta \simeq 20°$, results from all three

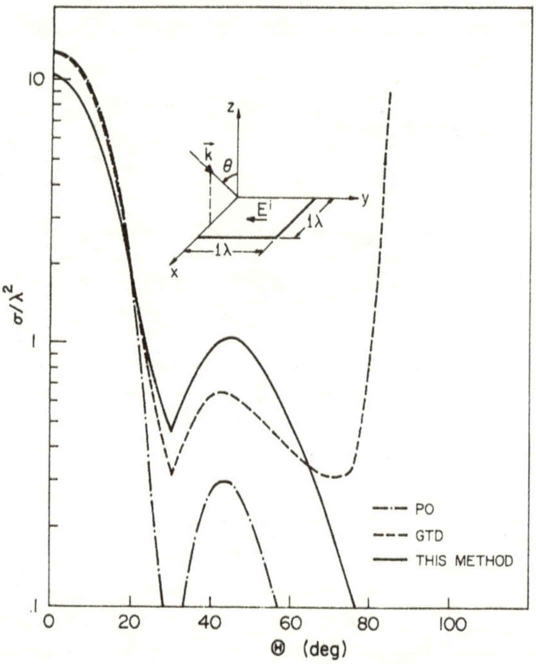

Figure 2.24 RCS of 1λ × 1λ plate *versus* aspect angle (from [49]).

methods are in reasonable agreement; for $20° \simeq \theta \simeq 65°$, that is, in the off-broadside region, PO results disagree significantly. Because the GTD expression [51] is inaccurate near the grazing incidence, the integral equation results seem to be reasonable here although no experimental results were given in [49].

Figure 2.25 compares the RCS results measured in the normal direction for a square plate with those determined by integral equation and variational methods. The integral equation method provides results that agree better with the experiment.

2.10.1.4 Circular Disk

The RCS of a circular metallic disk can be analyzed by using PO, GTD, and integral equation methods. A large contribution to the scattered field and, hence, to the RCS occurs when the plate is illuminated by normal incidence. For disks that are large compared to wavelength, a significant amount of scattering may occur on edge-on incidence, particularly when the *E*-field is oriented parallel to the disk.

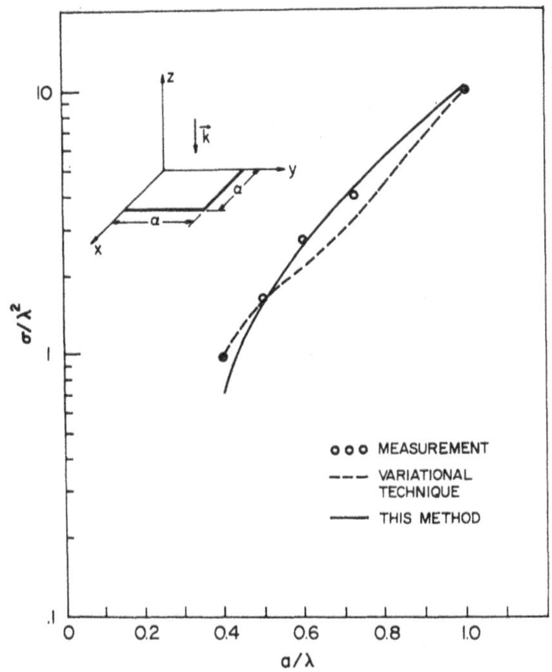

Figure 2.25 RCS of square plate for normally incident plane wave (from [49]).

For a large circular disk, oriented in the horizontal xy plane, the backscattering cross section in the θ direction as determined by the PO method is [7, Chap. 5]

$$\sigma = \frac{\pi a^2}{\tan^2\theta}\left[J_1\left(\frac{4\pi a \sin\theta}{\lambda}\right)\right]^2 \tag{2.67}$$

where a is the radius of the disk, and J_1 is the Bessel function of the first kind and order 1. Equation (2.67) is independent of polarization and accurate only near normal incidence.

Flammer [52] obtained an exact solution for disk scattering in terms of oblate spheroidal functions; such solutions are also discussed in [2]. Some selected exact results have been compared with experiment by Mattson [53].

GTD has been applied with success to analyze scattering from disks. The geometry of the problem considered in [54] is shown in Figure 2.26. In the first-order

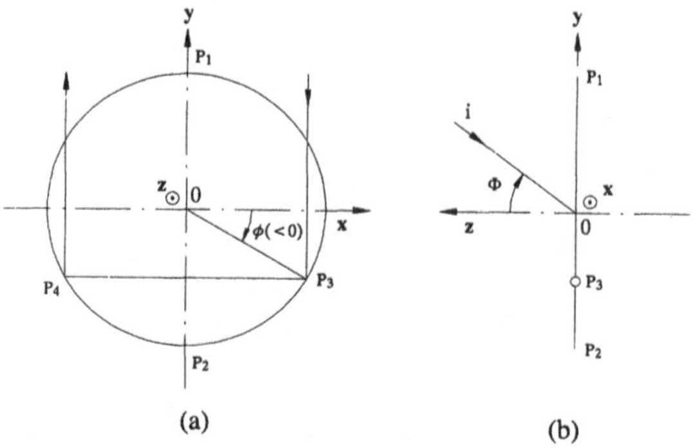

Figure 2.26 Geometry of electromagnetic scattering by a disk: (a) in the plane of the disk; (b) in the plane of incidence.

theory, the scattering function is [54]

$$S_E = -\frac{1}{4}\left(\frac{ka}{\pi \sin\phi}\right)^{1/2}\left[\left(1 \mp \frac{1}{\sin\phi}\right) e^{i2ka\sin\phi - i\pi/4} + \left(1 \pm \frac{1}{\sin\phi}\right) e^{i2ka\sin\phi + i\pi/4}\right] \tag{2.68}$$

where E and H refer to the polarization having E or H parallel to the plane of the disk, and φ and Φ are as described in Figure 2.26. S_E and S_H being the two orthogonal components of the scattering function, the backscattering cross section then can be obtained by using (1.29) and (2.68). Figures 2.27 and 2.28 illustrate the comparison between theoretical and experimental results. The first-order theory was

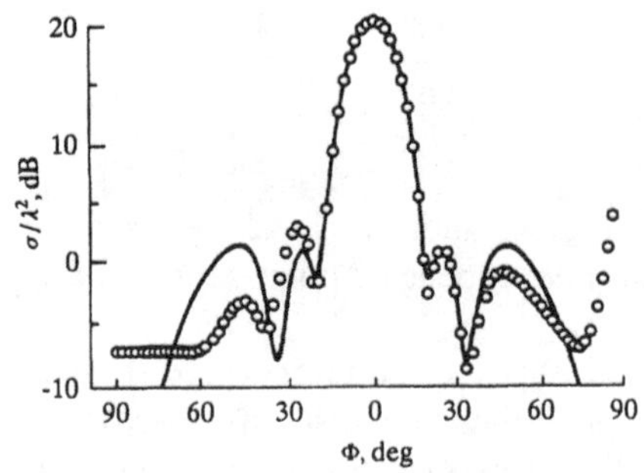

Figure 2.27 Comparison of theory and experiment for H-polarization with ka = 6: ∘∘∘ theory (first order) —— experiment (from [54]).

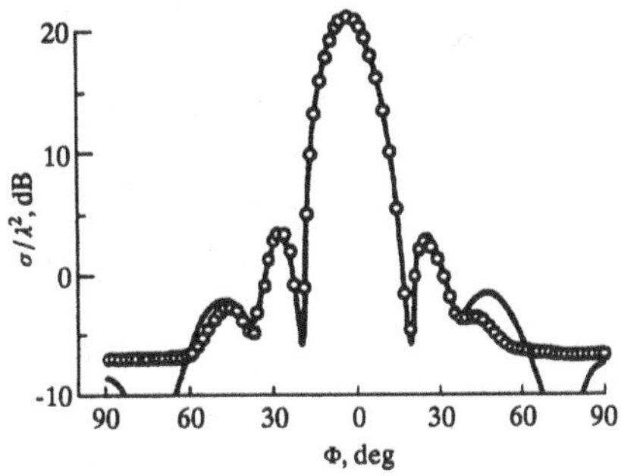

Figure 2.28 Comparison of theory with experiment for E-polarization with $ka = 6$: ∘∘∘ theory —— experiment (from [54]).

found to agree well with the results of the experiment for E-polarization; as indicated in Figure 2.27 a second-order theory is needed for H-polarization. Further details are given in [54].

Disk edge-on scattering may be of interest in many practical problems. The backscattering function under edge-on incidence with E-field parallel to the disk is given by the following asymptotic expression [55, 56]:

$$S = \left\{ 1/2\sqrt{ka/\pi}\left(1 - \frac{i}{4ka}\right) + B(ka)^{2/3} \exp i2ka \right.$$

$$\left. + i\pi ka[1 - 1/2(ka)]^{-2/3} \exp(-i\pi/3) \right\} \exp(-i3\pi/4) \quad (2.69)$$

where

$$B = \left(1 - \frac{\gamma}{\pi}\right)^{-1/2} \exp(i0.068\pi)$$

$$\gamma_1 = ka - 0.25(ka)^{1/3}$$

$$\gamma_2 = 0.43(ka)^{1/3}$$

Figure 2.29 shows a comparison between theoretical results obtained from (2.69) with the measured results. The agreement is found to be very good except for small disks where $|S|$ depends more on the volume than its shape.

Scattering from a planar array of small metal discs is of interest to radar engineers. An analysis based on the concept of dipole moments is available in [57]. The

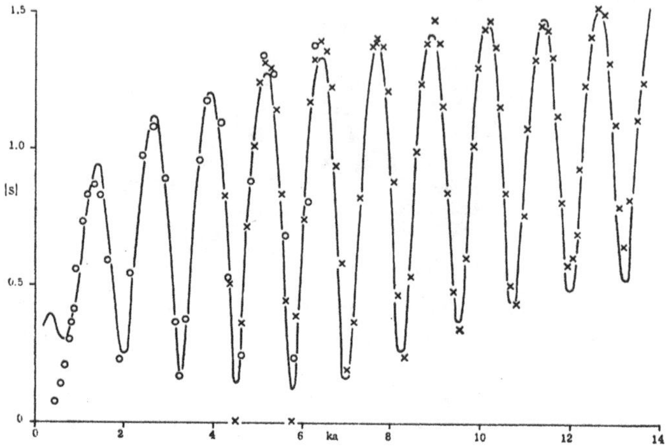

Figure 2.29 Comparison of theoretical and measured scattered field for disks, $\lambda = 0.8\pi$ $\cdots\cdot$ 0.0036λ in thickness × × × × 0.0027λ in thickness (from [55]).

complementary case of a planar array of circular apertures on a conducting plane has been analyzed in [58].

2.10.2 Targets of Intermediate Complexity

2.10.2.1 Finite Circular Cylinder

Scattering of electromagnetic waves by conducting and dielectric cylinders of infinite length are well documented in the literature [2; 6, Chap. 4; 7, Chap. 4; 8; 13]. Earlier papers on infinite cylinders are mentioned in [13], which also gives 18 references under the heading of circular cylinders and several under dielectric cylinders.

The geometry of the bistatic scattering of electromagnetic waves by a conducting circular cylinder of finite length is shown in Figure 2.30, where four significant scattering centers, S_1, S_2, S_3, and S_4, are identified. Assuming that the individual scattering center contribution to the bistatic scattering function is

$$\sqrt{\pi/\lambda^2}\,\sqrt{\sigma_i}\,e^{i\alpha_i}, \quad \text{with } i = 1,2,3,4$$

the total bistatic scattering cross section is given by (see 1.29)

$$\sigma = \left| \sum_{i=1}^{4} \sqrt{\sigma_i}\,e^{i\alpha_i} \right|^2 \tag{2.70}$$

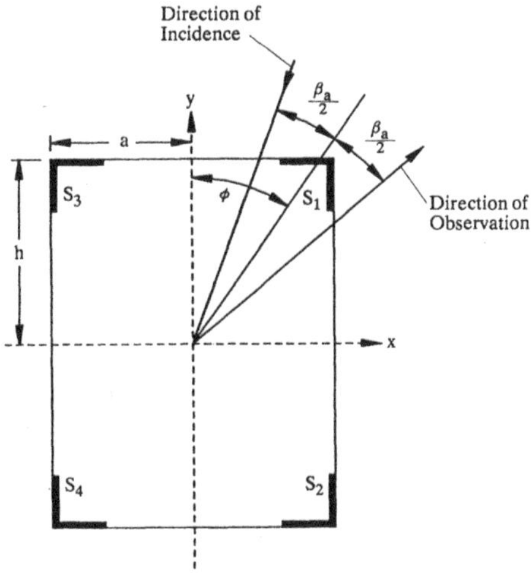

Figure 2.30 Geometry of the scattering from a finite cylinder.

GTD expressions for $\sigma_i'^s$ appropriate for a conducting finite cylinder have been obtained in [59, 60], and they are

$$\sqrt{\sigma_1} = \frac{2}{3}\sin\left(\frac{2\pi}{3}\right)\left(\frac{a \operatorname{cosec}\varphi}{k\cos\beta_{a/2}}\right)^{1/2}\left\{\left[\cos\left(\frac{2\pi}{3}\right) - \cos\frac{2}{3}(\pi + 2\varphi)\right]^{-1}\right.$$

$$\left.\mp\left[\cos\left(\frac{2\pi}{3}\right) - \cos\left(\frac{2\beta_a}{3}\right)\right]^{-1}\right\} \tag{2.71a}$$

$$\sqrt{\sigma_2} = \sin\left(\frac{2\pi}{3}\right)\left(\frac{a \operatorname{cosec}\varphi}{k\cos\beta_a/2}\right)^{1/2}\left\{\left[\cos\left(\frac{2\pi}{3}\right) - \cos\left(\frac{4\varphi}{3}\right)\right]^{-1}\right.$$

$$\left.\mp\left[\cos\left(\frac{2\pi}{3}\right) - \cos\left(\frac{2\beta_a}{3}\right)\right]^{-1}\right\},$$

$$\text{for } \varphi \geq \frac{\beta_a}{2}, \quad \sqrt{\sigma_2} = 0 \quad \text{for } \varphi < \frac{\beta_a}{2} \tag{2.71b}$$

$$\sqrt{\sigma_3} = \frac{2}{3}\sin\left(\frac{2\pi}{3}\right)\left(\frac{a \operatorname{cosec}\varphi}{k\cos\beta_a/3}\right)^{1/2}\left\{\left[\cos\left(\frac{2\pi}{3}\right) - \cos\frac{2}{3}(\pi - 2\varphi)\right]^{-1}\right.$$

$$\left.\mp\left[\cos\left(\frac{2\pi}{3}\right) - \cos\left(\frac{2\beta_a}{3}\right)\right]^{-1}\right\},$$

$$\text{for } \varphi \le \frac{\pi}{2} < \frac{\beta_a}{2} = 0,\ \varphi > \frac{\pi}{2} - \frac{\beta_a}{2} \tag{2.71c}$$

$$\sqrt{\sigma_4} = \sin\left(\frac{2\pi}{3}\right)\left(\frac{a \operatorname{cosec}\varphi}{k\cos\beta_a/2}\right)^{1/2}\left\{\left[\cos\left(\frac{2\pi}{3}\right) - \cos\frac{2}{3}(\pi + 2\varphi)^{-1}\right]\right.$$

$$\left.\mp\left[\cos\left(\frac{2\pi}{3}\right) - \cos\frac{2}{3}(2\pi - \beta_a)\right]^{-1}\right\}$$

$$\text{for } \left(\frac{\pi}{2} - \frac{\beta_a}{2}\right) \le \varphi \le \frac{\beta_a}{2} = 0,\ \frac{\beta_a}{2} < \varphi < \left(\frac{\pi}{2} - \frac{\beta_a}{2}\right) \tag{2.71d}$$

where the upper and lower signs are for vertical and horizontal polarizations assuming that the *xy*-plane in Figure 2.30 is vertically oriented. The corresponding phases α_i of the contributions by the scattering centers are [59]

$$\alpha_1 = -2\left(k\cos\frac{\beta_a}{2}\right)(a\sin\varphi + h\cos\varphi) + \frac{\pi}{4} \tag{2.72a}$$

$$\alpha_2 = -2\left(k\cos\frac{\beta_a}{2}\right)(a\sin\varphi - h\cos\varphi) + \frac{\pi}{4} \tag{2.72b}$$

$$\alpha_3 = 2\left(k\cos\frac{\beta_a}{2}\right)(a\sin\varphi - h\cos\varphi) - \frac{\pi}{4} \tag{2.72c}$$

$$\alpha_4 = 2\left(k\cos\frac{\beta_a}{2}\right)(a\sin\varphi + h\cos\varphi) - \frac{\pi}{4} \tag{2.72d}$$

Theoretical and experimental RCS results for a right circular cylinder for vertical and horizontal polarizations are shown in Figure 2.31. Similar results for horizontal and cross polarizations (i.e., horizontal incidence and vertical reception and *vice versa*) are shown in Figure 2.32. Relative contributions to the RCS by various scattering centers are shown in Figure 2.33.

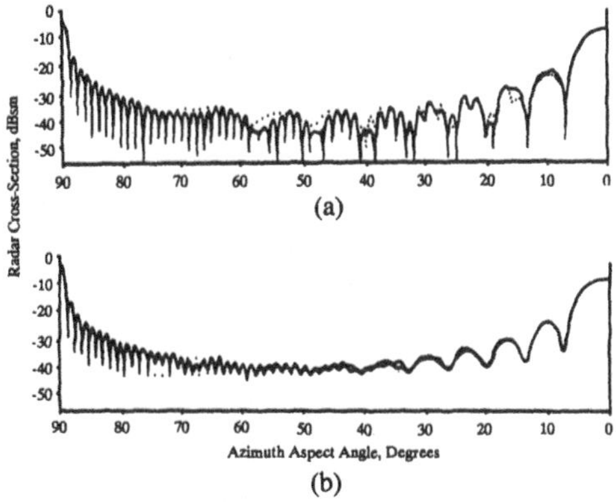

Figure 2.31 Radar cross section of right-circular cylinder for vertical and horizontal polarizations: $2h = 24.25\lambda$; $2a = 4.779\lambda$; $\beta_a = 0$ —— experimental; - - - GTD (with single diffraction); (a) horizontal polarization; (b) vertical polarization (from [59]).

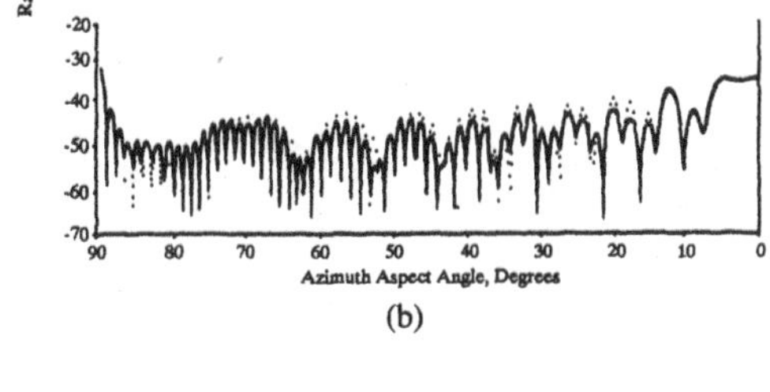

Figure 2.32 Radar cross section of right-circular cylinder for parallel and horizontal polarizations $2h = 24.51\lambda$; $2a = 4.961\lambda$; $\beta_a = 0.0$ parallel polarization (from [59]).

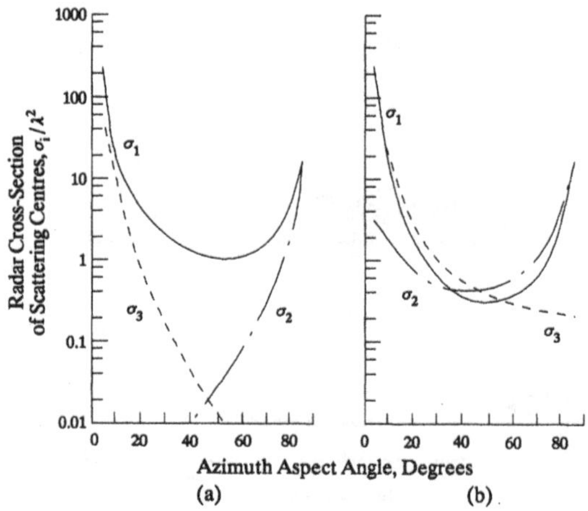

Figure 2.33 Radar cross section of scattering centers on cylinder for vertical and horizontal polarizations: $2h = 24.265\lambda$; $2a = 6.779\lambda$; $\beta_a = 0$; (a) horizontal polarization; (b) vertical polarization (from [59]).

The monostatic scattering geometry and the appropriate scattering centers for a finite metallic cylinder of rectangular cross section are shown in Figure 2.34, where it is assumed that xy-plane is the horizontal plane. Expressions for σ_i and α_i appropriate for the monostatic case are available [61]. They are where the upper and lower signs are for vertical and horizontal:

$$\sqrt{\sigma_1} = \frac{2b}{3\sqrt{\pi}} \sin\left(\frac{2\pi}{3}\right)\left\{\left[\cos\left(\frac{2\pi}{3}\right) - \cos\frac{2}{3}(\pi + 2\varphi)\right]^{-1} \mp \left[\cos\left(\frac{2\pi}{3}\right) - 1\right]^{-1}\right\} \tag{2.73a}$$

$$\sqrt{\sigma_2} = \frac{2b}{3\sqrt{\pi}} \sin\left(\frac{2\pi}{3}\right)\left\{\left[\cos\left(\frac{2\pi}{3}\right) - \cos\left(\frac{4\varphi}{3}\right)\right]^{-1} \mp \left[\cos\left(\frac{2\pi}{3}\right) - 1\right]^{-1}\right\} \tag{2.73b}$$

$$\sqrt{\sigma_3} = \frac{2b}{3\sqrt{\pi}} \sin\left(\frac{2\pi}{3}\right)\left\{\left[\cos\left(\frac{2\pi}{3}\right) - \cos\frac{2}{3}(\pi - 2\varphi)\right]^{-1} \mp \left[\cos\left(\frac{2\pi}{3}\right) - 1\right]^{-1}\right\} \tag{2.73c}$$

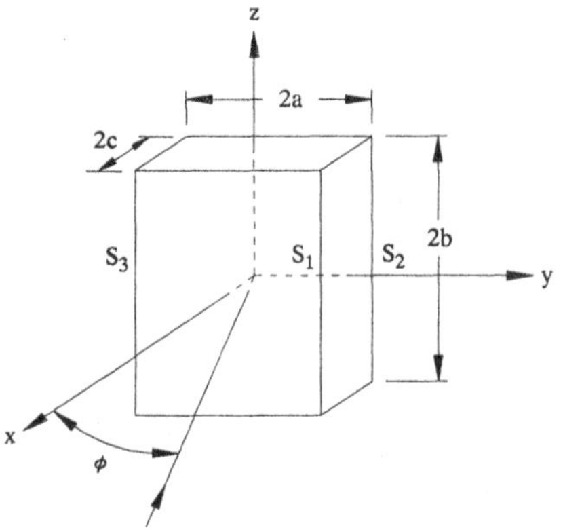

Figure 2.34 Scattering centers on a rectangular cylinder (from [58]).

polarizations, respectively; and

$$\alpha_1 = -2k(a \sin \varphi + c \cos \varphi) + \pi/4 \tag{2.74a}$$

$$\alpha_2 = -2k(a \sin \varphi - c \cos \varphi) + \pi/4 \tag{2.74b}$$

$$\alpha_3 = 2k(a \sin \varphi - c \cos \varphi) + \frac{\pi}{4} \tag{2.74c}$$

Theoretical and experimental results for the RCS of a finite cylinder are shown in Figure 2.35.

We mention here only the two key papers [62, 63] on scattering from finite dielectric cylinders. In [63] the integral equation and MOM methods have been used to obtain differential cross sections of dielectric cylinders. Typical results obtained are shown in Figure 2.36.

2.10.2.2 Finite Cones

Backscattering from a right circular cone shown in Figure 2.37 is of considerable practical interest. Depending on the incident angle, the backscattered field may consist of contributions from diffraction effects at the tip and the edge, formed by its curved surface and the base, and also from reflections off the curved surface and the base. GTD has been used extensively [64–67] to estimate the scattered fields and, hence, the RCS of finite length metallic cones. GTD expressions for the monostatic RCS of a finite

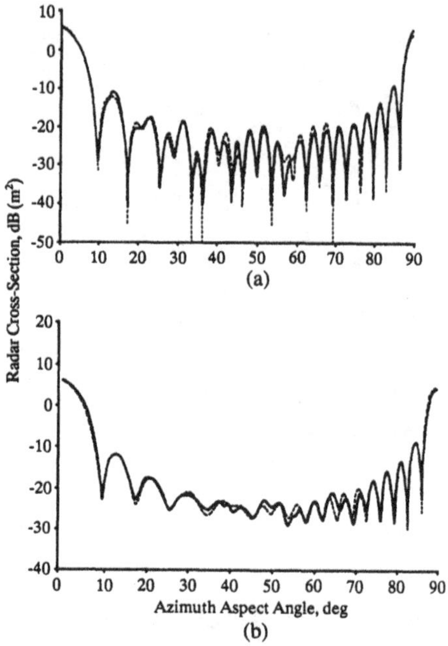

Figure 2.35 Radar cross section of rectangular cylinder for vertical and horizontal polarizations: $2a = 14.763\lambda$; $2b = 17.716\lambda$; $2c = 2.226\lambda$; ---- experimental; —— GTD (single-diffraction only); (a) horizontal polarization; (b) vertical polarization (from [60]).

metallic cone shown in Figure 2.37 are given by [67]

$$\sigma^{V,H} = \frac{ka}{4\pi^2}\left[\frac{\sin(\pi/n)}{n}\right]^2 \frac{1}{\sin\varphi}$$

$$\times \left|\exp\left[-i\left(2ka\sin\varphi - \frac{\pi}{4}\right)\right]\left[\left(\cos\frac{\pi}{n} - 1\right)^{-1}\right.\right.$$

$$\left.\mp\left(\cos\frac{\pi}{n} - \cos\frac{3\pi - 2\varphi}{n}\right)^{-1}\right] \tag{2.75}$$

$$+ \exp\left[i\left(2ka\sin\varphi - \frac{\pi}{4}\right)\right]\left[\left(\cos\frac{\pi}{n} - 1\right)^{-1}\right.$$

$$\left.\left.\mp\left(\cos\frac{\pi}{n} - \cos\frac{3\pi + 2\varphi}{n}\right)^{-1}\right]\right|^2, \quad 0 < \varphi < \gamma$$

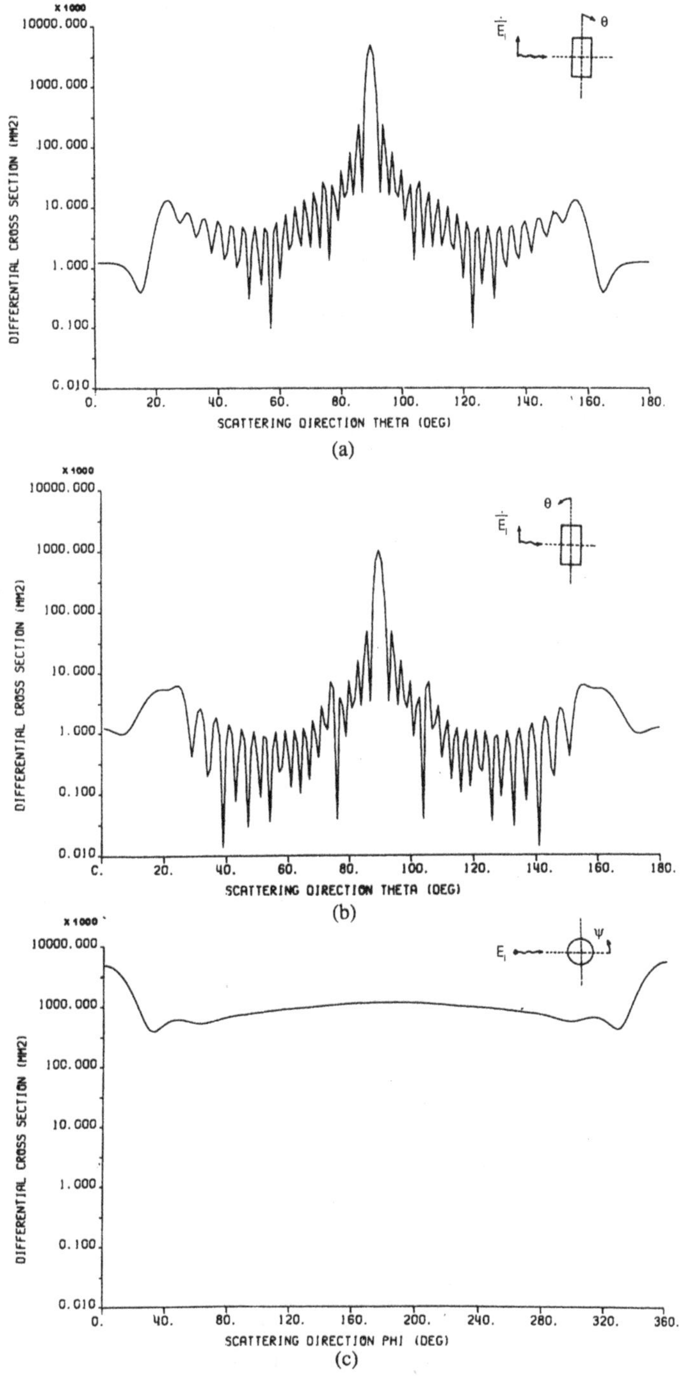

Figure 2.36 DCS of a dielectric cylinder *versus* θ: (a) forward direction $\theta = 0°$; (b) backward direction $\theta = 180°$; (c) intermediate direction $\theta = 45°$ (from [63]).

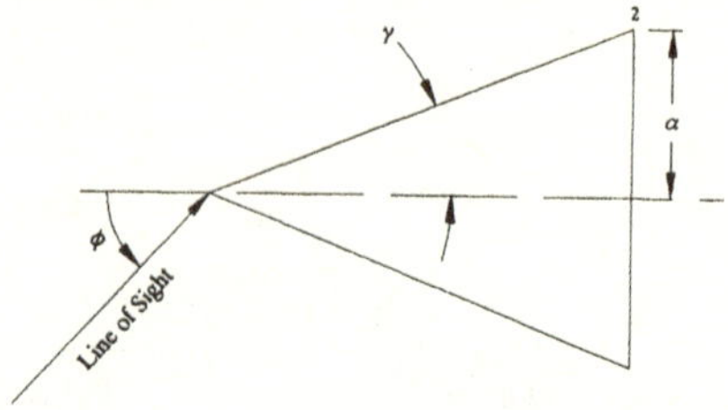

Figure 2.37 Longitudinal section of a right circular cone.

$$\sigma^{V,H} = \frac{ka}{4\pi^2}\left[\frac{\sin(\pi/n)}{n}\right]^2 \frac{1}{\sin\varphi}\left[\left(\cos\frac{\pi}{n} - 1\right)^{-1}\right.$$

$$\left.\mp\left(\cos\frac{\pi}{n} - \cos\frac{2\pi - 2\varphi}{n}\right)^{-1}\right]^2, \tag{2.76}$$

$$\gamma < \varphi < \frac{\pi}{2}, \ \varphi \neq \frac{\pi}{2} - \gamma$$

$$\sigma^{V,H} = \frac{ka}{4\pi^2}\left[\frac{\sin(\pi/n)}{n}\right]^2 \frac{1}{\sin\varphi}$$

$$\times \left|\exp[-i(2ka\sin\varphi - \pi/4)]\left[\left(\cos\frac{\pi}{n} - 1\right)^{-1}\right.\right.$$

$$\left.\mp\left(\cos\frac{\pi}{n} - \cos\frac{3\pi - 2\varphi}{n}\right)^{-1}\right] \tag{2.77}$$

$$+\exp[i(2ka\sin\varphi - \pi/4)]\left[\left(\cos\frac{\pi}{n} - 1\right)^{-1}\right.$$

$$\left.\left.\mp\left(\cos\frac{\pi}{n} - \cos\frac{2\varphi - \pi}{n}\right)^{-1}\right]\right|^2, \quad \frac{\pi}{2} < \varphi < \pi$$

$n = 2(1 - \gamma/\pi)$ and γ is the half-angle of cone (Figure 2.37). The upper and lower signs refer to the vertical and horizontal polarizations.

For $\varphi = 0°$, the nose-on monostatic RCS is independent of polarization and is given by [66]

$$\frac{\sigma}{\lambda^2} = \frac{1}{\pi}\left[\frac{ka\sin(\pi/n)}{n}\right]^2\left(\cos\frac{\pi}{n} - \cos\frac{3\pi}{n}\right)^2 \tag{2.78}$$

with n as just given.

For $\varphi = \pi/2 - \gamma$, that is, incidence is broadside on the curved surface [66],

$$\frac{\sigma}{\lambda^2} = \frac{8}{9}\left(\frac{a}{\lambda}\right)^2\left[\sin^2\gamma \cos\gamma\right]^{-1} \tag{2.79}$$

and for $\varphi = \pi$,

$$\frac{\sigma}{\lambda^2} = \frac{(ka)^4}{4\pi} \tag{2.80}$$

Results given by (2.79) and (2.80) have been obtained under PO approximation and are independent of polarization.

Calculated and measured RCS results for a cone under various conditions are shown in Figures 2.38 and 2.39.

2.10.3 Complex Targets

Analytically determining the scattering is extremely difficult, if not impossible, from complex targets such as an aircraft or aircraftlike structures. Rigorous analytical formulation is possible in the form of integral equations, the solution of which is significantly involved and not generally possible. Some computer codes capable of providing numerical solutions are available [68–72]. Depending on the directions of incidence and observation, high-frequency radar returns from a complex target may be thought of as composed of contributions from a number of scattering centers. Also, whenever possible a complex target is assumed to consist of a number of simple shapes whose scattering properties are known. Based on these observations a number of models have been developed to determine the RCS of complex targets. We shall briefly describe a few.

2.10.3.1 Combination Method

The following steps describe the procedure to determine the RCS of satellites, aircraft, and missiles [7, Chap. 10; 73; 74]:

1. The complex shape is divided into a number of component simple shapes. Geometrically simple shapes are now chosen to approximate the component shapes.

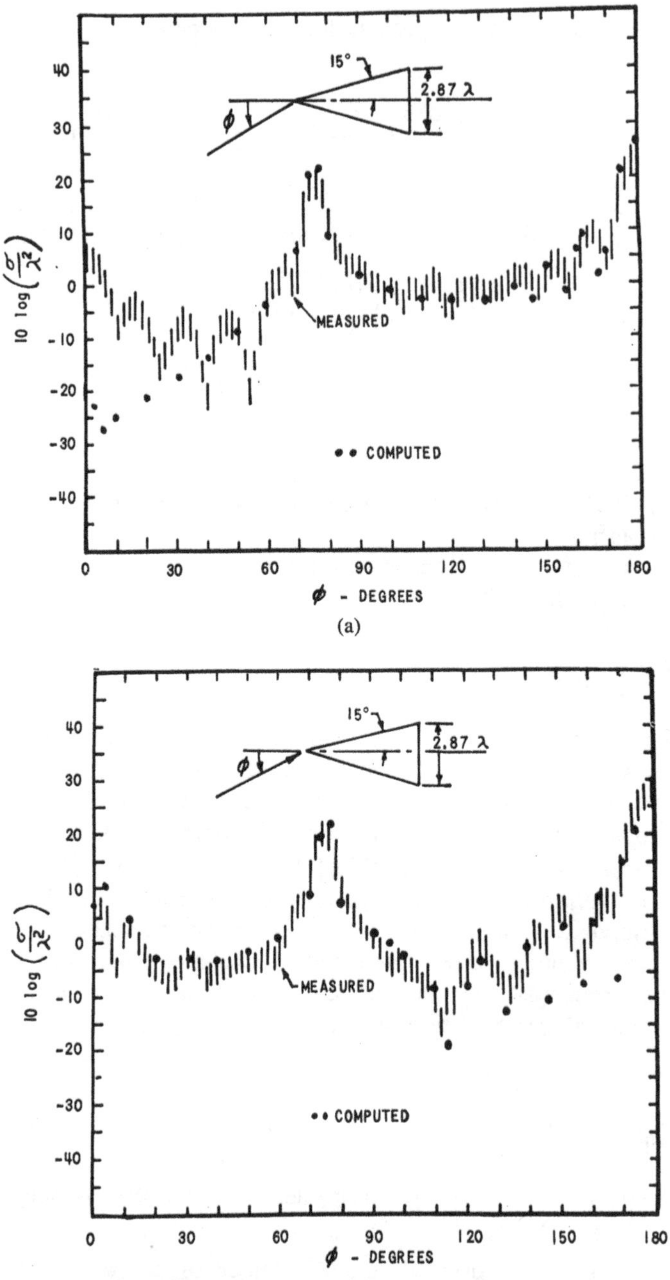

(b)

Figure 2.38 (a) Vertically polarized RCS for cones with $\gamma = 15°$ and $ka = 9.02$; (b) horizontally polarized RCS for cones with $\gamma = 15°$ and $ka = 9.02$ (from [65]).

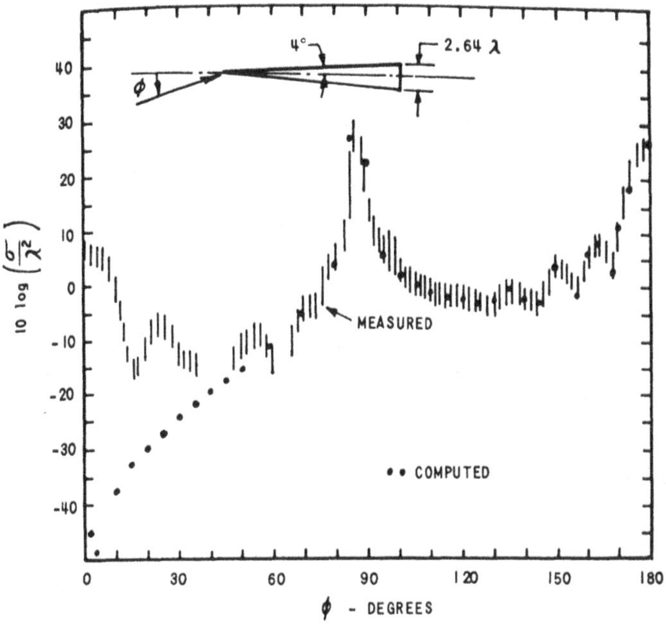

(a)

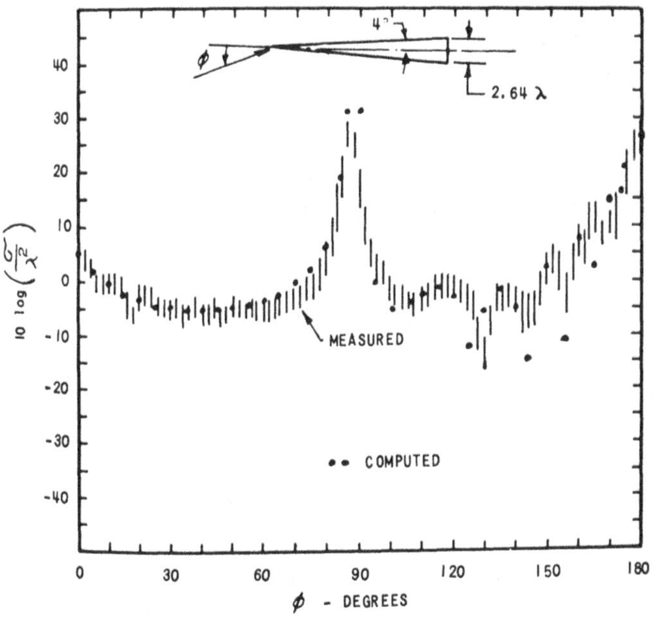

(b)

Figure 2.39 (a) Vertically polarized RCS for cones with $\gamma = 4°$, $ka = 8.28$; (b) horizontally polarized RCS for cones with $\gamma = 4°$, $ka = 8.28$ (from [65]).

2. The RCS is calculated for each of the simple shapes in step 1.
3. A combination of steps (1) and (2) is used to determine the RCS of the entire complex target in a manner that will be described in this section. Because of the existence of a large number of scattering centers with different scattering mechanisms (edge, tip, and corner diffraction; multiple diffraction; creeping waves, *et cetera*) the overall RCS pattern of a complex target generally will possess large number of sharp variations. This happens as a result of destructive and constructive interferences produced by the combinations from many scattering centers.

   Assuming that each component (or scattering center) contributes an amount

$$\sqrt{\frac{\pi}{\lambda^2}}\ \sqrt{\sigma_n}\, e^{i\phi_n}$$

With $i = 1, R \ldots N$, the effective RCS of a complex target consisting of N components can be written as

$$\sigma = \left| \sum_{n=1}^{N} \sqrt{\sigma_n}\ e^{i\phi_n} \right|^2 \tag{2.81}$$

where ϕ_n may be interpreted as the relative phase of the contribution σ_n. Thus, the maximum value of the RCS is

$$\sigma_{\max} = \left| \sum_{n=1}^{N} \sqrt{\sigma_n} \right|^2 \tag{2.82}$$

An alternative method for obtaining the overall RCS is to assume that the values of φ_n are randomly distributed. Under this condition, the RCS sought is the "average" RCS, given by

$$\sigma_{\text{av}} = \sum_{n=1}^{N} \sigma_n \Big/ N \tag{2.83}$$

In the analysis, a parameter called the *rms spread*, S, often has been found useful; it is defined as

$$S^2 = \left(\sum_{n=1}^{N} \sigma_n \right)^2 - \sum_{n=1}^{N} \sigma_n^2 \tag{2.84}$$

The purpose of an RCS calculation generally dictates the manner in which the contributions of the scattering center are to be combined. A theoretical method, based on the preceding discussion, for calculation of RCS of aircraft and missiles and some relevant experimental results are given in [7, Chap. 10; 75].

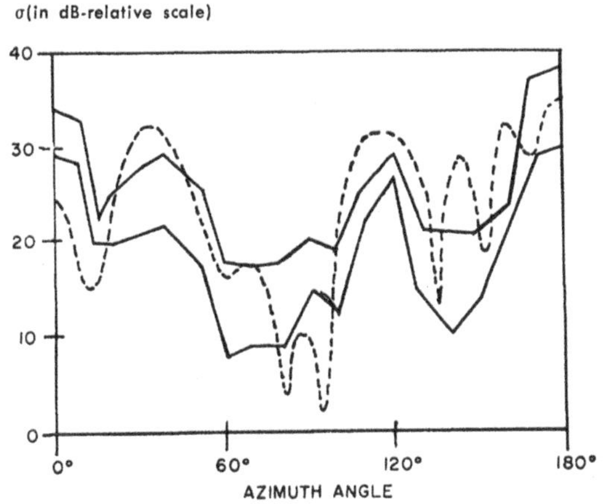

Figure 2.40 Radar cross section of a missile body: ---- experimental data; —— theoretical (rms spread) (from [72]).

Figure 2.40 compares the RCS results obtained by the rms spread equation (2.84) for a missile with those obtained by measurement. The model of a missile with fins is sketched in Figure 2.41. Theoretical and experimental RCS results for the missile shown in Figure 2.41, are given in Figure 2.42 [74]. The results were obtained with the E-vector slanted at 45° angle with the plane in which aspect is measured. Similar results obtained with vertical polarization for the same missile are shown in Figure 2.43. Experimental near broadside RCS results for the missile are shown in Figure 2.44, which also shows theoretical results obtained by relative phase method of combination discussed earlier.

The geometry of a large aircraft scattering is sketched in Figure 2.45; and its monostatic and bistatic RCS results are shown in Figure 2.46 (a) and (b), respectively. The upper and lower bounds for bistatic RCS results determined by using the monostatic-bistatic theorem [74] for a missile are shown in Figure 2.47 (a) and (b), respectively. More detailed information regarding various aspects of aircraft scattering and their RCS estimation may be found in [7, Chap. 10; 73; 74–76].

2.10.3.2 MI 105 Shell Model

The geometry of the MI 105 shell with dimensions in millimeters is shown in Figure 2.48; a metallic version of this model is utilized to determine RCS analytically. The GTD backscattering RCS of the shell is determined by using the analytical expressions given in Table 2.1 [77]. The expressions for RCS at angles $\theta = 0$, $(\pi/2 - \alpha_1)$, $\pi/2$,

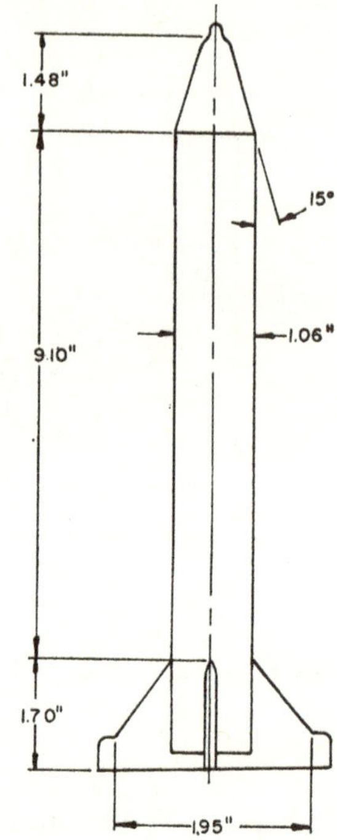

Figure 2.41 Sketch of a missile model.

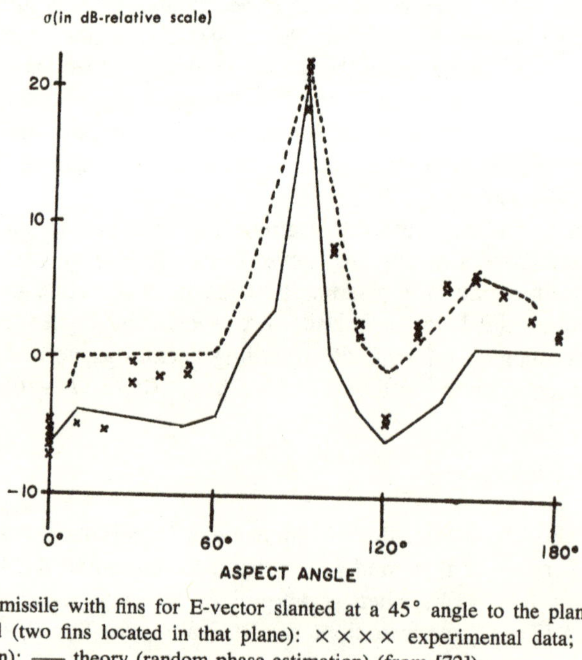

Figure 2.42 RCS of missile with fins for E-vector slanted at a 45° angle to the plane in which aspect is measured (two fins located in that plane): × × × × experimental data; --- theory (in-phase estimation); —— theory (random phase estimation) (from [72]).

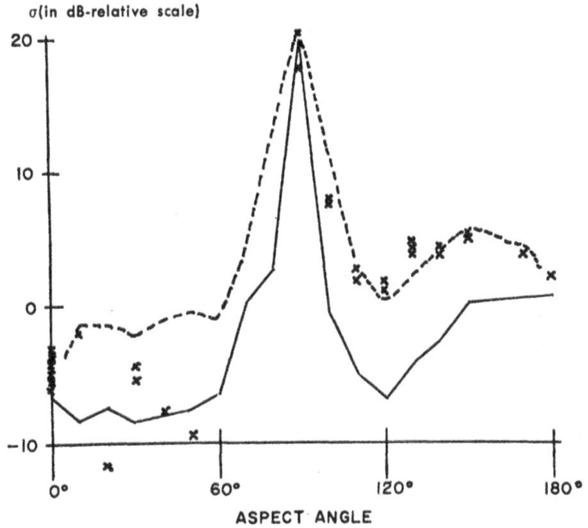

Figure 2.43 RCS of missile with fins for vertical polarization (E-vector normal to plane in which aspect is measured, two of the fins contained in that plane): × × × × experimental points; - - - theory (in-phase estimation); —— theory (random phase estimation) (from [72]).

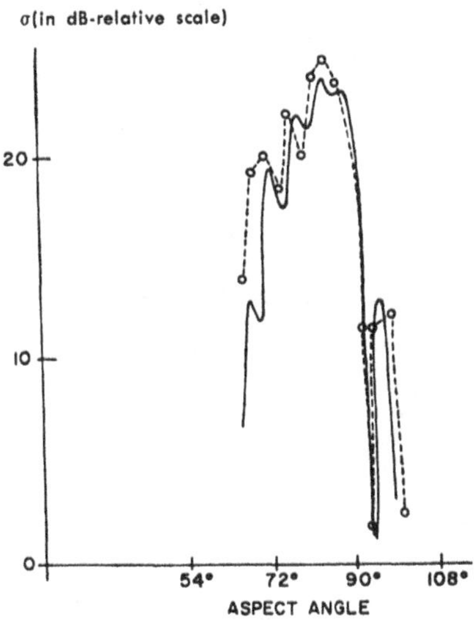

Figure 2.44 RCS of a missile near broadside: × × × × experimental data, - - - theory (relative phase method) (from [72]).

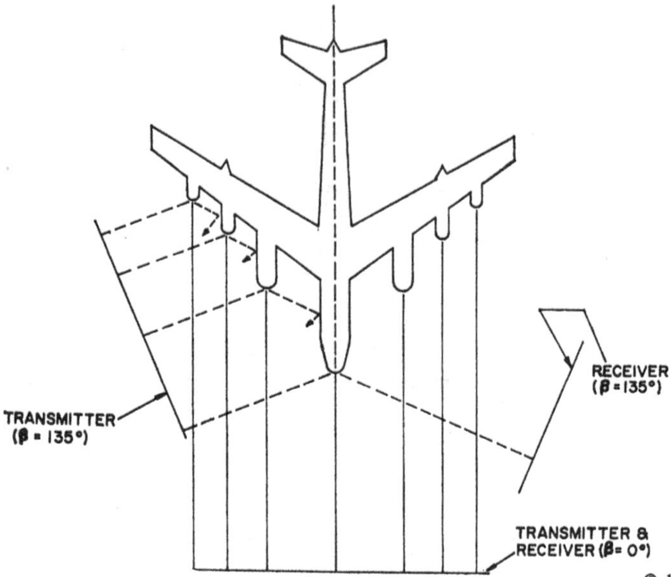

Figure 2.45 Geometry of the aircraft scattering (from [72]).

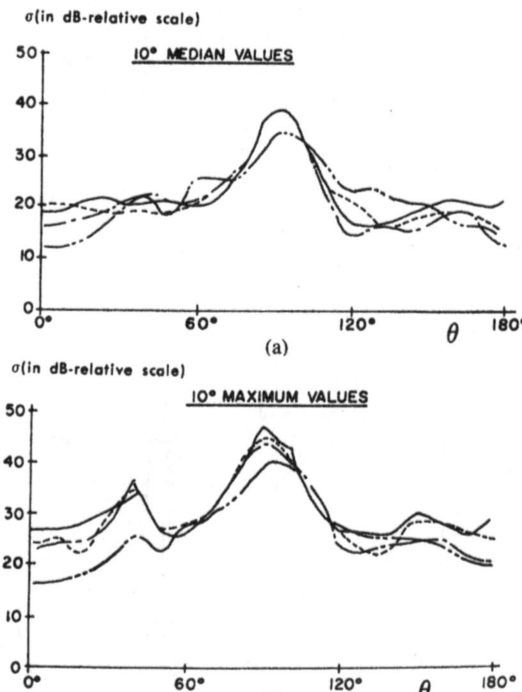

Figure 2.46 Monostatic and bistatic scattering from the nose of a large aircraft (a) monostatic; (b) bistatic (from [72]).

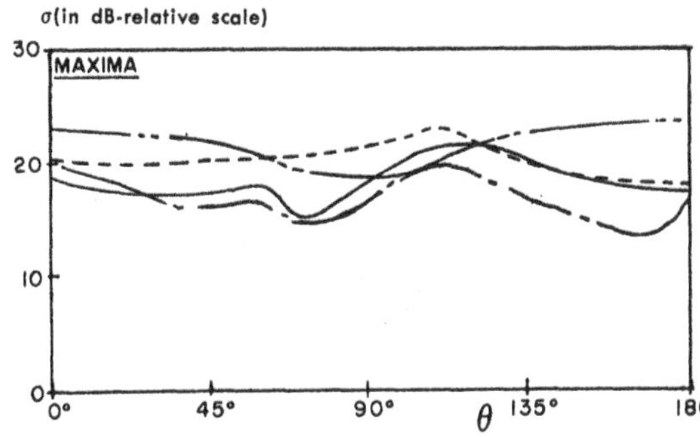

(a)

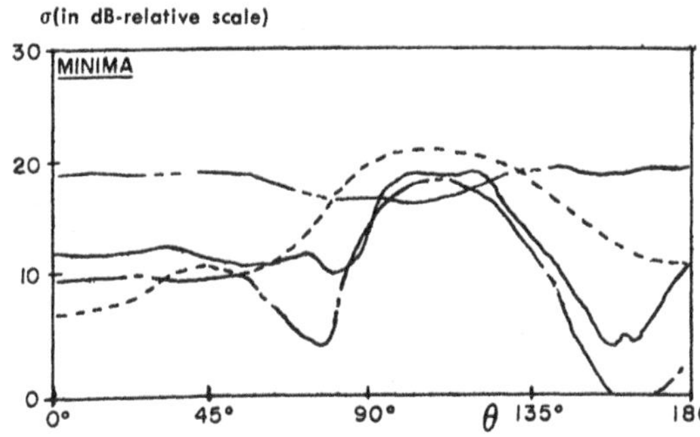

(b)

Figure 2.47 Monostatic and bistatic scattering from the nose of a large aircraft —— $\beta = 0°$; $\beta = 30°$; $\beta = 60°$; $\beta = 135°$; $f = 250$ MHz (from [72]): —— $\beta = 0°$; — · — · $\beta = 40°$; - - - - $\beta = 90°$; — · · — · · — $\beta = 150°$.

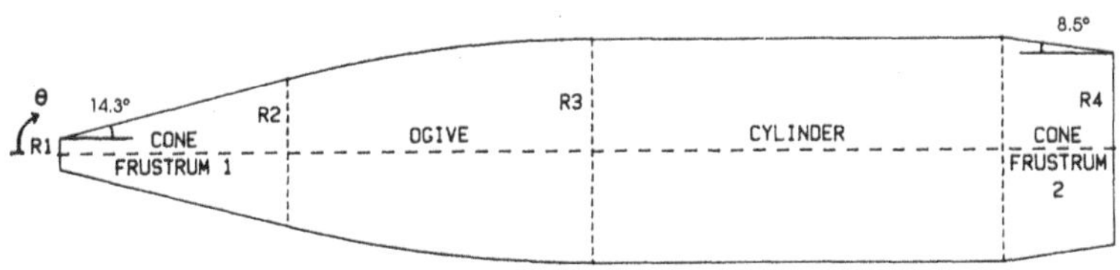

Figure 2.48 Sketch of the M105 shell model.

Table 2.1
Backscattering Cross Section for the Shell of Figure 2.48 as a Function of the Aspect Angle

$\underline{\theta = 0}$

$$\sigma(0) = \pi k_0^2 r_1^4$$

$\underline{\theta = \frac{\pi}{2} - \alpha_1}$

$$\sigma(\theta) = \frac{4}{9} k_0 \cos\alpha_2 \cot^2\alpha_1 \left(r_2^{3/2} - r_1^{3/2}\right)^2$$

$\underline{\theta = \frac{\pi}{2}}$

$$\sigma_{\perp,\parallel}\left(\frac{\pi}{2}\right) = k_0 r_3 (l_3 - l_2)^2 \left|1 + \frac{i}{2k_0 r_3}\left(-\frac{3}{8} \mp 1\right)\right|^2$$

$\underline{\theta = \frac{\pi}{2} + \alpha_2}$

$$\sigma(\theta) = \frac{4}{9} k_0 \cos\alpha_2 \cot^2\alpha_2 \left(r_3^{3/2} - r_4^{3/2}\right)^2$$

$\underline{\theta = \pi}$

$$\sigma(\pi) = \pi k_0^2 r_4^4$$

$\underline{0 < \theta < \theta_1}$

$$\sigma(\theta) = \pi r_1^2 \left[\frac{J_1(2k_0 r_1 \sin\theta)}{\tan\theta}\right]^2$$

$$\sigma_{\perp,\parallel}(\theta) = 4\pi \left| \frac{r_0}{2}\left[1 - \frac{r_0 - r_3}{r_0 \sin\theta}\right]^{1/2} \right.$$
$$\times \exp\left[2k_0 i\left(l_2\cos\theta + (r_0 - r_3)\sin\theta - r_0\right)\right]$$
$$\mp D_{\perp,\parallel}\left(\frac{\pi}{2} + \theta, n_1\right)\rho(r_1)^{1/2}\exp(-i\psi_1)$$
$$\mp D_{\perp,\parallel}\left(\frac{\pi}{2} - \theta, n_1\right)\rho(-r_1)^{1/2}\exp(i\psi_1)$$
$$\mp D_{\perp,\parallel}(\theta, n_2)\rho(r_3)^{1/2}\exp(i\psi_2)$$
$$\left. \mp D_{\perp,\parallel}(\theta - \alpha_2, n_3)\rho(r_4)^{1/2}\exp(i\psi_4) \right|^2$$

$\underline{\text{For } \frac{\pi}{2} < \theta < \pi - \alpha_1}$

$$\sigma_{\perp,\parallel}(\theta) = 4\pi \left| D_{\perp,\parallel}(\pi - \theta - \alpha_1, n_1)\rho(r_1)^{1/2}\exp(-i\psi_1) \right.$$
$$+ D_{\perp,\parallel}(\pi - \theta + \alpha_2, n_2)\rho(r_3)^{1/2}\exp(i\psi_2)$$
$$+ D_{\perp,\parallel}\left(\frac{3\pi}{2} - \theta, n_3\right)\rho(r_4)^{1/2}\exp(i\psi_4)$$
$$\left. + D_{\perp,\parallel}\left(\theta - \frac{\pi}{2}, n_3\right)\rho(-r_4)^{1/2}\exp(i\psi_5) \right|^2$$

$\theta_1 < \theta < \alpha_2$

$$\sigma_{\perp,\parallel} = 4\pi \left| D_{\perp,\parallel}\left(\frac{\pi}{2} + \theta, n_1\right)\rho(r_1)^{1/2}\exp(-i\psi_1) + D_{\perp,\parallel}\left(\frac{\pi}{2} - \theta, n_1\right)\rho(-r_1)^{1/2}\exp(1\psi_1) + D_{\perp,\parallel}(\theta, n_2)\rho(r_3)^{1/2}\exp(i\psi_2)\right|^2$$

For $\alpha_2 < \theta < \frac{\pi}{2} - \alpha_1$

$$\sigma_{\perp,\parallel}(\theta) = 4\pi \left| D_{\perp,\parallel}\left(\frac{\pi}{2} + \theta, n_1\right)\rho(r_1)^{1/2}\exp(-i\psi_1) + D_{\perp,\parallel}\left(\frac{\pi}{2} - \theta, n_1\right)\rho(-r_1)^{1/2}\exp(i\psi_1) + D_{\perp,\parallel}(\theta, n_2)\rho(r_3)^{1/2}\exp(i\psi_2) + D_{\perp,\parallel}(\theta - \alpha_2, n_3)\rho(r_4)^{1/2}\exp(i\psi_4)\right|^2$$

For $\frac{\pi}{2} - \alpha_1 < \theta < \frac{\pi}{2}$

For $\pi - \alpha_1 < \theta - \pi - \alpha_2$

$$\sigma_{\perp,\parallel}(\theta) = 4\pi \left| D_{\perp,\parallel}(\pi - \theta + \alpha_2, n_2)\rho(r_3)^{1/2}\exp(i\psi_2) + D_{\perp,\parallel}\left(\frac{3\pi}{2} - \theta, n_3\right)\rho(r_4)^{1/2}\exp(i\psi_4) + D_{\perp,\parallel}\left(\theta - \frac{\pi}{2}, n_3\right)\rho(-r_4)^{1/2}\exp(i\psi_5)\right|^2$$

For $\pi - \alpha_2 < \theta < \pi - \theta_2$

$$\sigma_{\perp,\parallel}(\theta) = 4\pi \left| D_{\perp,\parallel}(\pi - \theta + \alpha_2, n_2)\rho(r_3)^{1/2}\exp(i\psi_2) + D_{\perp,\parallel}(\theta - \pi + \alpha_2, n_2)\rho(-r_3)^{1/2}\exp(i\psi_3) + D_{\perp,\parallel}\left(\frac{3\pi}{2} - \theta, n_3\right)\rho(r_4)^{1/2}\exp(i\psi_4) + D_{\perp,\parallel}\left(\theta - \frac{\pi}{2}, n_3\right)\rho(-r_4)^{1/2}\exp(i\psi_5)\right|^2$$

For $\theta_2 < \theta < \pi$

$$\sigma(\theta) = \pi r_4^2 \left[\frac{J_1(2k_0 r_4 \sin\theta)}{\tan\theta}\right]^2$$

$(\pi/2 + \alpha_2)$, π, and $\theta < 0 < \theta$ (nonzero), are obtained by the PO expressions. For angles other than these, RCS is obtained by the GTD and superposition of contributions from individual scattering centers determined by the range of aspect angles of interest. The contribution to the scattered field by each scattering center is obtained by multiplying the incident field at the position of the scattering center by the diffraction coefficient and the associated phase contributions.

The quantities σ_1, σ_2, γ_0 in Table 2.1 are defined as

$$\alpha_1 = \tan^{-1}\frac{\gamma_2 - \gamma_1}{l_1} = 14.3°$$

$$\alpha_2 = \tan^{-1}\frac{\gamma_3 - \gamma_4}{l_4 - l_3} = 8.5°$$

$$\gamma_0 = \frac{\gamma_3 - \gamma_2}{2} + \frac{(l_2 - l_1)^2}{8(\gamma_3 - \gamma_2)} = 568.2 \text{ mm}$$

where r_0 is the radius of curvature of the ogive, $n_1 = 3/2 - \alpha_1/\pi$, $n_2 = 1 + \alpha_2/\pi$, and $n_3 = 3/2 - \alpha_3/\pi$. The phases of the diffracted rays, referred to a reference plane, are given by

$$\psi_1 = 2k_0 r_1 \sin\theta$$

$$\psi_2 = -2k_0 r_3 \sin\theta + 2k_0 l_3 \cos\theta$$

$$\psi_3 = +2k_0 r_3 \sin\theta + 2k_0 l_3 \cos\theta$$

$$\psi_4 = 2k_0 r_4 \sin\theta + 2k_0 l_4 \cos\theta$$

$$\psi_5 = 2k_0 r_4 \sin\theta + 2k_0 l_4 \cos\theta$$

The diffraction coefficients listed in Table 2.1 are

$$D_{\perp/\parallel},(\varphi, n) = \frac{\exp(i\pi/4)\sin(\pi/n)}{n\sqrt{2\pi k_0}}\left[\left(\cos\frac{\pi}{n} - 1\right)^{-1} \mp \left(\cos\frac{\pi}{n} - \frac{2\varphi}{n}\right)^{-1}\right]$$

where the upper and lower signs on the right-hand side of the equations are to be used with $\perp$ and $\parallel$, respectively, on the right-hand side. Numerical RCS results obtained with this model are discussed in [77].

2.10.3.3 Summary of Other Approaches

Lin and Richmond [76] used a wire grid model to simulate an aircraft at low frequencies and Galerkin's method to obtain numerical results. They also studied the effects of increasing the number of segments in wire grid model. Specific models used for the MG19 aircraft, and the numerical and experimental results obtained for the aircraft at two different frequencies, are shown in Figure 2.49.

A number of publications describe various approaches used for complex target modeling and the computer codes developed to obtain specific results [69, 70, 78, 79]. Electromagnetic surface patch modeling of perfectly conducting structures using polygonal plates is described and illustrated by examples in [70, 80]; some results obtained by this technique are given in Figures 1.9–1.11.

A summary of the computer code RECOTA (*Return from complex target*), developed and tested at Boeing Aerospace, is described in [71]. The code is based on a computer-aided design package that carries out target modeling in terms of facets and wedges and uses PO, PTD, and semiempirical formulations; it also can account for the effects of shadowing, multiple scattering, and discontinuities for monostatic calculations. Figure 2.50 gives a missile model and its RECOTA predicted and measured RCS at 12 GHz. The RECOTA is designed to provide the following advantages:

1. No limitation on target geometry;
2. No limitation on the shape of the components;
3. Flexibility in changing the shape;
4. Phase calculations already incorporated;
5. Ability to consider multiple scattering;
6. Ability to take into account the shadow effects;
7. Can take into account the effect of dielectric materials placed at any portion of the target surface.

Recently, more realistic models of aircraft have been proposed by fairing the junctions between basic shapes [80, 81, 82]. Figures 2.51 (a) and (b) show typical faired and unfaired test bodies, and Figure 2.52 shows the model of an airplane composed of basic shapes with faired junctions. Computed X-band total RCS maps for the faired and unfaired test bodies are shown in Figure 2.53. Computed and measured RCS results for the unfaired body are given in Figure 2.54. These results indicate that the envelope techniques described in [80, 82] appear to be useful for efficient computation and display of full volumetric RCS patterns.

2.11 CROSS-POLARIZED RCS

In this section, we discuss certain aspects of the cross-polarization phenomenon and models developed to estimate the cross-polarized RCS of targets; and we illustrate them with selected results. In radar terminology the topic comes under the general subject named *polarimetric scattering*, where the scattered field having polarization different than that of the incident fields also is of interest. In particular, we shall discuss the phenomenon of cross polarization that generally has different definitions [84]. We shall

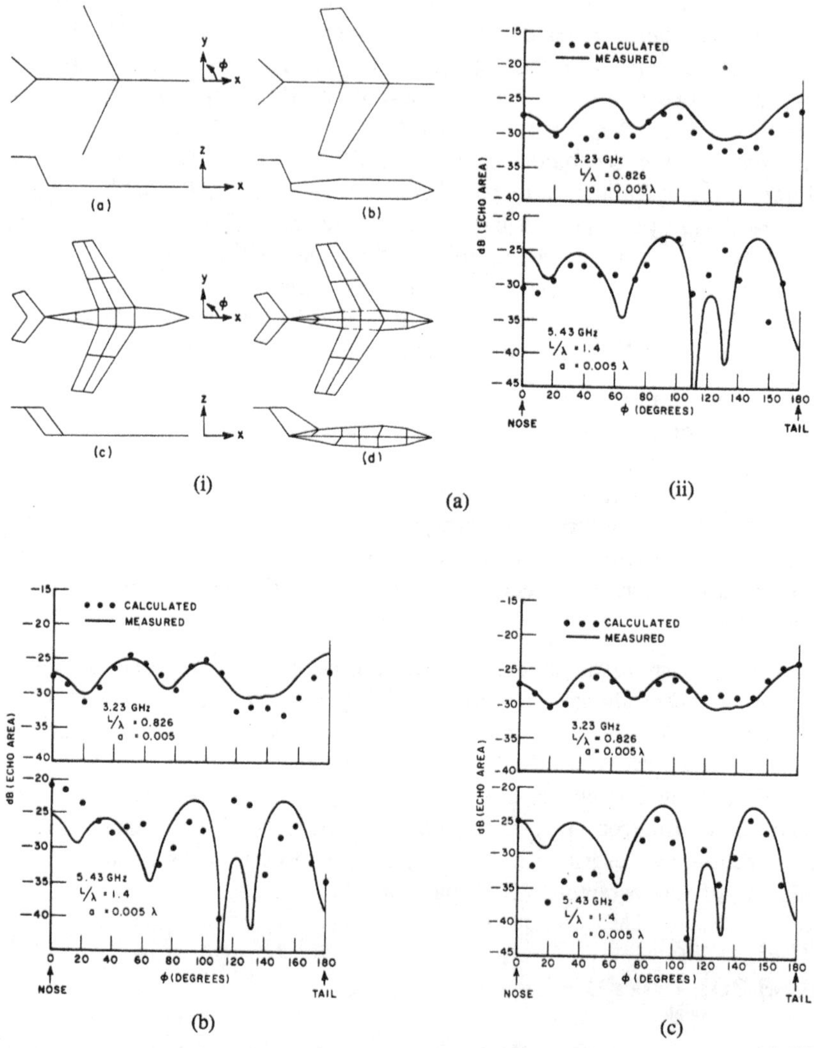

Figure 2.49 Backscattering echo area *versus* aspect angle for a MIG19 aircraft. (i) Several wire-grid models: (a) 8-segment model, (b) 18-segment model, (c) 45-segment model, (d) 70-segment model; (ii) backscattering characteristics: (a) 8-segment model, (b) 18-segment model, (c) 45-segment model (from [76]).

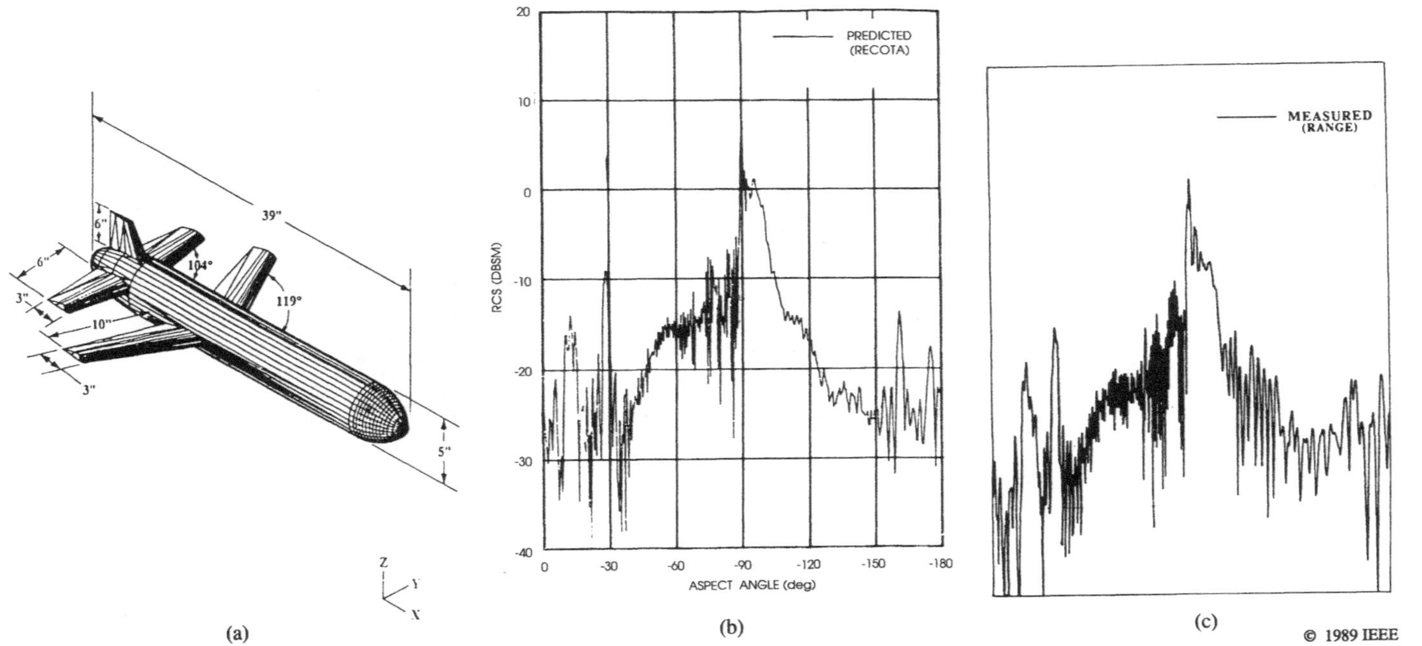

Figure 2.50 RECOTA predicted results for a missile model: (a) the missile model; (b) RECOTA predicted theoretical values; (c) measured values; $f = 12$ GHz; pitch = $-10.5°$; roll = $0°$; horizontal polarization (from [71]).

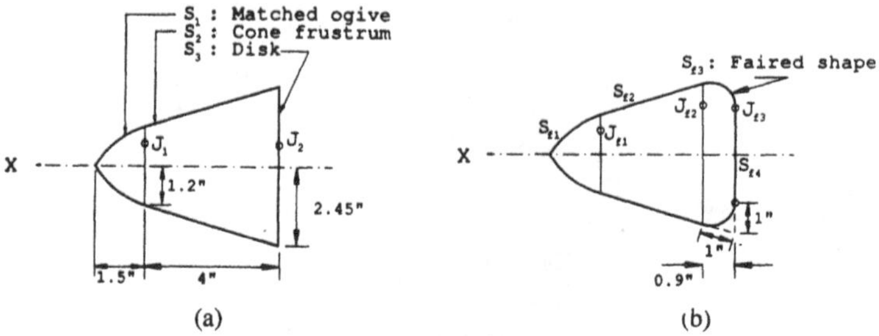

Figure 2.51 Test body (a) unfaired (b) faired (from [81]).

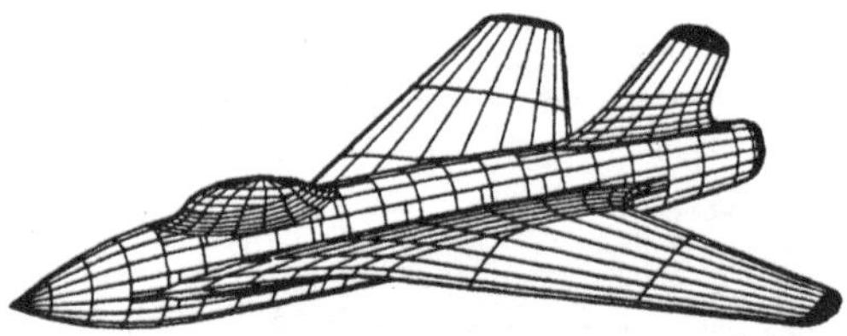

Figure 2.52 Airplane composed of basic shapes with faired junctions (from [81]).

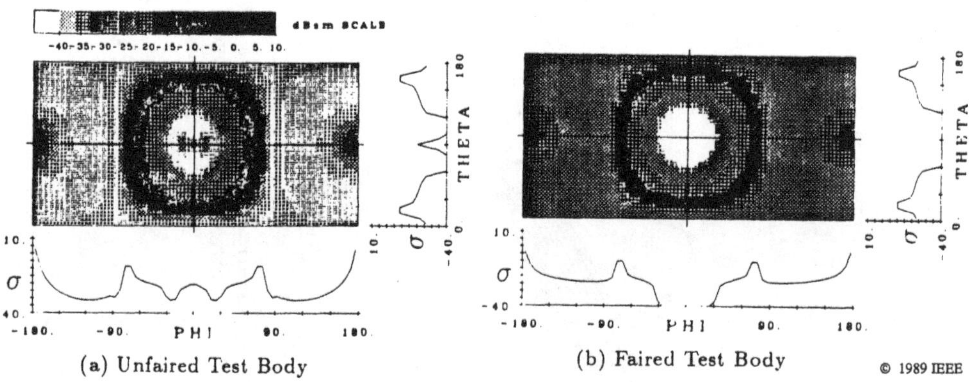

Figure 2.53 Computed total RCS maps of the faired and unfaired test bodies with ϕ-polarization at 10 GHz (from [81]).

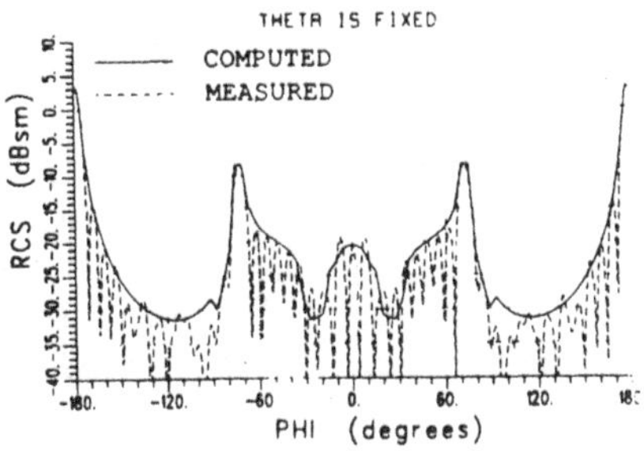

Figure 2.54 Results for the unfaired body at 10 GHz with ϕ-polarization (from [81]).

use the IEEE definition: "In a specified plane containing the reference polarization ellipse, the polarization orthogonal to a specified reference polarization is called the cross polarization." The reference polarization is called the *copolarization* (or copol).

Early designers and users of radar systems were satisfied with the capability of the radar to process only the copolarized returns from targets. Because any geometrical or material constraint on the target or absence of symmetry in the target can cause depolarization in the scattered field, we expect the cross-polarized component of the radar return to contain more detailed and different information about the target than the co-polarized returns. Hence, a need has arisen for knowledge of cross-polarization characteristics of targets in a variety of applications; for example, terrain mapping, terrain classification, and target identification. Current radars therefore are required to be polarimetric, which in turn requires that scattering models developed for estimation of scattering by targets also be capable of providing their depolarization characteristics.

Not much information about the depolarization phenomenon in radar target scattering is available in the open literature. An early theoretical work on the subject was by Knott and Senior [84, 85] who showed that the parallel (i.e., parallel to the incident plane) and perpendicular parts of the complex scattering matrix for a conducting target can be represented by

$$S_{\parallel} = 1/2(S_V + S_H) + 1/2(S_V - S_H)\cos 2\beta \tag{2.85}$$

$$S_{\perp} = -1/2(S_V - S_H)\sin 2\beta \tag{2.86}$$

where β is the angle between the incident polarization and the vertical direction, and S_V and S_H are the complex scattering matrices for vertical and horizontal polarizations, respectively, given by

$$S_V = \frac{\mathrm{i}k^2 z_0}{4\pi} \int_S \hat{\mathbf{y}} \cdot \mathbf{J}(\mathbf{r}) \mathrm{e}^{-\mathrm{i}k\hat{r}\cdot\mathbf{r}'}\,\mathrm{d}s' \tag{2.87}$$

$$S_H = \frac{\mathrm{i}k^2 z_0}{4\pi} \int_S \hat{r} \cdot \hat{y} X \mathbf{J}(\mathbf{r}')\,\mathrm{e}^{-\mathrm{i}k\hat{r}\cdot\mathbf{r}'}\,\mathrm{d}s' \tag{2.88}$$

where $\hat{y}$ is defined to be the vertical direction, $\hat{r}$ is the unit vector in the direction of the field point, $\mathbf{J}(\mathbf{r})$ is the surface current at $\mathbf{r}'$ on the target with surface S, and the other parameters are as described in Figure 2.5 above.

In general, S_V and S_H are different, implying thereby from (2.85) and (2.86) that depolarization occurs for all β except $\beta = 0$ and $\pi/2$; that is, where the incidences are out of the plane of symmetry of the scatterer. Note that to obtain cross-polarized behavior from these equations, $J(r')$ in (2.87) and (2.88) must be non-PO current because, as indicated earlier, the PO method does not predict depolarization even if the incidences are out of the symmetry plane of the scatterer. Figure 2.55 [84] gives the normalized cross-polarized RCS *versus* the normalized linear dimension in order of decreasing edge contributions for four types of targets: thin disks, right circular cones, cones with rounded edges, and cone-spheres. The values on the graphs are averages of the peak returns of each RCS pattern closest to, and on either side of, the axial incidence for a given *ka*. As expected, disks and sharp-edged cones cause strongest cross-polarized RCS. From Figure 2.55 also note that the difference between the cone

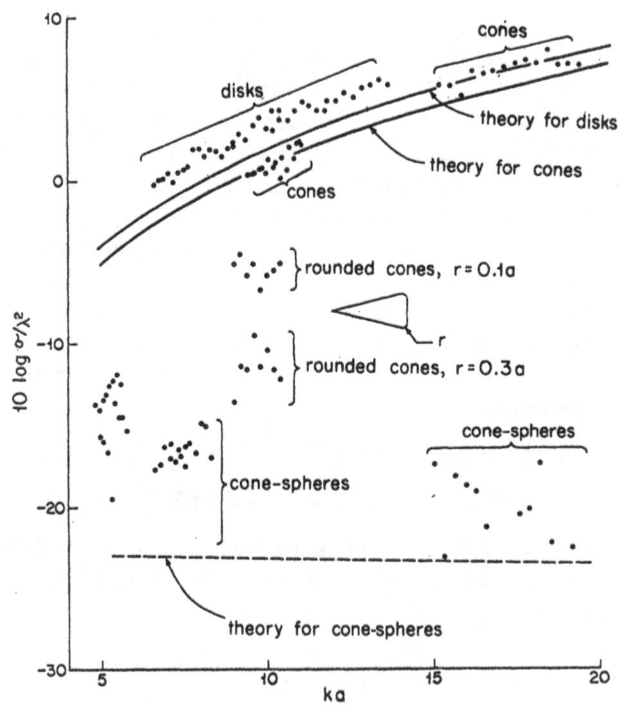

Figure 2.55 Cross-polarized radar cross sections for different targets in order of decreasing edge diffractions (from [84]).

and cone-sphere returns is as large as 30 dB for $ka = 20$; hence, the cone-sphere is the weakest depolarizing target among the four considered in Figure 2.55.

Depolarization properties of an infinite (conducting) cylinder illuminated by circularly polarized plane waves have been discussed in [86], which gives the backscattered fields from the conducting cylinder (oriented along the z-direction of a cylindrical coordinate system ρ, φ, z):

$$\mathbf{E}^S = \left(\frac{2i}{\pi k\rho}\right)^{1/2} \mathrm{e}^{-ik\rho}\left[-\hat{v}E_v \mathrm{e}^{i\psi_v} + i\hat{h}E_h \mathrm{e}^{i\psi_h}\right] \tag{2.89}$$

where

$$E_v \mathrm{e}^{i\psi_v} = E_0 \sum_{n=0}^{\infty} \epsilon_n(-1)^n \frac{J_n'(ka)}{H_n^{(2)\prime}(ka)} \tag{2.90}$$

$$E_h \mathrm{e}^{i\psi_n} = E_0 \sum_{n=0}^{\infty} \epsilon_n(-1)^n \frac{J_n(ka)}{H_n^{(2)}(ka)} \tag{2.91}$$

$$\epsilon_n = 2 \quad \text{if } n = 0$$

$$= 1 \quad n \neq 0$$

E_0 is the amplitude of the incident electric field polarized, $\hat{v}(\varphi)$ and $\hat{h}(z)$ are unit vectors in the vertical and horizontal directions, respectively, which also are transverse and parallel to the cylinder.

The scattered field given by (2.89) generally is elliptically polarized. Let v_c and l_c refer to the unit vectors in the *right circular* (RC) and *left circular* (LC) polarization directions of the scattered field. These can be obtained by using the transformation given by (1.40) and (1.41). An expression for the complex circular depolarization is now obtained by taking the ratio of the RC to the LC scattered field, given

$$D\,\mathrm{e}^{i(\psi_r-\psi_e)} = \left(\frac{E_r}{E_e}\right)\mathrm{e}^{i(\psi_r-\psi_e)} = \frac{-E_h \mathrm{e}^{i\psi_h} - E_r \mathrm{e}^{i\psi_v}}{E_h \mathrm{e}^{i\psi_h} - E_v \mathrm{e}^{i\psi_v}} \tag{2.92}$$

where (E_r, E_l), (ψ_r, ψ_l) now represent the amplitude and phase, respectively, of the right and left circularly polarized scattered fields; D is the amplitude of the complex depolarization, and $(\psi_r - \psi_l)$ is phase associated with depolarization.

The backscattered cross section $\sigma_{\alpha\beta}$ for any two-dimensional target like the infinite cylinder is defined as

$$\sigma_{\alpha\beta}(\text{per meter}) = \lim_{\rho\to\infty} 2\pi\rho\left|\frac{E_\alpha}{E_\beta}\right|^2$$

where α and β stand for (r, l) or (v, h) respectively. Using (2.89)–(2.93) σ_{rr} and σ_{ll} are obtained as

$$\left.\begin{matrix}\sigma_{rr}\\ \sigma_{11}\end{matrix}\right\} = 1/2\left[\sigma_{vv} + \sigma_{hh} \pm 2(\sigma_{vv}\sigma_{hh})^{1/2}\cos(\psi_h - \psi_v)\right] \tag{2.94}$$

Expressions for ψ_r and ψ_l are given in [86].

Depolarization of circularly polarized waves by an infinite cylinder has been investigated in [87]. Figure 2.56 compares the results between theory and experiment for amplitude and phase of the complex depolarization due to an infinite cylinder of variable radius. An important observation from Figure 2.56 is that care must be taken to calibrate circularly polarized radars with circular cylinders as references because the amplitude and phase of circular complex depolarization is constant for $ka < 0.1$, where a is the radius of the cylinder (or wire). Figures 2.57 and 2.58 show radar widths and normalized phases, respectively, for various cases.

Cross polarization due to scattering by metallic disks have been studied in [88, 89]. Figure 2.59 gives measured and calculated values of cross-polarized RCS for a disk illuminated by a plane wave with polarization angle 45°. Figures 2.60 and 2.61 [16] give a comparison of theory and experiment for two different sizes of disks.

Figure 2.62(a) shows the geometry and axis of rotation with additional surfaces that generally cause strong depolarization. Figure 2.62(b) depicts theoretical and experimental cross-polarized RCS of a cube with shadowing surface.

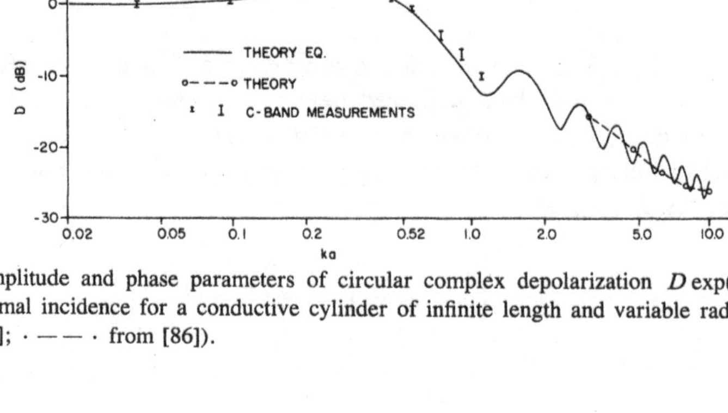

Figure 2.56 Amplitude and phase parameters of circular complex depolarization $D\exp(j(x\psi_r - \psi_l))$ at normal incidence for a conductive cylinder of infinite length and variable radius (—— from [87]; · —— · from [86]).

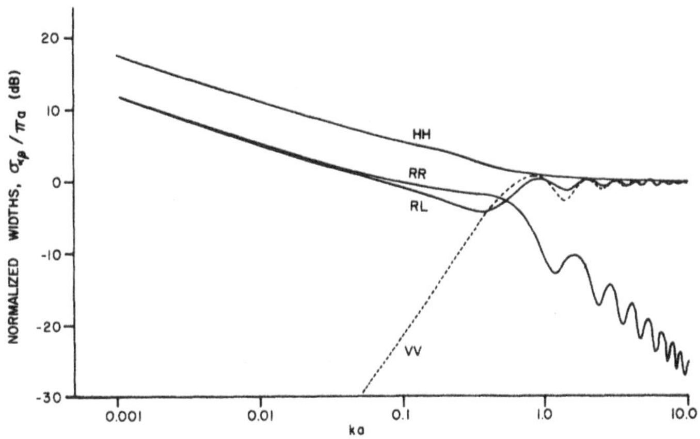

Figure 2.57 Normalized radar widths for linear and circular polarizations at normal incidence for a conductive cylinder of infinite length and variable radius (from [87]).

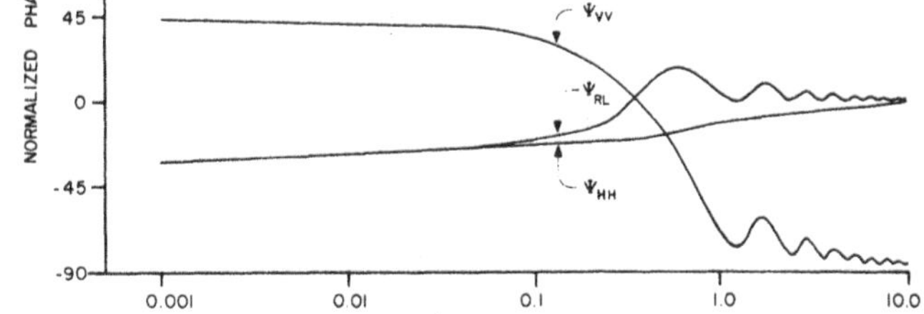

Figure 2.58 Normalized field phases $\psi_{\alpha\beta}$ for linear and circular polarizations at normal incidence for a conductive cylinder of infinite length and variable radius (from [87]).

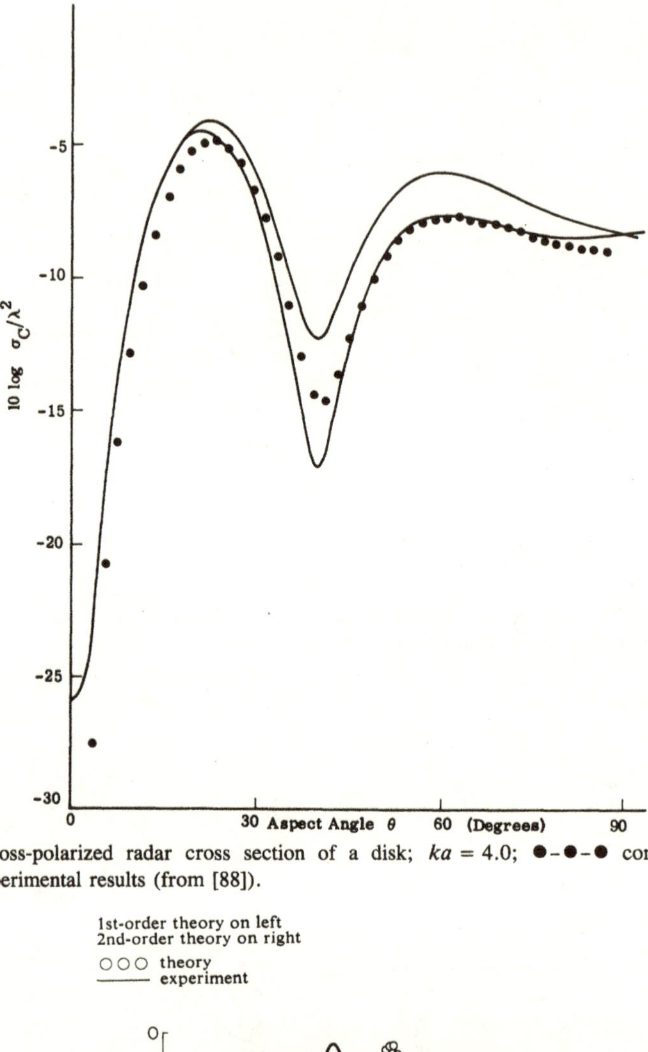

Figure 2.59 Cross-polarized radar cross section of a disk; ka = 4.0; ●-●-● computed values, —— experimental results (from [88]).

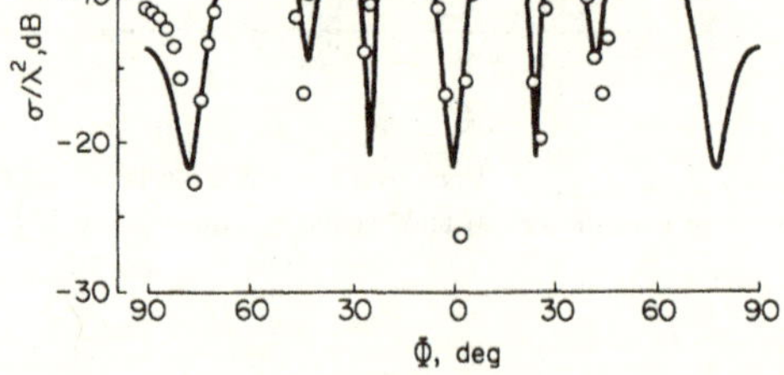

Figure 2.60 Comparison of theory with experiment for cross-polarization with ka = 6 for a disk, first-order theory on left; second-order theory on right; theory, ○-○-○; experiment, —— (from [54]).

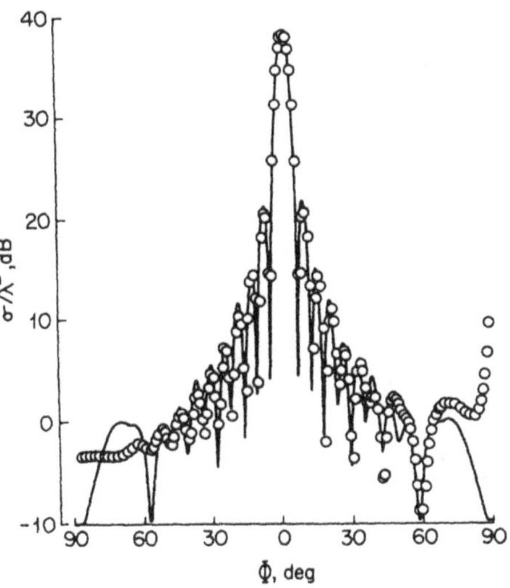

Figure 2.61 Comparison of theory with experiment for cross-polarization with $ka_{\mathrm{eff}} = 17.155$ for a disk, first-order theory on left; second-order theory on right;——theory; O-O-O, experiment (from [54]).

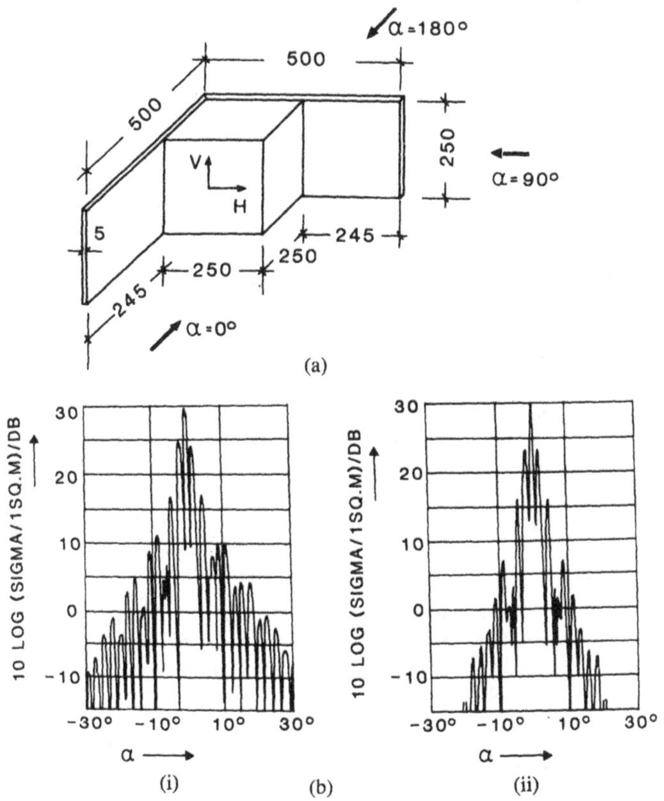

Figure 2.62 (a) Cubic with shadowing surfaces (all dimensions are in mm); (b) cross-polarized RCS of cube with shadowing surfaces: (i) experimental results; (ii) theoretical results (from [89]).

2.12 COMPARISON OF DIFFERENT METHODS

Different methods available for RCS analysis and their applications have been discussed in the previous sections. Table 2.2 now compares the advantages and disadvantages of these methods, and where a specific method should and should not be applied. Although the choice of a specific technique for a given practical problem is evident from the

Table 2.2
Comparison of Methods of RCS Analysis

Method	Advantages and Applications	Limitations and Comments	Recommendation
Geometrical optics (GO)	Works well for calculations of (1) ray optical field in a ray tube (convergent or divergent); (2) reflected field by tangent plane approximation for curved and plane surfaces. Good for ray tracings that do not encounter shadow boundaries and caustics.	Cannot be applied to surfaces whose radii of curvatures are not very large compared to wavelength. Cannot take care of phase of the field, diffraction effects, caustic regions, shadow boundaries, or shadow region fields.	Good for (1) ray field estimation in a beam in uniform medium; (2) plane wave reflection from an infinite plane or curved surface; (3) ray tracing for a given scenario.
Physical optics (PO)	Useful for calculation of near- or far-zone scattered field from a surface or antenna. Gives good results for $\pm 40°$ from normal incidence by thumb rule. It can treat caustics and shadow region fields without singularities.	A severe limitation is that classical PO method cannot predict currents due to a discontinuity on a body; cannot handle depolarization for a monostatic single scatterer; cannot predict diffraction effects and, hence, shadow region fields. Inaccurate for wide-angle lobe predictions.	Use anywhere where we need an accurate field over a relatively small angular section from normal to the source current, with the limitations and, hence, modifications in mind. First-order cross-pol estimation is possible.
Physical theory of diffraction (PTD)	No diffraction coefficients need be postulated. Gives very good results once the fringe current is available. Fields are finite in and around the shadow boundaries.	Expression for fringe current is difficult to achieve in many cases.	Can be applied to a large number of cases including wide-angle and cross-pol estimations.

Table 2.2 *Continued*

Method	Advantages and Applications	Limitations and Comments	Recommendation
Integral equation method and methods of moments (MOM)	Any problems can be formulated using this method in terms of the unknown field quantities. Can be solved by MOM. ISB and RB caustics and shadow regions are taken care of.	Normally, for electrically large structures, the computer memory requirement and processing time may be large. Special fast algorithms to be used for such cases.	Apply where the diffraction coefficient is not available. Recommended when the body is not electrically large.
Hybrid method (HM)	Reduces computer time and memory requirements. Does not matter if the diffraction coefficient is not known. Finds a numerical diffraction coefficient for hybrid method involving GTD.	The combination of different techniques may vary with the type and complexity of the problems.	Has been tested for a limited number of cases, and the results seem to be promising.
Keller geometrical theory of diffraction (Keller's GTD)	First time a diffraction coefficient for scattered field calculation was provided for a user. Very good for monostatic calculations.	Cannot handle incident and reflection shadow boundaries or caustic regions, cannot predict cross-polarization, violates edge condition.	Use for monostatic calculations with accuracy over wide angles of observation. Do not use if in a region where ISB, RB, or caustic are present. Not accurate near edge.
Uniform theories of diffraction (UTD, UAT)	The restrictions of Keller's GTD at and around ISB and RB have been removed. UTD has a diffraction coefficient; and UAT uses asymptotic fields. Scope of the GTD has been widely extended.	Cannot handle caustics. For a two-edge problem, (example, a flat plate for grazing incidence) when one edge is in the SB of the other, the incident field in the first edge is a nonoptical field. Cannot be applied directly. Needs special attention. UTD needs a slope diffraction correction when the edge is close to the source to accommodate nonuniformity of edge illumination. Treatment of cross polarization is not possible.	Can be applied to a large variety of radiation, scattering, and coupling problems.

discussions given earlier, a comparative evaluation of their applicability, advantages and limitations is summarized in Table 2.2.

REFERENCES

1. R. F. Harrington, *Time-Harmonic Electromagnetic Fields*, McGraw-Hill, 1968.
2. J. J. Bowman, T. B. A. Senior, and P. L. E. Uslenghi, *Electromagnetic and Acoustic Scattering by Simple Shapes*, North-Holland Publishing, Amsterdam, 1969.
3. M. Kline and I. W. Kay, *Electromagnetic Theory and Geometrical Optics*, John Wiley and Sons, New York, 1965.
4. W. D. Burnside, C. L. Yu, and R. J. Marhefka, "A Technique to Combine the Geometrical Theory of Diffraction and the Moment Method," *IEEE Trans. on Antennas and Propagation*, Vol. AP-23, No. 4, July 1975, pp. 551–558.
5. J. N. Saholos and G. A. Thiele, "On the Applications of the GTD-MM Techniques and Its Limitations," *IEEE Trans. on Antennas and Propagation*, Vol. AP-29, No. 5, September 1981, pp. 780–786.
6. G. T. Ruck, ed., *Radar Cross Section Handbook*, Plenum Press, New York, pp. 28–139.
7. J. W. Crispin, ed., *Methods of Radar Cross Section Analysis*, Academic Press, New York, 1968.
8. E. F. Knott *et al.*, *Radar Cross Section*, Artech House, Dedham, MA, Chap. 5, 1985.
9. R. E. Kell and R. A. Ross, "Radar Cross Section of Targets," Chap. 27 in M. I. Skolnik, ed., *Radar Handbook*, McGraw-Hill, New York, 1970.
10. E. F. Knott, "A Progression of High Frequency RCS Prediction Techniques," *Proc. IEEE*, Vol. 73, No. 2, February 1985, pp. 252–264.
11. E. F. Knott and T. B. A. Senior, "Comparison of Three High-Frequency Diffraction Techniques," *Proc. IEEE*, Vol. 62, No. 11, November 1974, pp. 1468–1474.
12. J. B. Keller, R. M. Lewis, and B. D. Seckler, "Asymptotic Solution of Some Diffraction Problem," *Comm. Pure Applied Math.*, Vol. 9, 1956, pp. 207–265.
13. P. C. Fritch, ed., "Special Issue on Radar Reflectivity," *Proc. IEEE*, Vol. 53, No. 8, August 1965.
14. D. L. Sengupta, "The Sphere," Chap. 10; T. B. A. Senior and P. L. E. Uslenghi, "The Circular Cylinder," Chap. 2; J. J. Bowman, "The Cone," Chapter 18 in [2].
15. S. W. Lee *et al.*, "Twelve Versions of Physical Optics: How Do They Compare?" 1988 *IEEE AP-5 Symp. Dig.*, Syracuse, New York, pp. 408–411.
16. D. Klement, J. Preissner, and V. Stein, "Special Problems in Applying the Physical Optics Method for Backscatter Computations of Complicated Objects," *IEEE Trans. on Antennas and Propagation*, Vol. AP-36, No. 2, February 1988, pp. 228–237.
17. J. B. Keller, "Geometrical Theory of Diffraction," *J. Optics Soc. of America*, Vol. 52, 1962, pp. 116–130.
18. J. B. Keller, "Backscattering from a Finite Cone-Theory and Experiment," *IEEE Trans. on Antennas and Propagation*, Vol. AP-8, No. 2, March 1960, pp. 411–412.
19. M. E. Bechtel and R. A. Ross, "Radar Scattering Analysis," CAL Report No. ER/RIS-10, Cornell Aeronautical Laboratory, Buffalo, New York, August 1966.
20. D. S. Ahluwalia, R. M. Lewis, and J. Boersma, "Uniform Asymptotic Theory of Diffraction by a Plane Screen," *SIAM J. Applied Math.*, Vol. 16, 1968, pp. 783–807.
21. D. S. Ahluwalia, "Uniform Asymptotic Theory of Diffraction by the Edge of a Three-Dimensional Body," *SIAM J. Applied Math.*, Vol. 18, 1970, pp. 287–301.
22. S. W. Lee and G. A. Deschamps, "A Uniform Asymptotic Theory of Electromagnetic Diffraction by a Curved Wedge," *IEEE Trans. on Antennas and Propagation*, Vol. AP-24, No. 1, January 1976, pp. 25–34.
23. G. A. Deschamps, J. Boersma, and S. W. Lee, "Three-Dimensional Half Plane Diffraction: Exact

Solution and Testing of Uniform Theories," *IEEE Trans. on Antennas and Propagation*, Vol. AP-32, No. 2, March 1984, pp. 264–271.

24. R. G. Kouyoumjian and P. H. Pathak, "A Uniform Geometrical Theory of Diffraction for an Edge in a Perfectly Conducting Surface," *Proc. IEEE*, Vol. 62, No. 11, November 1974, pp. 1448–1461.
25. D. A. McNamara, C. W. I. Pistrorius, and J. A. G. Malherbe, *Introduction to the Uniform Geometrical Theory of Diffraction*, Artech House, Norwood, MA, 1990.
26. J. D. Cashman, Comments on "A Uniform Geometrical Theory of Diffraction for an Edge in Perfectly Conducting Surface," *IEEE Trans. on Antennas and Propagation*, Vol. AP-25, No. 3, May 1977, pp. 447–449.
27. J. A. Stratton, *Electromagnetic Theory*, McGraw-Hill, New York, 1941.
28. S. Silver, *Microwave Antenna Theory and Design*, McGraw-Hill, New York.
29. A. Sommerfeld, *Optics*, Academic Press, New York, 1954, pp. 325–328.
30. E. K. Miller and A. J. Poggio, "Moment Method Techniques in Electromagnetics from an Application Viewpoint," Chap. 9 in *Electromagnetic Scattering*, Academic Press, New York, 1978, P. L. E. Uslenghi, ed.
31. J. Van Bladel, *Electromagnetic Fields*, McGraw-Hill, New York, 1964.
32. R. F. Harrington, *Field Computation by Moment Methods*, Macmillan, New York, 1968.
33. S. W. Lee, "Comparison of Uniform Asymptotic Theory and Ufimtsev's Theory of Electromagnetic Edge Diffraction," *IEEE Trans. on Antennas and Propagation*, Vol. AP-25, No. 2, March 1977, pp. 162–170. See also P. Y. Ufimtsev, "Comments on Comparison of Three High-Frequency Diffraction Techniques," *Proc. IEEE*, Vol. 63, 1975, pp. 1736–1737.
34. P. Y. Ufimtsev, "Method of Edge Waves in the Physical Theory of Diffraction," Air Force System Command, Document ID No. FTD-HC-23-259-71, 1971.
35. J. B. Keller, "Diffraction by an Aperture," *J. Applied Physics*, Vol. 28, April 1957, pp. 626–666.
36. C. E. Ryan, Jr. and L. Peters, Jr., "Evaluation of Edge-Diffracted Fields Including Equivalent Currents for the Caustic Regions," *IEEE Trans. on Antennas and Propagation*, Vol. AP-18, No. 2, p. 275.
37. W. D. Burnside and L. Peters, "Axial Radar Cross Section of Finite Cones by the Equivalent-Current Concept with High-Order Diffraction," *Radio Science*, Vol. 7, No. 10, October 1972, pp. 943–948.
38. G. A. Thiele and T. H. Newhouse, "A Hybrid Technique for Combining Moment Methods with the GTD." *IEEE Trans. on Antennas and Propagation*, Vol. AP-23, No. 1, January 1975, pp. 62–69.
39. G. A. Thiele and T. H. Newhouse, "A Hybrid Technique for Combining Moment Methods with the GTD," *IEEE Trans. on Antennas and Propagation*, Vol. AP-23, No. 1, January 1975, pp. 62–69.
40. W. D. Burnside, C. L. Yu and R. J. Marhefka, "A Technique to Combine the Geometrical Theory of Diffraction and the Moment Method," *IEEE Trans. Antennas and Propagation*, Vol. AP-23, No. 4, July 1975, pp. 551–558.
41. S. Chang and V. V. Liepa, "Measured Backscattering Cross Section of Thin Wires," University of Michigan Radiation Laboratory, Rep. No. 8077-4-T, May 1967.
42. Richard H. Turpin, "Average RCS for Arbitrary Transmitter and Receiver Polarizations with Application to Wire Scatterers," *IEEE Trans. on Antennas and Propagation*, Vol. AP-19, No. 1, January 1971, pp. 139–161.
43. S. L. Borison, "Bistatic Scattering Cross Section of a Randomly Oriented Dipole," *IEEE Trans. on Antennas and Propagation*, Vol. AP-15, No. 2, March 1967, pp. 320–321.
44. C. J. Palerno and L. H. Baner, "Bistatic Scattering Cross Section of Chaff Dipoles with Applications to Communications," *Proc. IEEE*, Vol. 53, No. 8, August 1965, pp. 1119–1121.
45. J. W. Burns and T. B. A. Senior, "The Backscattered Field of a Thin Wire Loop for H-Polarization." *IEEE Trans. on Antennas and Propagation*, Vol. AP-35, No. 9, September 1987, pp. 1049–1057.
46. G. A. Thiele, "Radar Cross Section of Open Circular Loops," *IEEE Trans. on Antennas and Propagation*, Vol. AP-16, May 1968, pp. 373–375.
47. J. L. Lin, W. L. Curtis, and M. C. Vincent, "Radar Cross Section of a Rectangular Conducting Plate by Wire Mesh Modeling," *IEEE Trans. on Antennas and Propagation*, Vol. AP-22, September 1974, pp. 718–720.

48. C. H. Chan and R. Mittra, "Comparative Study of Iterative Techniques, Moment Method and Spectral Galenkin Approach for Solving the Problem of Electromagnetic Scattering by a Flat Plate," 1987 IEEE AP-S/URSI Meeting, Blacksburg, VA.
49. Y. Rahmat Sàmii and R. Mittra, "Integral Equation Solution and RCS Computation of a Thin Rectangular Plate," *IEEE Trans. on Antennas and Propagation*, Vol. AP-23, No. 2, March 1975, p. 302.
50. E. H. Newton and M. R. Schrote, "An Open Surface Integral Formulation for Electromagnetic Scattering by Material Plates," *IEEE Trans. on Antennas and Propagation*, Vol. AP-32, No. 7, July 1984, pp. 672–678.
51. R. A. Ross, "Radar Cross Section of Rectangular Flat Plates as a Function of Aspect Angle," *IEEE Trans. on Antennas and Propagation*, Vol. AP-14, No. 3, May 1966, pp. 329–335.
52. C. Flammer, "The Vector Wave Function Solution of the Diffraction of Electromagnetic Waves by Circular Discs and Apertures," Part 1. Oblate Spheroidal Vector Wave, Function: Part 2. The Diffraction Problem, *J. Applied Physics*, Vol. 24, September 1953, pp. 1218–1231.
53. G. R. Mattson, "Backscattering from a Thin Metallic Disc," *IEEE Trans. on Antennas and Propagation*, Vol. AP-19, No. 1, January 1971, pp. 143–145.
54. E. F. Knott, T. B. A. Senior, and P. L. E. Uslenghi, "High-Frequency Scattering from a Metallic Disc," *Proc. IEE*, Vol. 118, No. 12, December 1971, pp. 1736–1762.
55. T. B. A. Senior, "Disk Scattering at Edge-On Incidence," *IEEE Trans. on Antennas and Propagation*, Vol. AP-17, No. 6, November 1969, pp. 751–756.
56. T. B. A. Senior, "Scattering by Resistive Strips," *Radio Science*, Vol. 14, No. 5, September 1979, pp. 911–924.
57. R. K. L. Poon, "The Scattering of a Planar Array of Small Disc at Arbitrary Angle of Incidence," *IEEE Trans. on Antennas and Propagation*, Vol. AP-25, No. 3, May 1977, pp. 438–441.
58. A. K. Bhattacharyya and T. K. Sen, "Electromagnetic Scattering by a Thin Conducting Flat Plate with Donbly, Periodic Punched Circular Apertures," *Proc. IEEE*, Vol. 126, No. 11, November 1979, pp. 1149–1154.
59. R. A. Ross, "Scattering by a Finite Cylinder," *Proc. IEE*, Vol. 114, 1967, pp. 864–868.
60. R. A. Ross, "Bistatic Scattering Matrix for a Finite Right-Circular Cylinder," *Electronics Letters*, 25 March 1968.
61. R. A. Ross, H. R. Witt, and J. R. Graham, "Radar Cross Section of a Finite Rectangular Cylinder," *Electronic Letters*, 21 June 1967.
62. A. G. Papayiannakis and E. E. Kriezis, "Scattering from a Dielectric Cylinder of Finite Length," *IEEE Trans. on Antennas and Propagation*, Vol. AP-31, No. 5, September 1983, pp. 725–731.
63. S. S. Sekhar and A. Schneider, "Electromagnetic Scattering from a Dielectric Cylinder of Finite Length," *IEEE Trans. on Antennas and Propagation*, Vol. AP-36, No. 2, February 1988, pp. 303–307.
64. R. A. Ross, "Small-Angle Scattering by a Finite Cone," *IEEE Trans. on Antennas and Propagation*, Vol. AP-17, No. 2, March 1969, pp. 241–242.
65. M. E. Bechtel, "Application of Geometric Diffraction Theory to Scattering from Cones and Disks," *Proc. IEEE*, Vol. 53, 1965, pp. 877–882.
66. T. B. A. Senior and P. L. E. Uslenghi, "High-Frequency Backscattering from a Finite Cone," *Radio Science*, Vol. 6, No. 3, March 1971, pp. 393–406.
67. M. E. Bechtel, "Vertically Polarized Radar Backscattering from the Rear of a Cone or Cylinder," *IEEE Trans. on Antennas and Propagation*, Vol. AP-8, No. 3, March 1960, pp. 244–246.
68. G. J. Burke and A. J. Poggio, "Numerical Electromagnetic Code (NEC)—Method of Moments," Naval Ocean Syst. Cont. Tech. Doc., 116, AFWL-TR-76-320, July 1977.
69. E. H. Newman, "A Users Manual for Electromagnetic Surface Patch Code (ESP)," OSUESL, Dept. Elect. Eng., Rep. 713602-1, prepared under Contract DAAG 39-81-K-0020 for Dept. Army, U.S. Army Research Office, Research Triangle Park, NC, July 1981.

70. E. H. Newman and P. Alexandropoulous, "A Users Manual for Electromagnetic Surface Patch Code: Version II—Polygonal Plates and Wires," OSUESL, Dept. Elec. Eng., prepared under Contract N0016-18-C-0049 for the Dept. of Navy, Office of Naval Res., Arlington, VA, September 1983.
71. N. N. Youssef, "Radar Cross Section of Complex Targets," *Proc. IEEE*, Vol. 77, No. 5, May 1989, pp. 722–734.
72. J. W. Crispin, Jr. and A. L. Moffatt, "Radar Cross Section Estimating for Complex Targets," *Proc. IEEE*, Vol. 53, August 1965, pp. 972–982.
73. S. M. Correa, D. L. Sengupta, and W. J. Anderson, "Inflight Aircraft Vibration Modes and Thin Effect on Aircraft Radar Cross Section," *J. Aircraft*, Vol. 18, No. 4, April 1981, pp. 318–319.
74. J. W. Crispin, Jr., R. F. Goodrich, and K. M. Siegel, "A Theoretical Method for the Calculation of the Radar Cross Section of Aircraft and Missiles," University of Michigan Radiation Laboratory, Ann Arbor, Rep. 2591-1-H, July 1959.
75. J. W. Crispin, Jr., R. E. Hiatt, F. B. Sleator, and K. M. Siegel, "The Measurement and Use of Scattering Matrices," University of Michigan Radiation Laboratory, Ann Arbor, Rep. 2952-1-F, May 1960.
76. Y. T. Lin and J. H. Richmond, "EM Modeling of Aircraft at Low Frequencies," *IEEE Trans. on Antennas and Propagation*, Vol. AP-23, No. 1, January 1975, pp. 53–56.
77. G. Y. Delisle, M. Lecours, G. Gonthier, G. Ratte, and C. Deschenes, "RCS of Projectiles—Analytical and Experimental Results," *Proc. of the National Radar Conf.*, Los Angeles, March 12–13, 1986, pp. 83–88.
78. W. R. Stone, ed., Special Issue on RCS of Complex Objects, *Proc. IEEE*, Vol. 77, No. 5, May 1989.
79. W. R. Stone, ed., Special Issue on RCS of Complex Objects, *IEEE Trans. on Antennas and Propagation*, Vol. AP-37, No. 5, May 1989.
80. E. H. Newman, P. Alexandropoulous, and E. K. Walton, "Polygonal Modeling of Realistic Structures," *IEEE Trans. on Antennas and Propagation*, Vol. AP-32, No. 7, July 1984, pp. 742–747.
81. B. J. E. Tante, W. D. Burnside, and I. J. Gupte, "Envelope Radar Cross Section Analysis of Faired Composite Bodies," 1989 IEEE Conf. Antennas and Propagation Society, San Jose, CA, pp. 860–863.
82. B. J. E. Tante, "Envelope RCS Analysis of Faired Composite Bodies," Ph.D. Dissertation, Ohio State University, 1988.
83. A. C. Ludwig, "The Definition of Cross Polarization," *IEEE Trans. on Antennas and Propagation*, Vol. 21, No. 1, January 1973, pp. 116–119. (See comments by G. H. Knittel *ibid.*, Vol. 21, No. 11, November 1973, pp. 917–918.
84. E. F. Knott and T. B. A. Senior, "Polarization Characteristics of Scattered Fields," *Electronics Letters*, Vol. 7, April 22, 1971, pp. 183–184.
85. E. F. Knott and T. B. A. Senior, "Cross Polarization Diagnostics," *IEEE Trans. on Antennas and Propagation*, Vol. AP-20, No. 2, March 1972, pp. 223–224.
86. L. J. Kaplan, J. E. A. Ormsby, E. N. Fowle, K. R. Johnson, R. H. Bates, and S. H. Buckel, "Radar Calibration Test Satellite," *IEEE Trans. on Aerospace Elec. System*, Vol. AES-5, No. 4, July 1969, pp. 654–660.
87. T. M. Smith and S. L. Borison, "Depolarization of a Circularly Polarized Wave by an Infinite Cylinder," *IEEE Trans. on Antennas and Propagation*, Vol. AP-22, No. 6, November 1974, pp. 796–798.
88. G. R. Mattson, "Backscattering from a Thin Metallic Disk," *IEEE Trans. on Antennas and Propagation*, Vol. AP-19, No. 1, January 1971, pp. 143–145.
89. D. Klement, J. Preissner and V. Stein, "Special Problems in Applying the Physical Optics Methods for Backscatter Computations of Complicated Objects," *IEEE Trans. on Antennas and Propagation*, Vol. AP-36, No. 2, February 1988, pp. 228–237.

Chapter 3
Radar Cross-Section Enhancement

3.1 INTRODUCTION

Some practical applications require enhancement of the RCS. For example, when an aircraft or satellite is tracked, we want a constant return during the tracking period, thereby necessitating augmentation of the RCS of the vehicle in some preselected range of angles. Passive RCS augmentation devices are used to enhance the returns from small vehicles and deception devices; for example, light aircraft, buoys, small boats, or missile targets. For scattering and remote-sensing measurements RCS augmentation is used for calibration targets like corner reflectors.

An RCS augmenter is an integrated device with or without a lens and a retro-reflecting surface; and it can be a monostatic or a bistatic device. In general, there are three methods for achieving RCS enhancement. First is the proper shaping of the target to present a large echo area over the ranges of parameters of interest; this method is least useful for airborne targets. The second is the use of impedance loading at selected points on the target to disturb the induced current so as to achieve the goal; this method generally is limited to the narrow band. The third is the addition of a predesigned RCS augmentation device to the target to meet the specific requirements.

In this chapter we present different methods of RCS enhancement and describe different practical devices for RCS augmentation, including their key design considerations. Important characteristics of some standard devices also are presented.

3.2 CORNER REFLECTORS

Corner reflectors are structures formed with two or three metal plates. Figures 3.1(a) and (b) shows some dihedral and trihedral corner reflectors, respectively. Though these devices are passive, due to multiple scattering effects they have considerably large monostatic and bistatic RCS over a relatively wide range of aspect angles; one typical multiple scattering path ABCD is shown in Figure 3.1(a). A dihedral corner reflector gives large RCS only in the plane perpendicular to the reflector; for example, in the

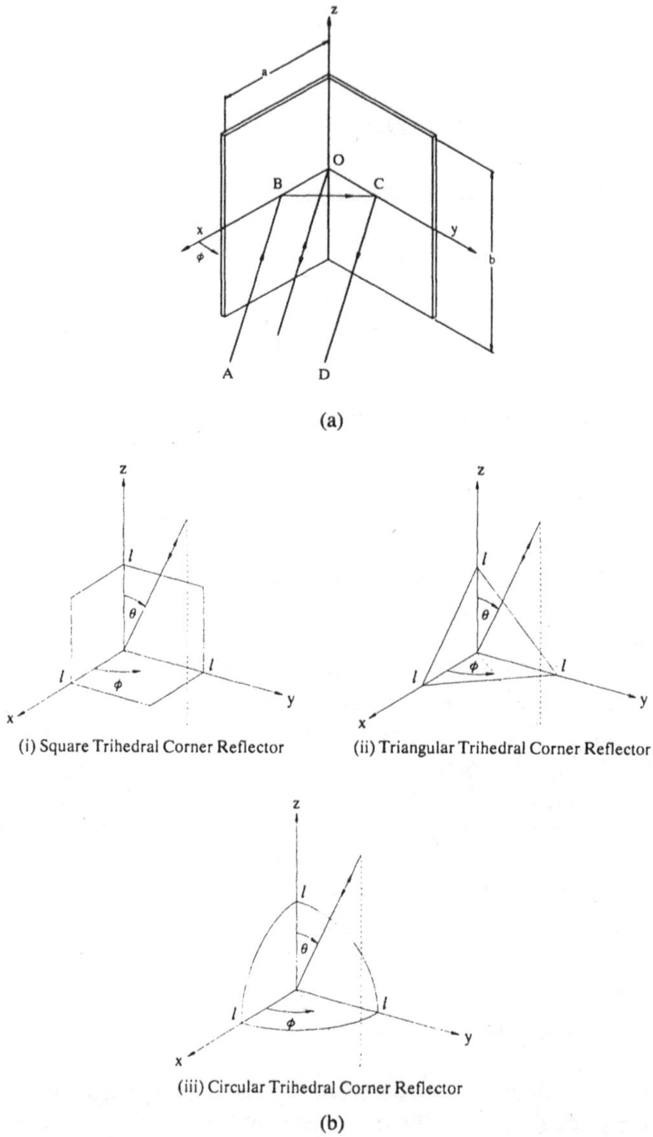

Figure 3.1 Corner reflectors: (a) dihedral; (b) trihedral, including (i) square, (ii) triangular, and (iii) circular.

xy-plane in Figure 3.1(a). This can be improved, and the structure can be made to provide large RCS in other planes by addition of a third plate, as shown in Figure 3.1(b). Depending on the shape of the plates, the trihedral corner reflectors are called *square trihedral*, *triangular trihedral*, or *circular trihedral corner reflectors*. A good account of different types of reflectors is given in [1, 2].

3.2.1 Dihedral Corner Reflectors

Recently, several authors reported extensive work performed on dihedral corner reflectors [1, Section 4.1, pp. 264–266; 3–10]. Corona *et al.* [3] discussed a mathematical model using the physical optics considerations and the effects of illumination of one face by the rays diffracted by the edge of the other face. Table 3.1 lists the key RCS formulas and angular coverage for dihedral corner reflectors.

The physical optics backscattered field due to the dihedral corner reflector, shown in Figure 3.1(a), illuminated by a plane wave with perpendicular polarization (i.e., E-field in the z-direction, xy-plane being the plane of incidence) is given by [3]

$$E_{\perp}^{\mathrm{PO}} = -\frac{ik}{4\pi}\frac{\mathrm{e}^{-ikr}}{r}2abE_{\perp}^{i}$$

$$\left[\Gamma_{\perp}^{x}\sin\varphi\frac{\sin(ka\cos\varphi)}{ka\cos\varphi}\mathrm{e}^{ika\cos\varphi}\right.$$

$$\left.+\Gamma_{\perp}^{y}\cos\varphi\frac{\sin(ka\sin\varphi)}{ka\sin\varphi}\mathrm{e}^{ika\sin\varphi}+2\Gamma_{\perp}^{x}\Gamma_{\perp}^{y}\sin\varphi\right] \tag{3.1}$$

Table 3.1
Dihedral Corner Reflectors

Particulars	Radar Cross Section
Dihedral reflector Monostatic Broadside incidence	$\sigma \approx \dfrac{4\pi A_e^2}{\lambda_0^2}$ (A_e = area of projected aperture on the incident ray)
Dihedral reflector Monostatic Angular coverage ± 30°	$\sigma = \dfrac{4\pi a^2 b^2 \sin^2(\pi/4+\phi)}{\lambda_0^2}$ (see Figure 3.1(a)) First-order PO formula. Polarization effects not accounted for.
Dihedral reflector Bistatic	$\sigma(\beta) = \sigma(0)\,\mathrm{e}^{-2\nu\beta}$ β = bistatic angle ν = design parameter given by Peter's design curve [1]

where $E^i_\perp$ is the amplitude of the incident electric field,

$$\Gamma^x_\perp = \frac{Z \sin \varphi - 1}{Z \sin \varphi + 1} \tag{3.2}$$

$$\Gamma^y_\perp = \frac{Z \cos \varphi - 1}{Z \cos \varphi + 1} \tag{3.3}$$

Z is the surface impedance of the reflector surface normalized to the intrinsic impedance of free space, superscripts x and y refer to the plates $y = 0$ and $x = 0$, respectively.

The PO field for parallel polarization (i.e., E-field in the xy-plane) is obtained from (3.1) by replacing $E^i_\perp$, $\Gamma^x_\perp$, and $\Gamma^y_\perp$ with $E^i_\parallel$, $\Gamma^x_\parallel$, $\Gamma^y_\parallel$, respectively; and the two reflection coefficients are given by

$$\Gamma^x_\parallel = \frac{\sin \varphi - Z}{\sin \varphi + Z} \tag{3.4}$$

$$\Gamma^y_\parallel = \frac{\cos \varphi - Z}{\cos \varphi + Z} \tag{3.5}$$

The scattered field expressions just given do not consider the effects of diffraction. Detailed considerations of the diffraction effects are given in [3]. We can show that for parallel polarization the total scattered field is given by [3]

$$\begin{aligned} E_\parallel = E^{\text{PO}}_\parallel &+ \frac{ik}{4\pi} \frac{\mathrm{e}^{-ikr}}{r} 2abE^i_\parallel \\ &\times \left\{ \frac{\cos \varphi - Z}{2a} \mathrm{e}^{ika\cos\varphi} \mathrm{e}^{-i\pi/4} \int_0^a \frac{1}{\sqrt{2\pi k d_1}} \mathrm{e}^{-ikd_1} \mathrm{e}^{iky\sin\varphi} \right. \\ &\times D_h(\beta_1, \gamma_1) \left[\frac{T_\parallel(\alpha_1)}{Z} \right] \mathrm{d}y + \frac{\sin \varphi - Z}{2a} \mathrm{e}^{ika\sin\varphi} \mathrm{e}^{-i\pi/4} \\ &\times \left. \int_0^a \frac{\mathrm{e}^{-ikd_2}}{\sqrt{2\pi k d_2}} \mathrm{e}^{ikx\cos\varphi} D_h(\beta_2, \gamma_2) \left[\frac{T_\parallel(\alpha_2)}{z} \right] \mathrm{d}x \right\} \end{aligned} \tag{3.6}$$

where

$$T_\parallel(\alpha) = \frac{2Z \cos \alpha}{Z + \cos \alpha} \tag{3.7}$$

and $D_h(\beta, \gamma_1)$, $D_h(\beta_2, \gamma_2)$ are hard diffraction coefficients for parallel polarization; the diffraction process involved for this case is shown in Figure 3.2(a), which may

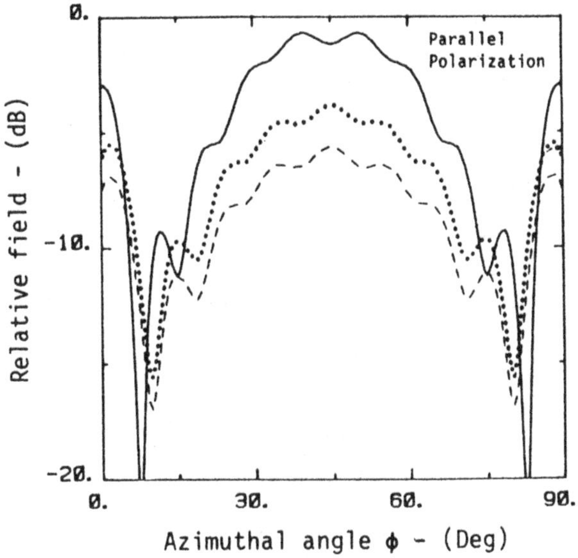

(a)

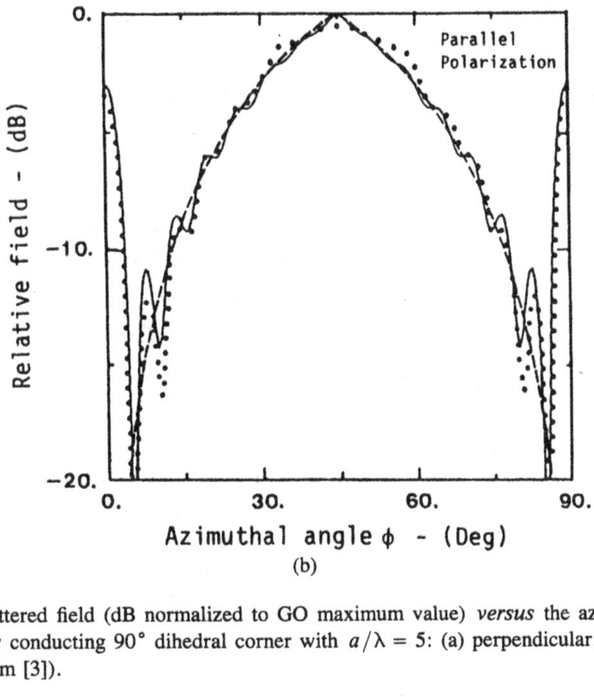

(b)

Figure 3.2 Backscattered field (dB normalized to GO maximum value) *versus* the azimuthal angle ϕ by a perfectly conducting 90° dihedral corner with $a/\lambda = 5$: (a) perpendicular; (b) parallel polarization (from [3]).

represent the top or bottom view of the dihedral reflector shown in Figure 3.1(a). Figure 3.2(b) and (c) show the backscattered field patterns for perpendicular and parallel polarizations, respectively, for a perfectly conducting corner reflector with $a/\lambda = 5$. The fields shown are normalized with respect to the GO maximum value.

Physical optics analyses of dihedral corner reflectors are given in [4–6]; PTD analysis is discussed in [5], and the effect of higher order reflections and diffractions also are described in [5]. Details of these considerations will not be discussed here. We shall present only a few important experimental results and their relationship with appropriate theoretical results. Figure 3.3 gives the experimental and theoretical RCS (normalized to 1 m^2) patterns for a 90° dihedral corner obtained under a variety of conditions. Researchers have found [5, 6] that a 90° dihedral corner has wide beamwidth of RCS pattern if ripples in excess of 5 dB only are of concern, particularly for horizontal polarization. When the reflector angle is not 90°, the pattern splits into several beams giving rise to large variations in the RCS with aspect angle.

Effects of the bistatic angle on the relative echo signal from a normal corner reflector and a corner reflector with 5°-error are shown in Figure 3.4(a) and (b), respectively. The results indicate that for larger bistatic angles the reflector error generally increases the beamwidth of response [7]. Normally, for the bistatic case, a large beamwidth is difficult to achieve without reduction of the cross section. However, the results of Figure 3.4 tend to indicate that it may be possible to obtain fairly large monostatic and bistatic RCS by using a cluster of reflectors because here the degradation of the reradiated pattern will be confined to one plane.

To understand this, we must briefly examine the polarization characteristics of dihedral corner reflectors. A linearly polarized wave incident on the reflector, with the electric vector either parallel or perpendicular to the seam of the reflector, is reflected with the same polarization, provided that the wave suffers an even number of bounces at the reflector. Similarly, a circularly polarized incident wave will also be reflected with the same sense of circular polarization. Polarization sensitivity of the dihedral corner reflector can be utilized for the calibration of the cross-polarization receiver of a linear dual-polarized radar. Further details of this important practical application are discussed in [8].

3.2.2 Trihedral Corner Reflectors and Widebanding

As indicated earlier, in a plane perpendicular to the plates the dihedral corner reflector provides maximum RCS over a range of aspect angles. In other planes (i.e., for $\theta \neq \pi/2$) the performance deteriorates. Use of a third plate is necessary to maintain a nominal RCS for wide range of the aspect angle φ for variable θ. This is how the trihedral reflectors were developed. To the best of our knowledge, no accurate analysis for the RCS of a trihedral reflector is currently available. However, some experimental

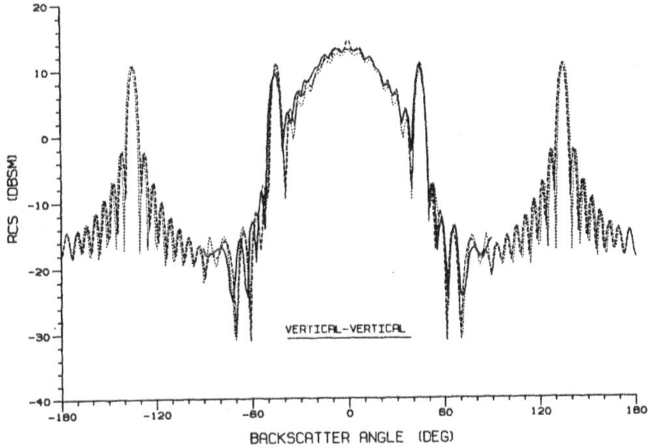

(a)

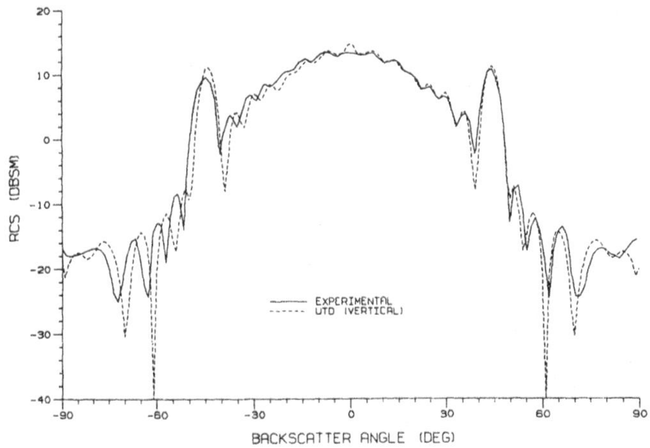

(b)

Figure 3.3 (a) Theoretical and experimental azimuthal patterns for backscattering RCS of a 90° dihedral corner reflector $a = b = 5.6088\lambda$; $f = 9.4$ GHz; polarization is vertical (from [5]). (b) Experimental and UTD RCS using higher-order diffraction of the 90° dihedral corner reflector in (a), where $a = b = 5.6088\lambda$; $f = 9.4$ GHz; polarization is vertical. (c) UTD cross section using higher-order rays for vertical and horizontal polarizations for the 90° dihedral, where $a = b = 5.6088\lambda$; $f = 9.6$ GHz; $R = 200\lambda$ (from [5]).

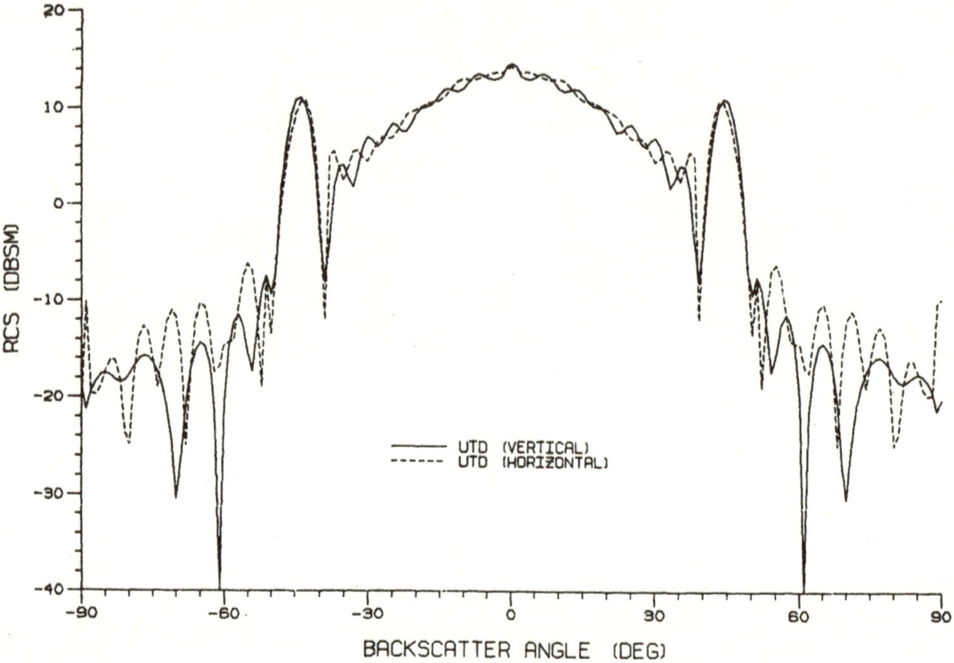

(c)

Figure 3.3 *Continued*

results and approximate expressions are available as guides for designing such reflectors. These are given in Table 3.2. Figure 3.5 shows the experimental backscattering cross-section pattern of a triangular trihedral corner reflector for different values of θ. In Figure 3.5, β is the bistatic angle, therefore $\beta = 0$ means the case of backscattering; the angle α, interpreted as the angle of incidence, is given by $\alpha = (\varphi - 45°)\sin\theta$ where θ and φ are as shown in Figure 3.1. Figure 3.4 indicates that peaks on the two sides of the broad pattern are attributed to the scattering effects of the corners. To reduce these effects, a compensated triangular trihedral corner reflector has been found useful; one is sketched in Figure 3.6. Monostatic RCS patterns for compensated and uncompensated reflectors are shown in Figure 3.7, which clearly show the effects of compensation. Analyses of such compensated structures are not yet available. Recently, some theoretical results have appeared for the trihedral corner reflector [9].

Figure 3.8(a) shows the trihedral corner geometry, and a comparison between the backscattered cross-section values obtained by using finite-difference time domain; and the shorting and bouncing ray codes are shown in Figure 3.8(b).

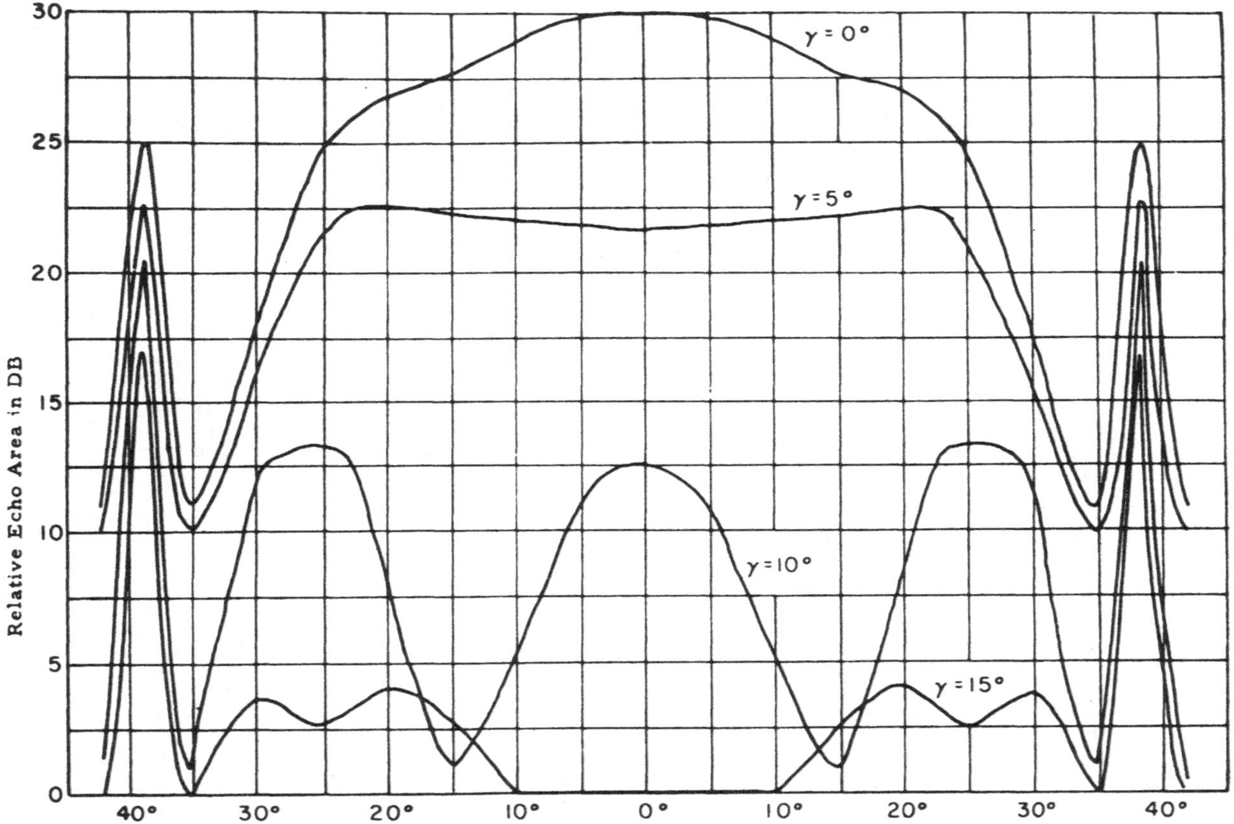

(a)

Figure 3.4 Effect of bistatic angle on response of (a) a normal corner reflector and (b) a corner reflector with 5° error (from [7]).

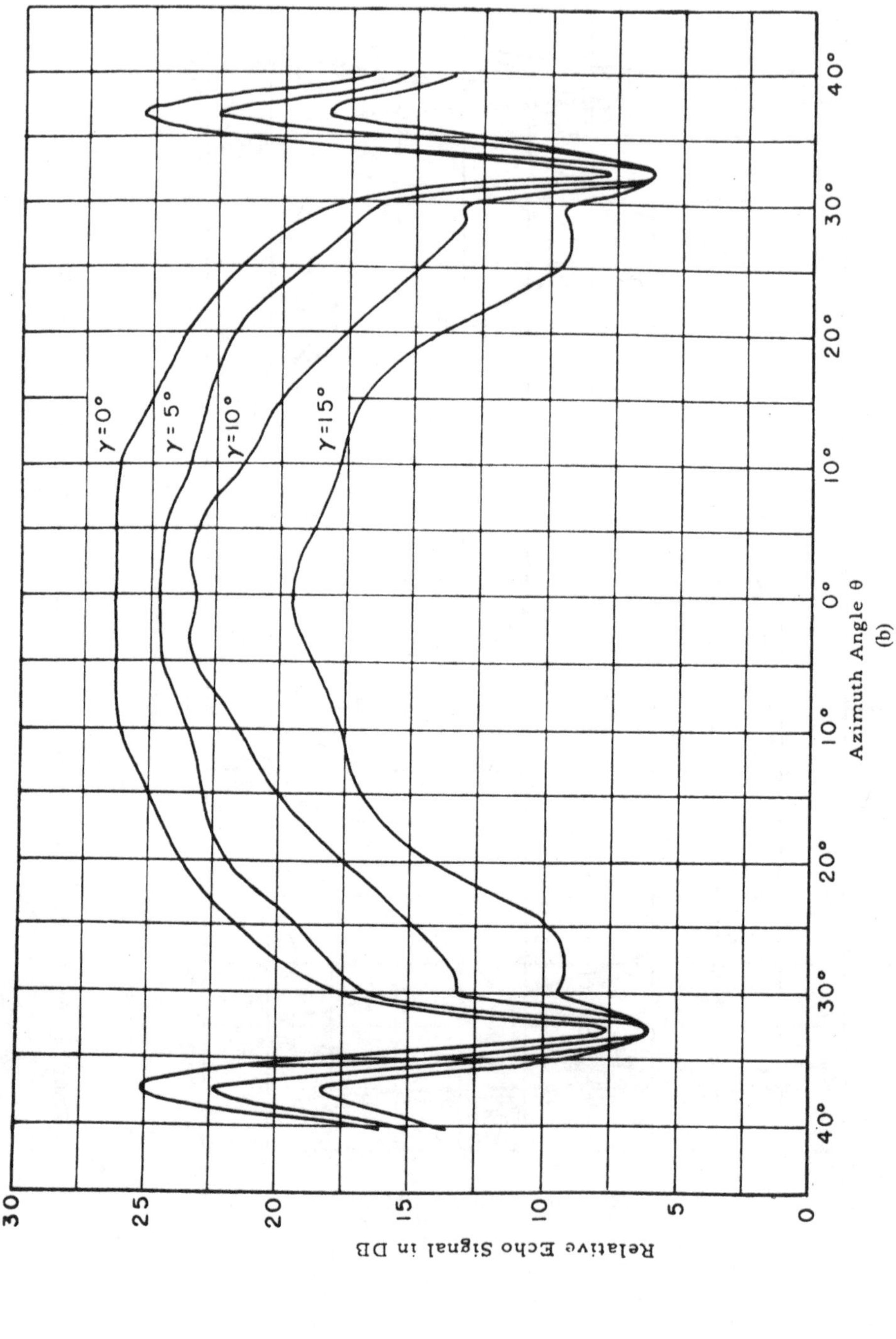

Figure 3.4 *Continued*

Table 3.2
Comparison of Trihedral Corner Reflectors

Trihedral Corner Reflector Type	Maximum RCS	Average RCS	Angular Coverage	RCS as a Function of θ, ϕ
Square	$\sigma = 12\pi a^4/\lambda_0^2$	$\langle\sigma\rangle = \dfrac{0.7a^4}{\lambda_0^2}$	25° cone about symmetry axis	(a)
Triangular	$\sigma = \dfrac{4\pi a^4}{\lambda_0^2}$	$\langle\sigma\rangle = \dfrac{0.17a^4}{\lambda_0^2}$	40° cone about symmetry axis	(b)
Circular	$\sigma = \dfrac{15.6a^4}{\lambda_0^2}$	$\langle\sigma\rangle = \dfrac{0.47a^4}{\lambda_0^2}$	32° cone about symmetry axis	—

Notes: The PO formulas are from [1].
(a) = The RCS with θ, ϕ near symmetry axis (or, $\theta \approx 54° \cdot 74$, $\phi \approx 45°$):

$$\sigma(\theta, \phi) \approx \frac{4\pi}{\lambda_0^2} \cos^2\theta (4 - \cot\phi)^2 a^4$$

(b) = The term a represents the side. The RCS with θ, ϕ for angles near grazing symmetry axis:

$$\sigma(\theta, \phi) \approx \frac{4\pi}{\lambda_0^2}\left[\cos\theta + \sin\theta(\sin\phi + \cos\phi) - 2\left(\cos\theta + \sin\theta(\sin\phi + \cos\phi)^{-1}\right)\right]^2$$

3.3 DIELECTRIC LENSES

By utilizing the properties of dielectric lenses it is possible to enhance the monostatic and bistatic RCS over a wide range of frequencies and aspect angles. Different types of lenses—for example, Kay's spherical lenses [10], the Eaton-Lippman lens [1, Section 4.1, pp. 264–266], and more popularly the Luneberg lens—have been used for such purposes; compared to metal targets of similar size, they provide larger monostatic and bistatic RCS. The cylindrical Eaton-Lippman lens has a normal incidence monostatic RCS equal to [1, Section 4.1]

$$\sigma(0) \simeq 16\pi a^2 \left(\frac{l}{\lambda}\right)^2 \tag{3.8}$$

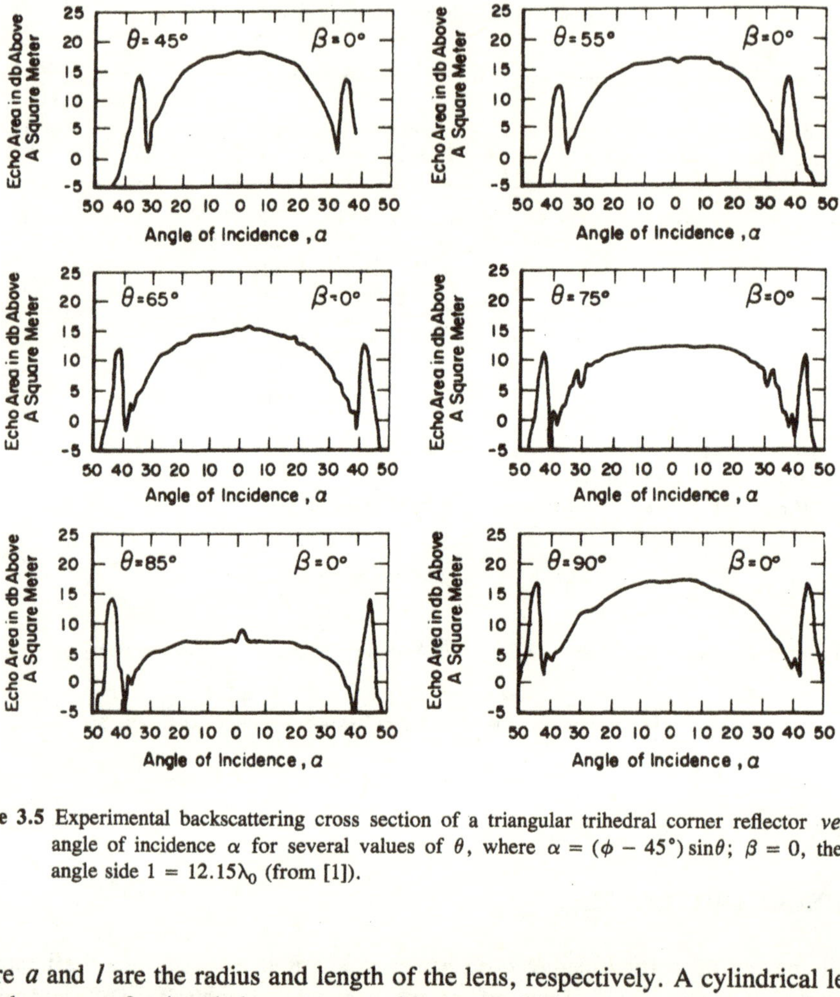

Figure 3.5 Experimental backscattering cross section of a triangular trihedral corner reflector *versus* the angle of incidence α for several values of θ, where $\alpha = (\phi - 45°)\sin\theta$; $\beta = 0$, the bistatic angle side 1 = $12.15\lambda_0$ (from [1]).

where a and l are the radius and length of the lens, respectively. A cylindrical lens has the advantage of azimuthal symmetry without requiring a metal reflector. The limitations of such lenses are their poor elevation plane performance and their requirement for an infinite or very high-dielectric constant at the center. The latter requirement is very difficult to satisfy in practice.

3.3.1 Luneberg Lenses

An ideal structure for antenna scanning applications and RCS augmentation is a symmetric configuration like a sphere; and the more popular Luneberg lens [10] has this

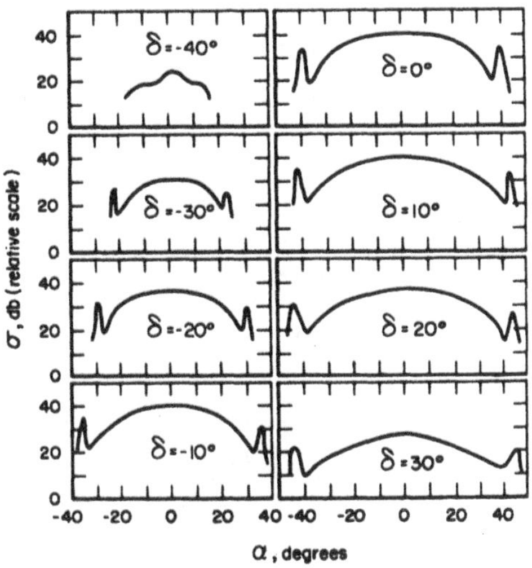

Figure 3.6 A compensated triangular corner reflector.

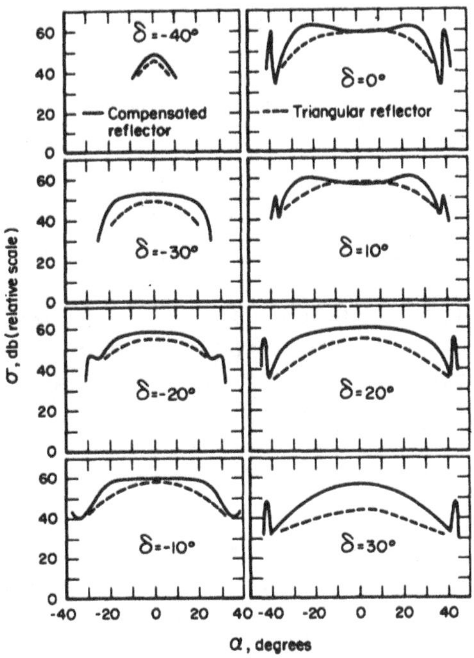

Figure 3.7 Experimental RCS of compensated and uncompensated triangular reflectors *versus* angle α for different δ, where $\delta = \theta - 54 \cdot 7°$ (from [1]).

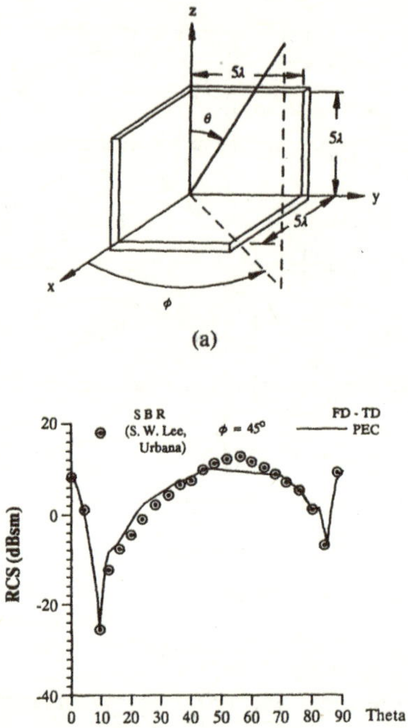

(b)

Figure 3.8 (a) A trihedral corner reflector and (b) its RCS with aspect angle θ, where $\phi = 45°$ and the dimensions are as shown in Figure 3.7.

structure. A Luneberg lens, shown in Figure 3.9, is a spherically symmetric inhomogeneous structure with the property that parallel rays incident on it are focused at a diametrically opposite point on its surface, and a point source located on its surface gives rise to parallel rays going out from its surface on the other side (Figure 3.9). The required dielectric constant of the lens radially varies from 2 at the center to unity at the surface. Typical ray paths in the lens are shown in Figure 3.9. The refractive index $n(r)$ at any point, r distance from the center, is given by

$$n^2(r) = 2 - (r/a)^2 \tag{3.9}$$

A practical Luneberg lens consists of several concentric spherical layers of dielectric materials with decreasing dielectric constants. The maximum backscattering cross section of a Luneberg lens of radius a is given by [8, Table 8.4, p. 599]

$$\sigma_{max} = \frac{4\pi^3 a^4}{\lambda_0^4} \tag{3.10}$$

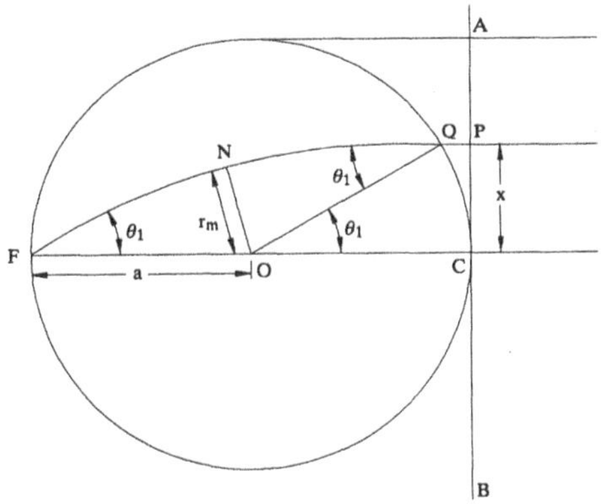

Figure 3.9 Path of a ray inside a Luneberg lens.

As an example, with a = 12.7 cm, eq. (3.10) indicates σ_{max} = 31.76 m^2 at λ_0 = 3.19 cm (9.4 GHz), whereas the measured value for $\sigma_{max} \approx$ 31.7 m^2 [9]. Figure 3.10 shows the measured backscattering cross sections *versus* aspect angle for a number of commercially available Luneberg lenses at 9.375 GHz [8, Table 8.4].

3.3.2 Luneberg Lens with a Reflector

Parallel rays incident on the Luneberg lens with a reflector, shown in Figure 3.11, generally are reflected back toward the transmitter. This then augments the backscattered RCS. The reflector may be connected separately or metallized on the lens surface. The RCS may be enhanced over a large range of aspect angles by varying the size of the reflector used. Table 3.3 gives some pertinent information about different lens models along with their backscattering cross sections [11].

Figure 3.12 shows the measured reflectivity pattern *versus* the aspect angle with different extents of the reflecting cap for the lens model C-04; for comparison, the reflectivity pattern for a 4-in. flat metallic plate is also given. Clearly, RCS enhancement over a wide range of angles is possible with a suitable choice of the reflector size.

3.4 PASSIVE AND ACTIVE DEVICES

Both passive and active devices can control the RCS of a target if the devices can be properly positioned on the target. In particular, the use of active diodes can significantly increase the cross section in certain directions. Very little information regarding active control of RCS is available in the open literature. The moment method solution for a strip dipole loaded with diodes is reported in [12]; we have reproduced the associated computer program in Appendix C.7. Figure 3.13(a) and (b) show the geometry of the

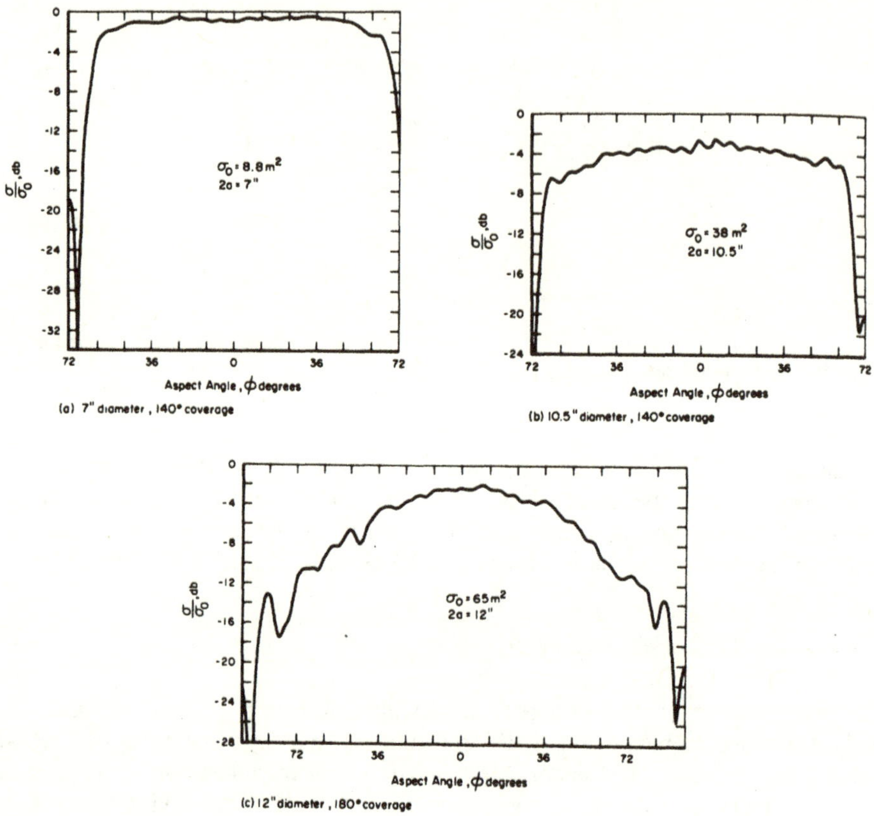

Figure 3.10 Measured backscattering cross sections *versus* aspect angle at 9.375 GHz for several commercially available Luneberg lenses (from [1, Section 4.1]).

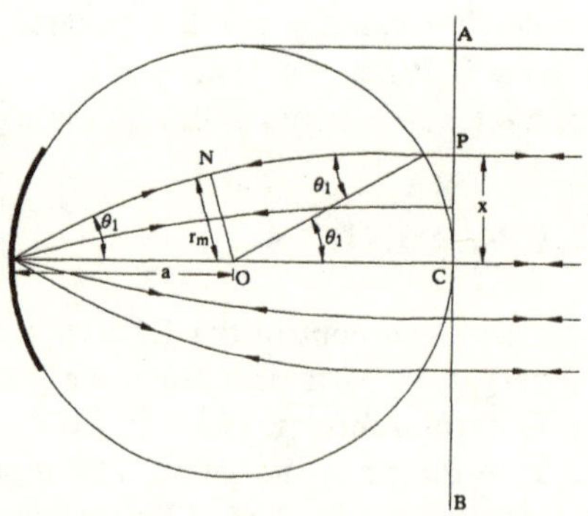

Figure 3.11 Luneberg lens with a reflector cap for RCS enhancement.

Table 3.3
Theoretical RCS for Different Lens Models (from [1, Section 4.1])

Model	Nominal Diameter Inches (cm)	Theoretical RCS in m^2			Weight* lbs (grams)
		S-Band	*X*-Band	*K*-Band	
C-03	3 (7.6)	0.03	0.25	1.16	0.6 (265)
C-04	4 (10.2)	0.08	0.81	3.67	1.6 (720)
C-05	5 (12.2)	0.20	2.20	9.00	3.3 (1500)
C-06	6 (15.2)	0.42	4.08	18.60	6.0 (2724)

*Does not include the mounting system.

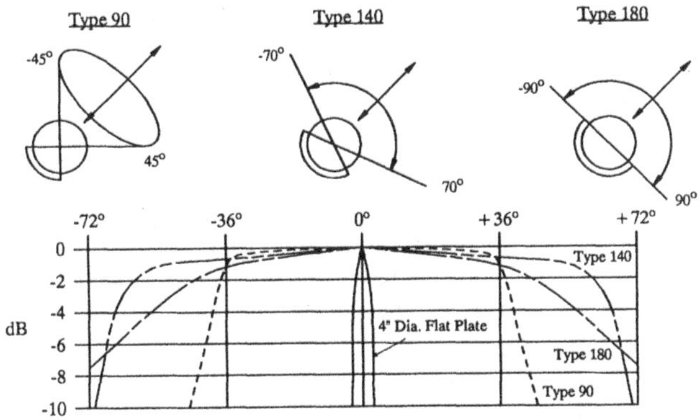

Figure 3.12 Reflectivity *versus* aspect angle for dielectric lenses with different sizes of reflector (from [11]).

scattering by a diode loaded strip dipole and its equivalent circuit, respectively. The equivalent circuits for a beam-lead PIN diode and a negative resistance diode are given in Figure 3.14(a) and (b), respectively. The backscattering cross section of a loaded dipole shown in Figure 3.13 is proportional to the square of the magnitude of the current I_L given by [12]

$$I_L = \frac{V_{\text{OC}}}{Z_{\text{in}} + Z_L} \tag{3.11}$$

and is inversely proportional to the square of the sum of the input impedance of the diode when fed as an antenna and the external load impedance Z_L. In [1, Section 4.1],

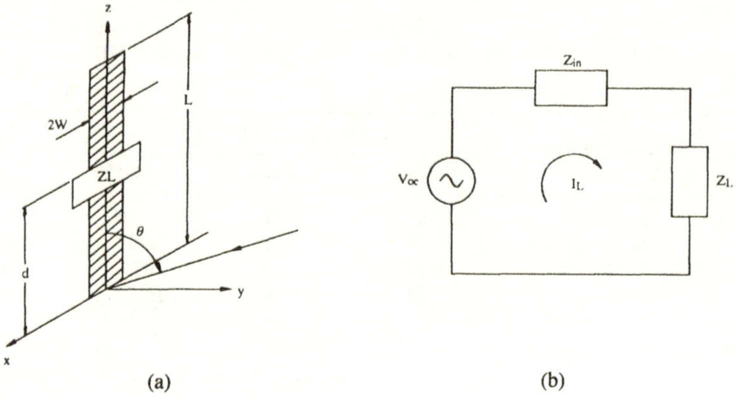

Figure 3.13 The (a) loaded strip dipole and (b) its equivalent circuit.

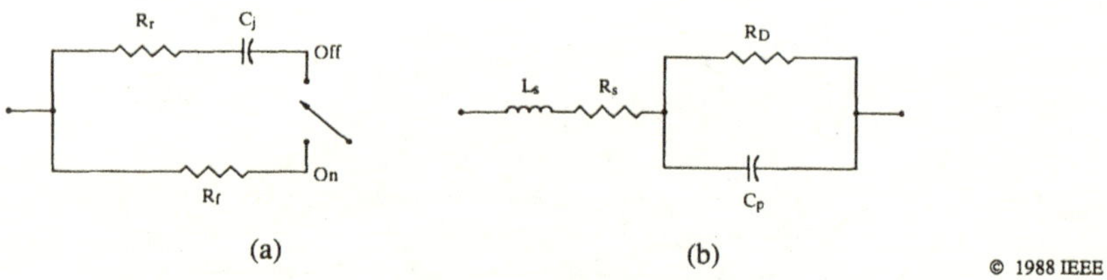

Figure 3.14 Equivalent circuits of x-band diodes: (a) beam-lead PIN, where $R_r = R_f = 7\ \Omega$, $C_j = 0.03$ pF; and (b) negative resistance diode, where $L_s = 0.25$ nH, $R_s = 1\ \Omega$, $R_D = -200\ \Omega$; $C_p = 0.25$ pF (from [12]).

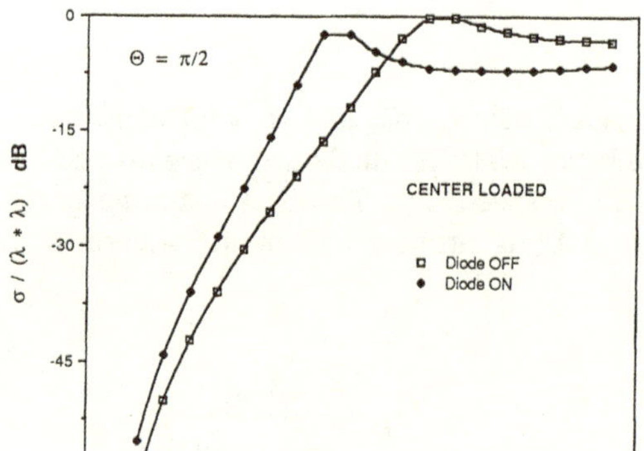

Figure 3.15 RCS of strip dipole loaded with beam-lead PIN diode, where $L = 1.5$ cm; $L / W = 75$.

three cases were considered: (1) a beam-lead PIN diode, (b) a *negative resistance diode* (NRD), and (3) a combination of PIN and NRD diodes.

The backscattering cross section *versus* frequency for a stripe dipole, center loaded with a beam-lead PIN diode, is shown in Figure 3.15, where results are shown for cases when the diode is off and on. Figure 3.16 gives the average RCS *versus* L/λ

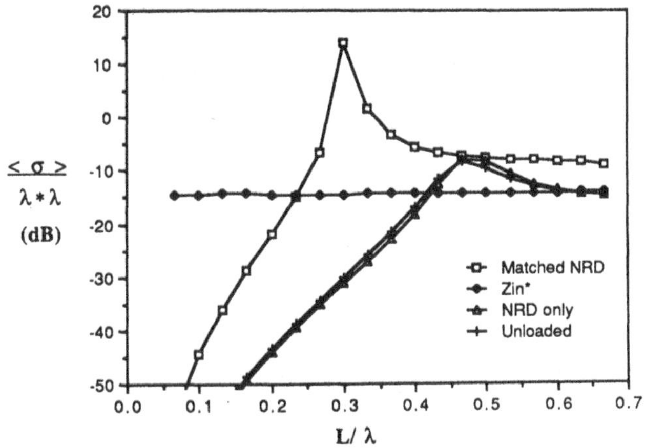

Figure 3.16 Average RCS of strip dipole loaded with NRD, where $L = 1$ cm, $L/W = 100$, and frequency variation is from 2 to 30 GHz (from [12]).

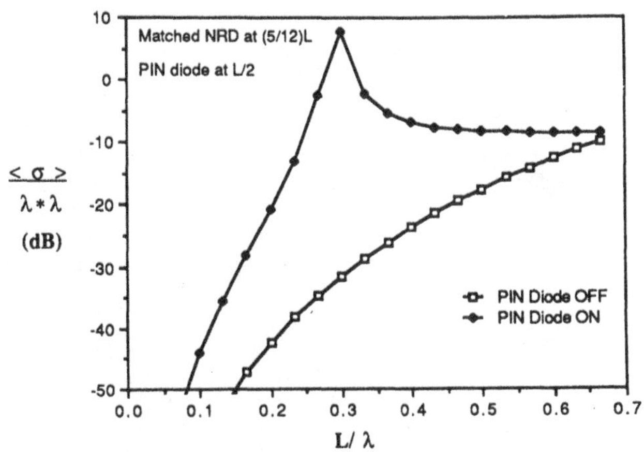

Figure 3.17 Average RCS of strip dipole loaded with PIN diode and NRD, where $L = 1$ cm, $L/W = 100$, and frequency variation is from 2 to 20 GHz (from [12]).

for a strip dipole of length $L = 1$ cm loaded with NRD; the case with matched NRD shows considerably increased RCS. With PIN diode on, there seems to be an increase in the RCS for the strip dipole loaded with both PIN diode and NRD as shown in Figure 3.17. The difference in RCS between the on and off conditions is increased appreciably by the presence of matched NRD. Table 3.4 lists RCS values obtained under a variety of conditions.

Table 3.4

Radar Cross Section of Strip Dipoles under Different Cases, where (left to right) $f = 13.5$ GHz, $L = 1.5$ cm, $L/W = 75$, and $L/\lambda_0 = 0.68$

Device	Z_{in} (Ω)	Z_{load} (Ω)	$Z_{in} + Z_{load}$ (Ω)	RCS (broadside)
Center-fed dipole	$350 + j300$	—	—	−7 dB
Packaged PIN diode	—	$4 + i250$ (off)	$354 + i550$	−16 dB
		$7 + i360$ (on)	$357 + i660$	−21 dB
Beam-lead PIN diode	—	$7 - i400$ (off)	$357 - i100$	0 dB
		$7 + i0$ (on)	$357 + i300$	−7 dB
Tunnel diode	—	$-1 + i40$	$369 + i340$	−8 dB
			$Z_{in} + Z_{load} + Z_M = 349$	−1 dB

Figure 3.18 RCS augmentation devices (courtesy Teledyne Ryan Electronics).

Table 3.5
Characteristics of Different RCS Augmentation Devices (*Source:* Teledyne Ryan Electronics, "Radar Augmentation Systems," Technical Manual)

Serial No.	Augmentater Type No.	Monostatic Bistatic	RCS Performance Frequency (GHz)	Bistatic Angle	RCS Specified Minimum	Typical	View Angles (degrees)	Pattern (Figure) Comments
1	SP-2-70M	Monostatic	10.0	0.0	2.0 m^2	6.0^2	120	3.10
2	SP-2-70M-CP*	Monostatic	16.0	0.0	+6 dB · 5 m	+8 dB · 5 m	100	3.11
3	SP-3-825M	Monostatic	10.0	0.0	6 m^2	10 m^2	120	3.12
					Minimum RCS (m^2) 100% of view angle	80% of viewing angle		
4	SP-4-140M-CP*	Monostatic	10.0	0.0	25	32	110	3.13
			5.0	0	25	30	100	
5	SP-5-160M	Monostatic	10.0	0	40	50	100	3.14
			15.0	0	25	40	100	

*These have the design capability of changing the direction of rotation of a circular polarization.

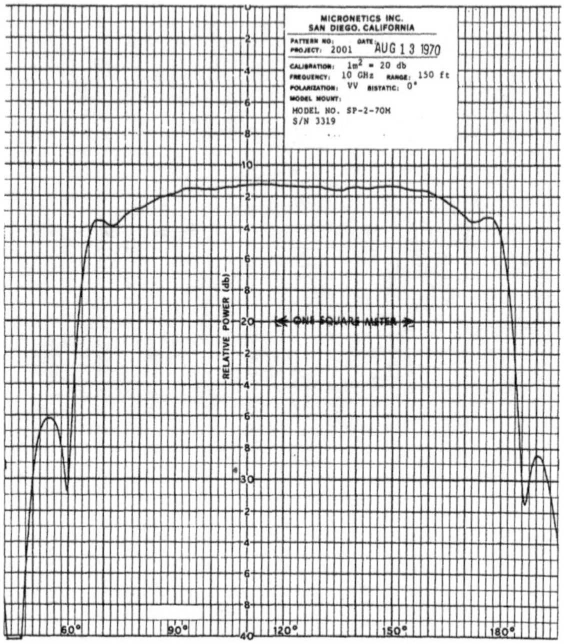

Figure 3.19 Power pattern of SP-2-70M augmenter, characteristics given in Table 3.4 (courtesy: Teledyne Ryan Electronics).

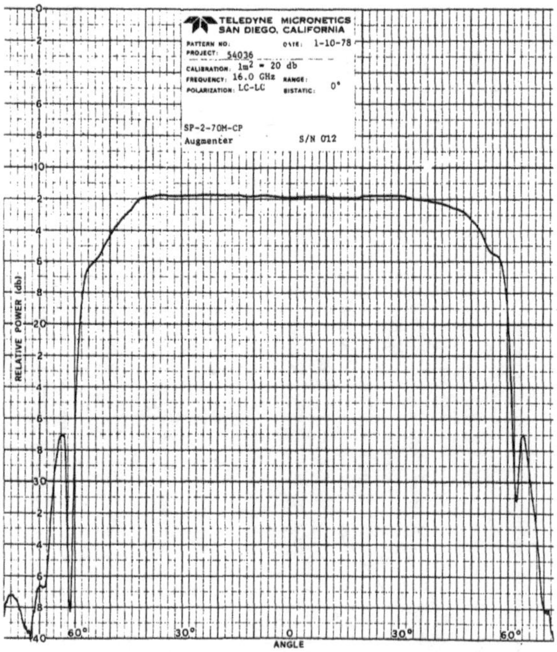

Figure 3.20 Power pattern of SP-2-70M-CP augmenter, characteristics given in Table 3.4 (courtesy: Teledyne Ryan Electronics).

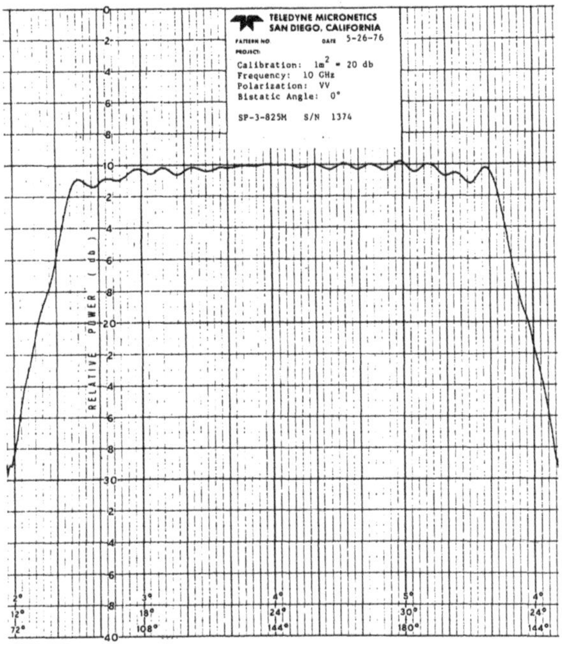

Figure 3.21 Power pattern of SP-3-825M, characteristics given in Table 3.4 (courtesy: Teledyne Ryan Electronics).

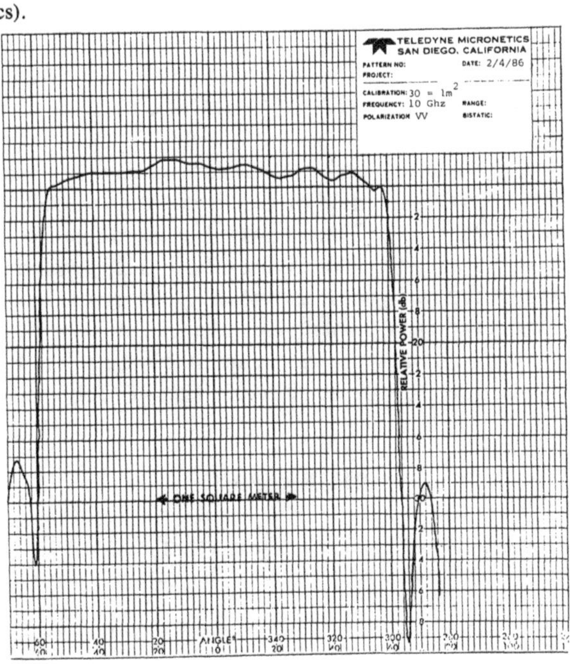

Figure 3.22 Power pattern of SP-4-140M-CP, characteristics given in Table 3.4 (courtesy: Teledyne Ryan Electronics).

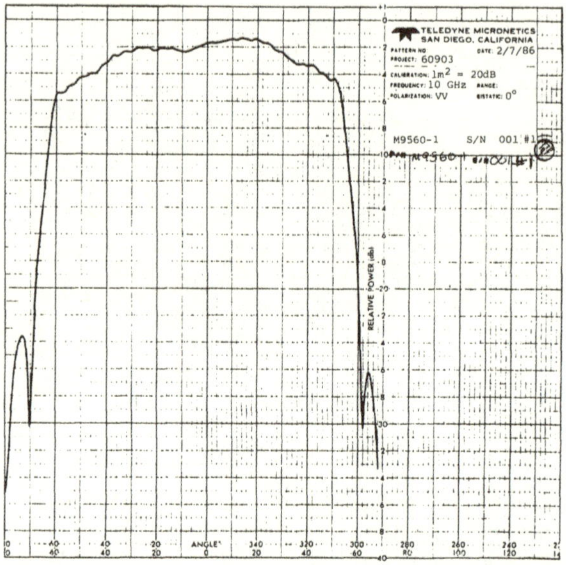

Figure 3.23 Power pattern of SP-5-160M, characteristics given in Table 3.4 (courtesy: Teledyne Ryan Electronics).

3.5 PRACTICAL RCS AUGMENTATION DEVICES

Some typical RCS augmentation devices that are currently available are shown in Figure 3.18. The characteristics claimed of the different augmentation devices shown in Figure 3.18 are given in Table 3.5. The measured backscattered power patterns for the devices listed in Table 3.5 are shown in Figures 3.19–3.23, which indicate that results are available only at selected (or spot) frequencies. Some of the devices can incorporate specially designed reflector caps, which can change the sense of incident circular polarization, and thereby can suppress rain clutter. Nothing has been reported in the literature on how to increase the frequency bandwidth of such structures.

REFERENCES

1. G. T. Ruck, ed., *Radar Cross Section Handbook*, Vol. 2, Section 8.2.2, pp. 588–597.
2. S. D. Robertson, "Targets for Microwave Radar Navigation," *B.S.T.J.*, Vol. 26, 1947, pp. 852–869.
3. P. Corona, G. Ferrara, and C. Gennarelli, "Backscattering by Loaded and Unloaded Dihedral Corners," *IEEE Trans. on Antennas and Propagation*, AP-35, No. 10, October 1987, pp. 1148–1153.

4. T. Griesser and C. A. Balanis, "Backscatter Analysis of Dihedral Corner Reflectors Using Physical Optics and the Physical Theory of Diffraction," *IEEE Trans. on Antennas and Propagation*, Vol. AP-35, No. 10, October 1987, pp. 1137–1167.
5. T. Griesser and C. A. Balanis, "Dihedral Corner Reflector Backscatter Using Higher Order Reflections and Diffraction," *IEEE Trans. on Antennas and Propagation*, Vol. AP-35, No. 11, November 1987, pp. 1235–1247. (Comments on and authors' reply in *IEEE Trans. on Antennas and Propagation*, Vol. AP-37, No. 10, October 1989, pp. 1339–1342).
6. K. Lin, C. A. Balanis, and T. Griesser, "An Integral Equation Solution for the RCS of Large Dihedral Corner Reflectors," 1989 IEEE AP-S Symp., San Jose, CA, pp. 852–855.
7. C. G. Bachman, *Radar Targets*, Lexington Books, Lexington, MA, 1982, Chaps. 2 and 3.
8. N. C. Currie, ed., *Radar Reflectivity Measurement: Techniques and Applications*, Artech House, Norwood, MA, 1989, Chap. 2.
9. A. Taflove and K. R. Umashankar, "The Finite Difference Time-Domain (FD-TD) Method for Numerical Modeling of Electromagnetic Scattering," *IEEE Trans. on Magnetics*, Special Issue on EM Field Computations, July 1989.
10. A. F. Kay, "Spherically Symmetric Lenses," *IEEE Trans. on Antennas and Propagation*, Vol. AP-7, No. 1, January 1959, pp. 32–38.
11. *Cuming Corporation Tech. Bulletin*, No. 510.
12. R. Janaswamy and S. W. Lee, "Scattering from Dipoles Loaded with Diodes," *IEEE Trans. on Antennas and Propagation*, Vol. 36, No. 11, November 1988, pp. 1669–1651.

Chapter 4
Reduction of Radar Cross Section

4.1 INTRODUCTION

An important problem often encountered in many military and civilian applications is how to reduce the radar return from a target to a specified minimum. In certain military applications, for obvious reasons, it is necessary to reduce the RCS of targets like aircraft, missiles, and spy satellites. In civilian applications, satisfactory performance of air traffic control radar requires that the scattering or reflections off the terminal buildings and aircraft hangars be reduced so that they cause less interference to the airport radars. RCS reduction finds important application also in the indoor measurements of antenna patterns and target scattering, where the required free space environment generally is simulated by reducing undesirable reflections from the surroundings.

Generally, there are four means of reducing (or controlling) the RCS of targets: target shaping, discrete loading, distributed loading, and active loading. In target shaping the shape of the target is modified to redirect the scattered energy from one angular region of interest in space to another region of little or no interest. For airborne targets, any shaping subject to aerodynamic considerations may be required to satisfy stringent requirements; however, static targets are more flexible to shaping.

In discrete loading the target is loaded at selected points with passive impedances to reduce the overall cross section. The method relies on phase cancellations in certain directions and, hence, it is severely limited in bandwidth.

In distributed loading the target is coated with lossy material, known as the radar absorbing material (RAM). The lossy material absorbs, in the form of heat, a considerable amount of incident energy and thereby reduces the RCS. For narrow bandwidths, a single coating of RAM may suffice, but broad bandwidth generally require multiple coating.

The active loading is similar to the discrete loading mentioned earlier, but uses active elements to achieve the necessary phase cancellation.

None of these techniques alone will reduce the RCS of a target below a certain minimum value under all circumstances; for example for a wide range of aspect angles, a wide band of frequencies, and both orthogonal polarizations. In most cases a judicious combination of two or more methods may be required to achieve the goal. For example, a ship or a vehicle may need the shaping criterion during the engineering design followed by a properly planned local application of RAM around the possible scattering centers like the edges and other discontinuities; generally, to coat the vehicle completely with RAM is not necessary.

Different methods of reducing the RCS are discussed in this chapter. Theoretical results supported by available experimental results are discussed to illustrate the status and promise of various RCS reduction techniques. Although detailed theoretical analyses are not given, appropriate references are cited wherever possible. A few practical designs with numerical examples also are described.

4.2 TARGET SHAPING

RCS of a target can be reduced by shaping it so that it does not project a large effective area to the radar. For flying objects, aerodynamic considerations severely restrict the choice of shapes for reduced RCS. Even for other objects, it seems impossible to shape a target so that it has a given minimum RCS for all radar parameters; for example, for all aspect angles. However, we can make some general comments. For moving objects like a ship or other vehicle, the RCS can be reduced considerably by shaping the vehicle to avoid dihedral angle corners with interior angles of 90° and so that the edges are smoothed. A mast can be designed with a polygonal cross section, which would give it an RCS lower than one with a circular cross section. When using foam columns for support, a square cross section can be used for low RCS. Generally, RCS reduction by shaping the target would assume some significance in situations where orienting the target to reduce the probability of its detection by radar is possible.

To help understand the effect of such shaping, we will briefly discuss the effect of body shape on RCS [1]. The aim here is to help the design of an object having a maximum volume but low (or minimum) RCS. Figure 4.1 shows the nose-on backscattering cross sections for a class of metallic objects having identical projected areas in all directions, length-to-diameter ratios, and volumes within a factor of two. The results shown in Figure 4.1 indicate that RCS of bodies can vary significantly depending on their shapes.

The effects of shaping the nose of a body is of considerable practical importance. A reduction in nose-on RCS may be achieved in general by using an ogival section instead of a spheroidal section for the nose of a body. Figure 4.2 shows the possible shape modifications applied to the nose cone of a target and their effects on the resultant RCS. An ogive-shaped nose cone can provide considerably lower RCS than a spheroidal one.

Various contributors to an aircraft RCS, their relative size, and the use of shaping to reduce the RCS are schematically represented in Figures 4.3, 4.4, and 4.5,

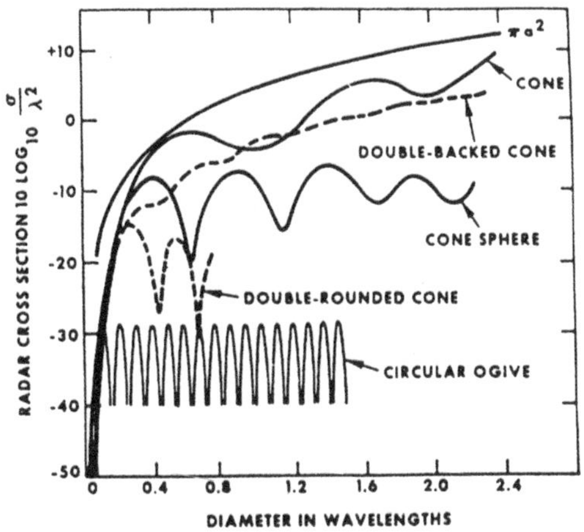

Figure 4.1 Nose-on backscattering cross sections of a class of objects (from [1]).

respectively. In aircraft, the use of shaping to achieve a low RCS usually involves a trade-off with aerodynamic performance. Some geometric modifications found to reduce RCS are shown in Figures 4.6–4.10. Note that fins and wings produce large radar returns only at normal and near-normal incidence, particularly at high frequencies. Figure 4.6 indicates that a flat plate has a large RCS under a variety of incident conditions. Curving the edges as shown in Figure 4.7 may cause less peak returns but the incident energy will be spread over a larger range of aspect angles. A second modification, shown in the lower diagram of Figure 4.7, uses a tapered dielectric material for a portion of the plate. A possible construction scheme for the reduction of RCS of a rotor blade is shown in Figure 4.8(a); contributions by various components to the RCS for both polarizations are shown in Figure 4.8(b). Figures 4.9 and 4.10 show some possible fin configurations. At selected aspect angles, the curved fins produce less RCS than the straight fins in Figure 4.8. The "good" fin design in Figure 4.10 is considered superior from the point of view of the "art" of smoothing radar returns.

4.2.1 Shaping Dihedral Corners

Conducting surfaces on ships and ground vehicles may form corner reflectors having increased RCS in some specific range of aspect angles. Their radar returns can be minimized by taking due care that the surfaces do not meet at right angles. A criterion has been developed in [2], giving the required angle between faces as a function of RCS reduction and the electrical size of the corner. The analysis first uses GO to find the ray and then uses physical optics to calculate the scattered far field. The geometry of the

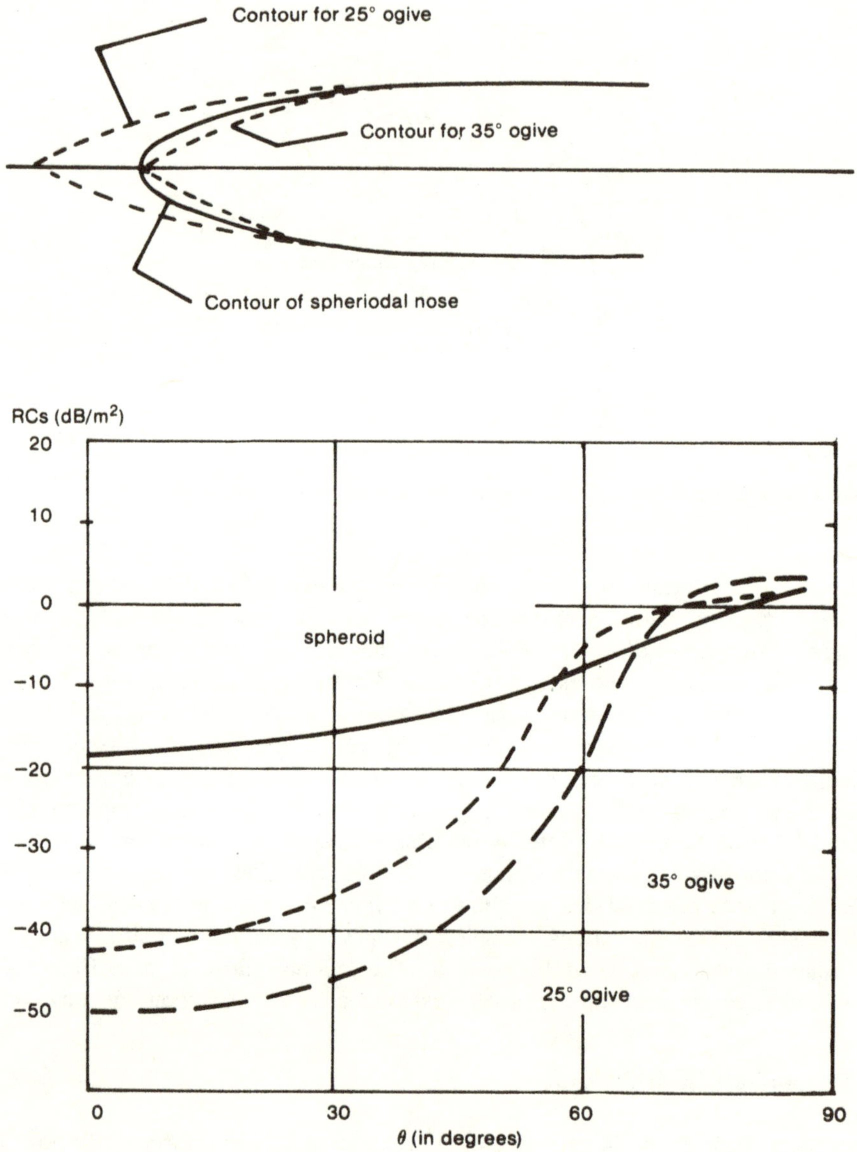

Figure 4.2 Possible modifications in the design of the nose section of the body (with RCS values at 3.0 GHz) (from [96]).

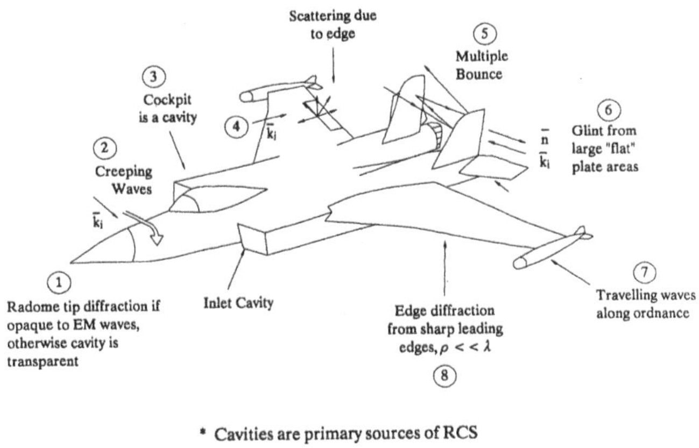

Figure 4.3 Contributors to an aircraft radar cross section (from [96]).

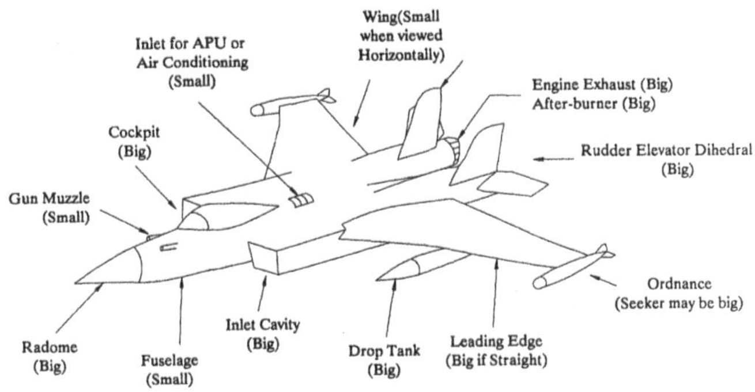

Figure 4.4 Relative size of contributors to aircraft RCS (from [96]).

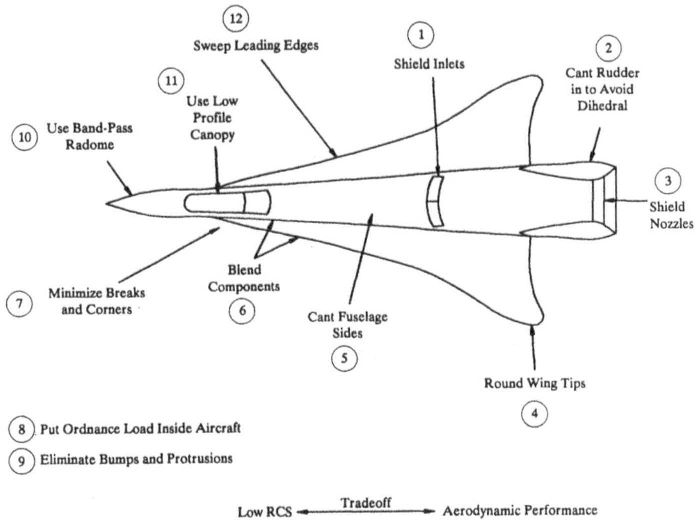

Figure 4.5 Shaping to reduce aircraft RCS (from [96]).

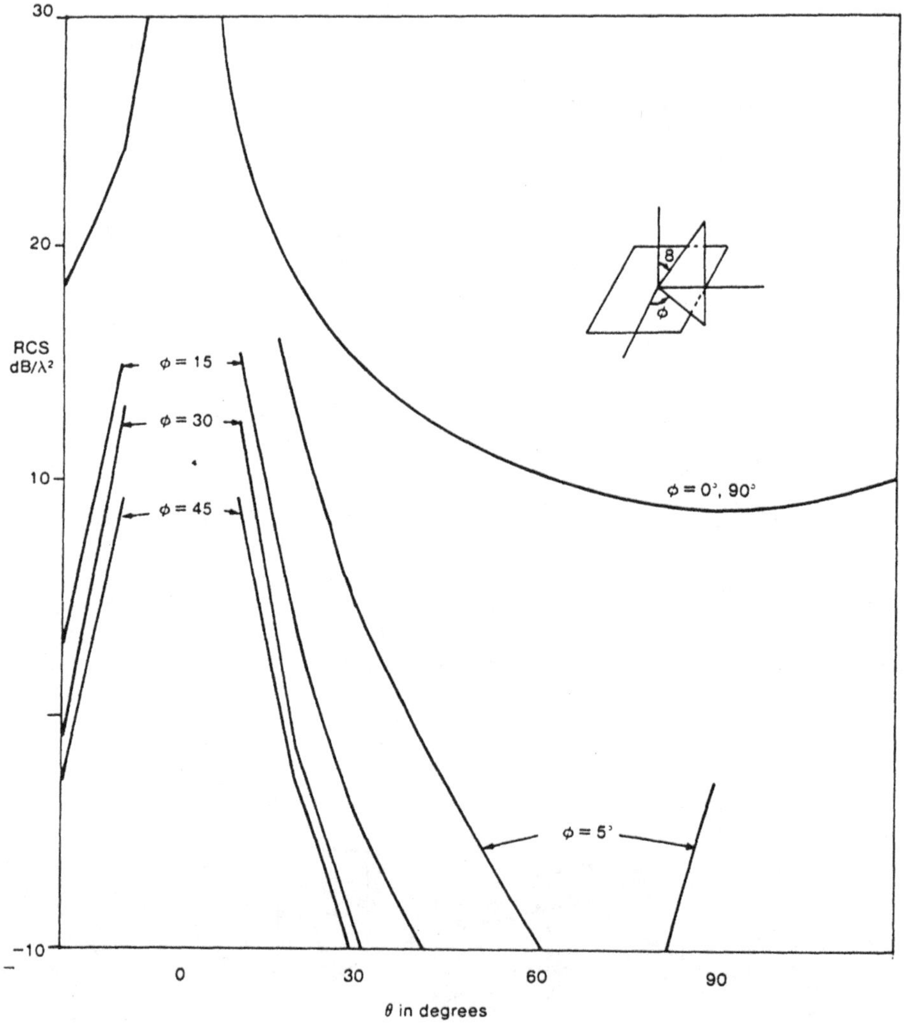

Figure 4.6 RCS of square flat plate 5λ on a side (physical optics) (from [96]).

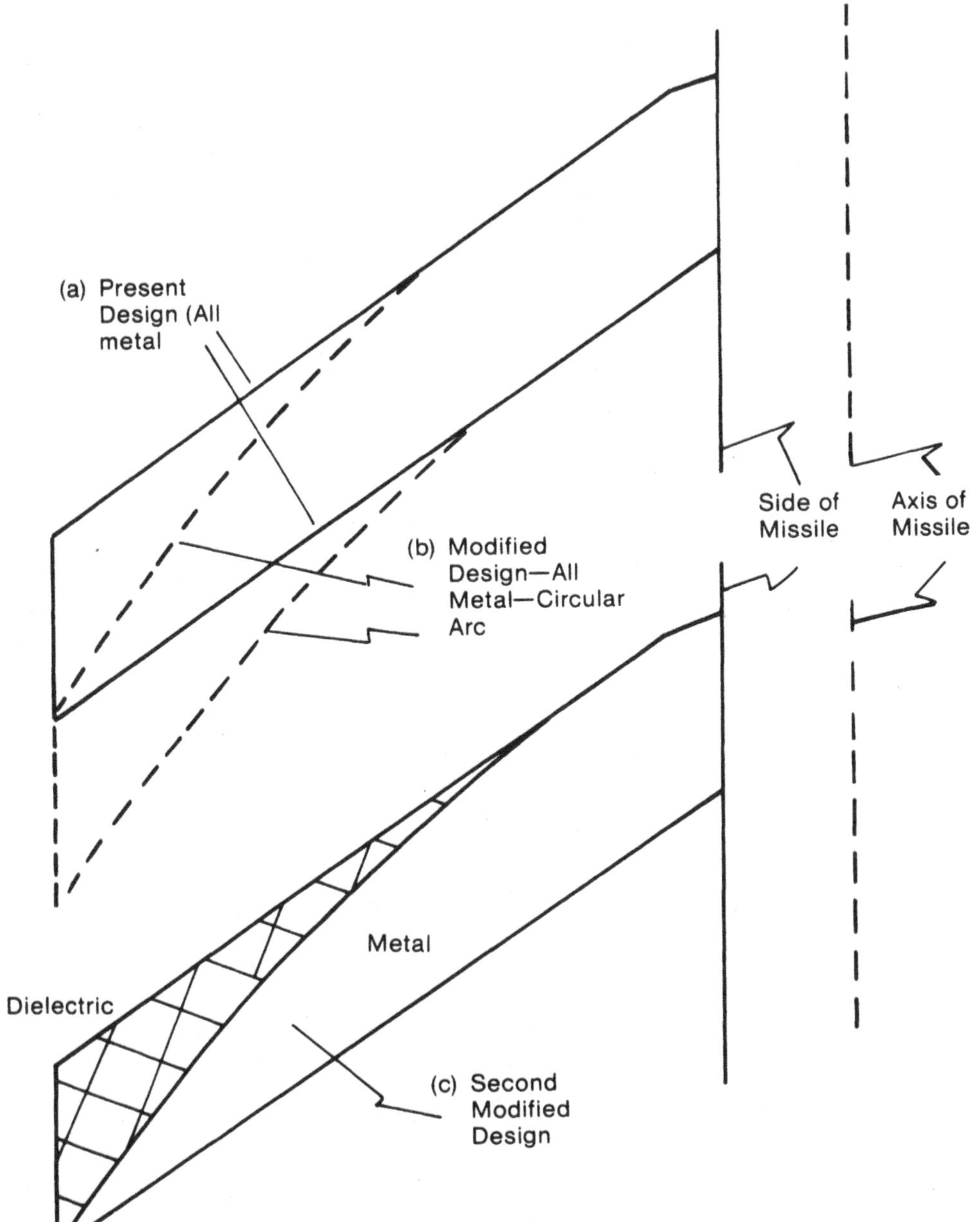

Figure 4.7 Modifications of the plate for RCS reduction (from [96]).

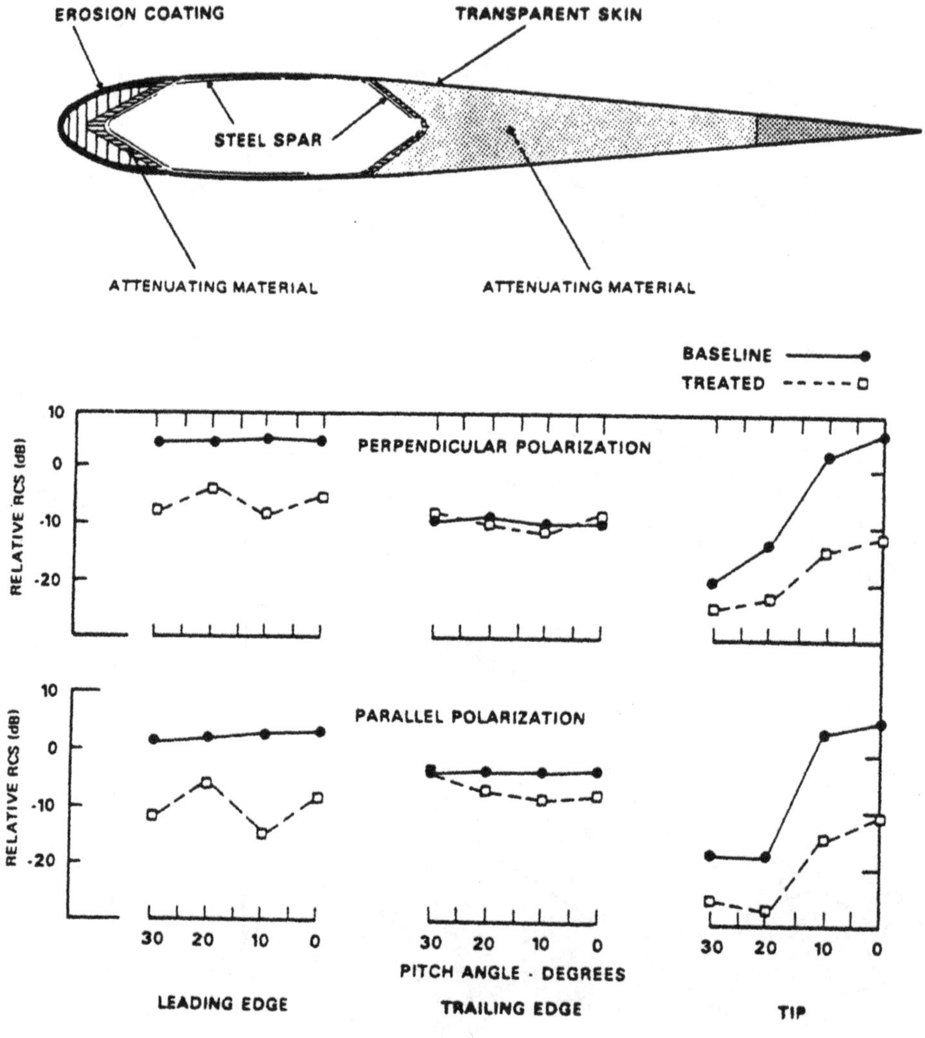

Figure 4.8 Rotor-blade RCS reduction (from [96]).

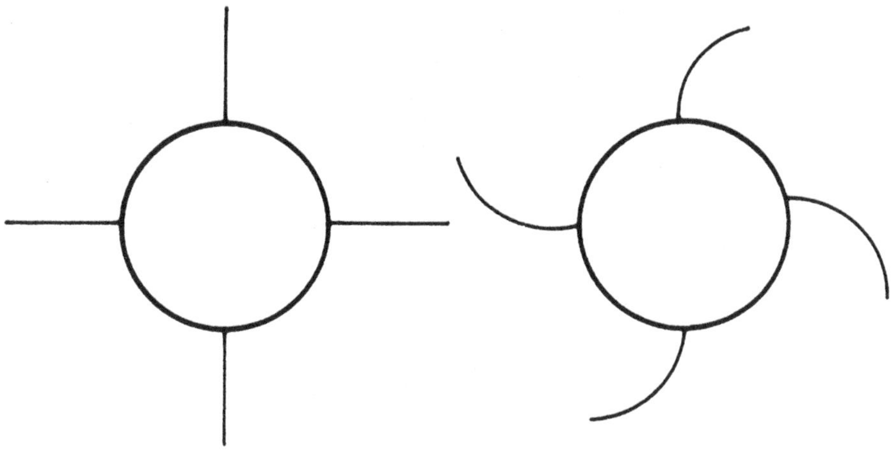

Figure 4.9 Fin designs.

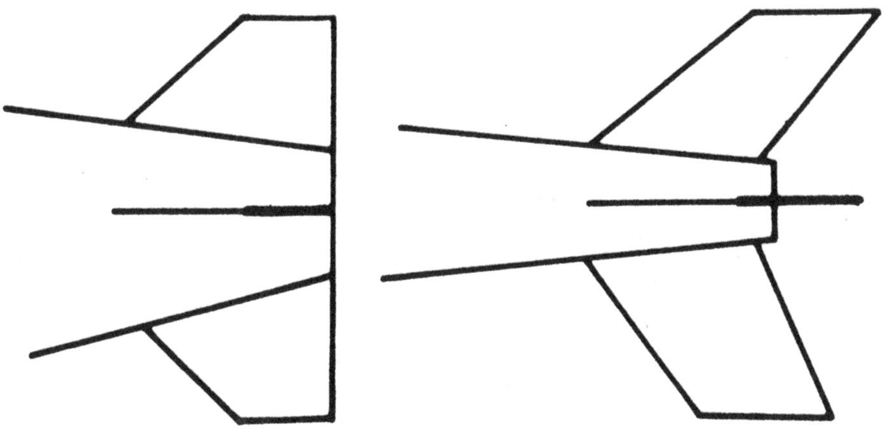

Figure 4.10 Typical fin designs.

dihedral corner reflector is shown in Figure 4.11. We assume that the dimension in a direction perpendicular to the plane of the figure is 1. The basic formula used for this purpose is the bistatic scattered field for a flat rectangular metal plate of length c and illuminated by plane waves as in Figure 4.12, where we assume the plate width is d (measured in a direction perpendicular to the incident plane). With the geometry given in Figure 4.12, we can show that the bistatic scattering function is [3]

$$\mathbf{S} = \frac{ikdc}{2\pi}(\hat{s} \times \hat{h}_r) \cdot (\hat{n} \times \hat{h}_i)\, e^{ik\mathbf{r}_0 \cdot (\hat{i} - \hat{s})} \cdot \sin\frac{[kc(\sin\theta_i - \sin\theta_s)/2]}{kc(\sin\theta_i - \sin\theta_s)/2} \tag{4.1}$$

where

$\hat{h}_i, \hat{h}_r$ = the unit vectors in the directions of the incident and reflected magnetic fields, respectively,
$\mathbf{r}_0$ = the position vector of the midpoint of the plate with respect to the origin,
$\hat{i}, \hat{s}$ = the unit vectors in the direction of incidence and scattering, respectively.

The RCS of the dihedral corner (Figure 4.11) can be obtained from

$$\sigma = \frac{\lambda^2}{\pi}|\mathbf{S}_a + \mathbf{S}_b + \mathbf{S}_{ab} + \mathbf{S}_{ba}|^2 \tag{4.2}$$

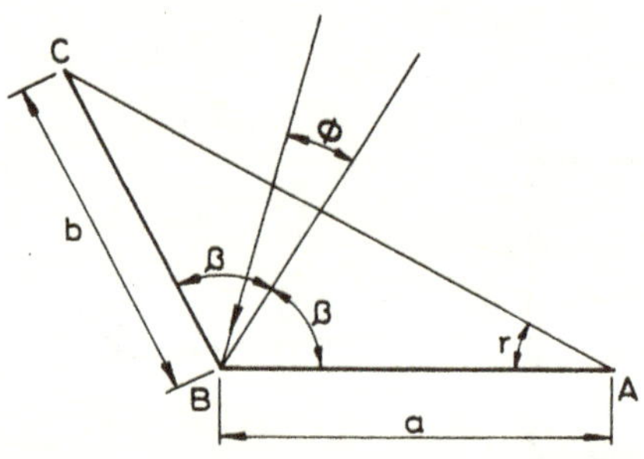

Figure 4.11 The dihedral corner reflector (from [2]).

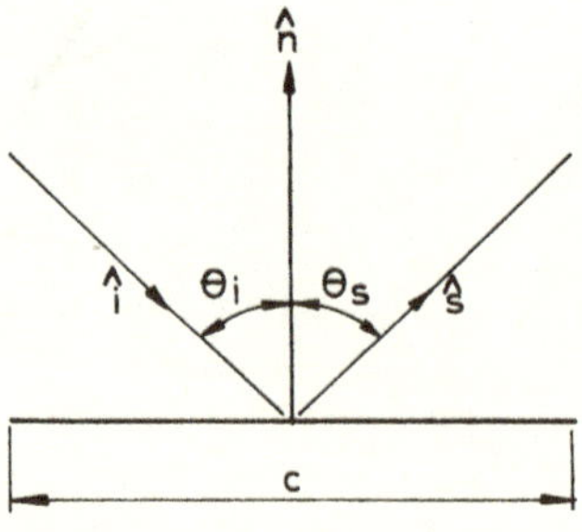

Figure 4.12 Incident and scattered directions of flat plate (from [2]).

Where $\mathbf{S}_a$ and $\mathbf{S}_b$ are direct contributions from the faces having dimensions a and b, and $\mathbf{S}_{ab}$, $\mathbf{S}_{ba}$ are the double-bounce contributions. With the apex of the dihedral corner placed at the origin, (4.1) can be used to determine the direct contributions for $-\beta \le \varphi \le \beta$. They are

$$S_a = i\frac{\mathrm{i}kad}{\lambda}\sin(\beta+\varphi)\exp[-\mathrm{i}ka\cos(\beta+\varphi)] \times \frac{\sin[ka\cos(\beta+\varphi)]}{ka\cos(\beta+\varphi)}, \quad \text{for } E\text{-polarization} \tag{4.3}$$

$$S_b = i\frac{kbd}{\lambda}\sin(\beta-\varphi)\exp[-\mathrm{i}kb\cos(\beta-\varphi)] \times \frac{\sin[kb\cos(\beta-\varphi)]}{kb\cos(\beta-\varphi)}, \quad \text{for } H \text{ polarization} \tag{4.4}$$

The secondary contributions can be determined after finding the illumination of one face due to the reflection from the other face. Because this illumination changes with φ and β, the illuminated sectors on the two faces must be determined for each φ and β. For E-polarization, the scattering functions for double bounce are given by [2]

$$S_{ab} = \frac{\mathrm{i}kb'd}{\lambda}\sin(3\beta+\varphi)\exp[-\mathrm{i}kb'\cos 2\beta\cos(\beta+\varphi)] \frac{\sin[kb'\cos 2\beta\cos(\beta+\varphi)]}{kn'\cos 2\beta\cos(\beta+\varphi)} \tag{4.5}$$

$$S_{ba} = -\frac{\mathrm{i}ka'd}{\lambda}\sin(3\beta-\varphi)\exp[-\mathrm{i}ka'\cos 2\beta\cos 2(\beta-\varphi)] \times \frac{\sin[ka'\cos 2\beta\cos(\beta-\varphi)]}{ka'\cos 2\beta\cos(\beta-\varphi)} \tag{4.6}$$

where a' and b' are the illuminated portions of the corner surfaces as shown in Figure 4.13. The corresponding scattering functions for H-polarization are obtained from (4.5) and (4.6) after replacing $\sin(3\beta \pm \varphi)$ with $\sin(3\beta \mp \varphi)$. The normalized RCS for the dihedral corner can now be determined by using (4.3)–(4.6) in (4.2), given by

$$\frac{\sigma}{d^2} = \frac{1}{4\pi}\left|\sum_{m=1}^{4} R_m \sin(P_m)[\exp(-\mathrm{i}2Q_m)-1]/Q_m\right|^2 \tag{4.7}$$

where P_m, Q_m, and R_m are as shown in Table 4.1.

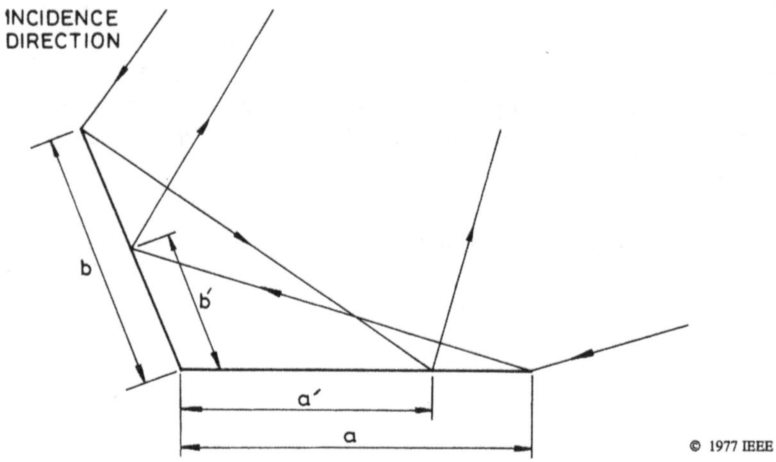

Figure 4.13 Incomplete illumination of face A by wave reflected from face B and *vice versa* (from [2]).

Table 4.1
Variables to Be Used in (4.7)

m	S	P_m E-polarization	P_m H-polarization	Q_m	R_m
1	S_a	$\beta + \varphi$	$\beta + \varphi$	$ka \cos(\beta + \varphi)$	ka
2	S_b	$\beta - \varphi$	$\beta - \varphi$	$kb \cos(\beta - \varphi)$	kb
3	S_{ab}	$3\beta + \varphi$	$-(\beta - \varphi)$	$kb' \cos 2\beta \cos(\beta + \varphi)$	kb'
4	S_{ba}	$3\beta - \varphi$	$-(\beta + \varphi)$	$ka' \cos 2\beta \cos(\beta - \varphi)$	ka'

We can show [2, 4] that the amplitude of the scattered far field due to a dihedral corner with $2\beta \neq \pi/2$ is less than that due to a 90° dihedral corner. The reduction R in the scattered field amplitude can be expressed as

$$R \simeq \left(kb \sin \tau \cos \gamma_0\right)^{-1} = \left(ka \sin \tau \sin \gamma_0\right)^{-1} \tag{4.8}$$

where $\tau = 2\beta - \pi/2$ and $\gamma_0 = \pi/4 + \varphi$.

The reduction factor for a metallic cylindrical segment compared to a rectangular segment shown in Figure 4.14 is given by

$$R = \frac{\theta}{2}\left(\frac{d}{h} + \frac{h}{d}\right), \quad \text{for } \Delta\theta < \gamma$$

$$= 1, \quad \text{for } \Delta\theta > \gamma \tag{4.9}$$

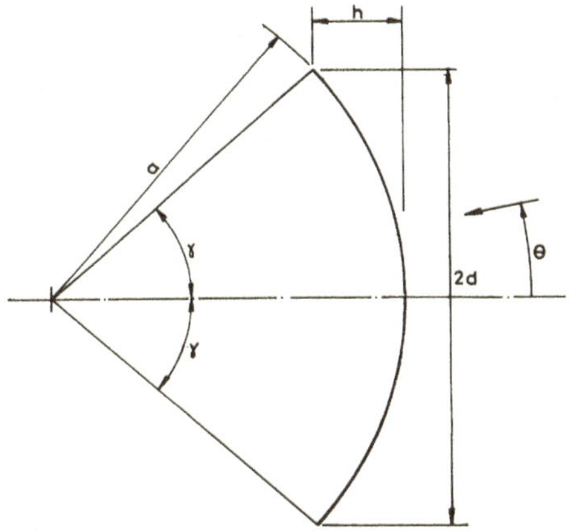

Figure 4.14 Geometry of the cylindrical segment.

where $\Delta\theta$ is the base interval in aspect angles, and d and h are as shown in Figure 4.14. Figure 4.15 shows the reduction R *versus* h/d based on mean returns averaged over the indicated base intervals.

The calculated RCS patterns for dihedral corners in $ka = 45$ and $kb = 30$, and $2\beta = 100°$ and $90°$, are shown in Figure 4.16, where we find that an RCS reduction of about 17.5 dB is possible around $\varphi = 0°$; that is, when the incidence is along the line bisecting the dihedral angle. There is a minimum reduction of about 12.7 dB in the central angular region bounded by $\pm 25°$. Also note from (4.9) that such reduction in RCS is not independent of the size of the corner faces. We therefore recommend that in using the present method we must achieve a specific amount of reduction in RCS at the lowest frequency in the band of frequencies of interest, which will translate to more reduction at the higher frequencies of the band. Further reduction in RCS then can be obtained by using RAM in localized regions.

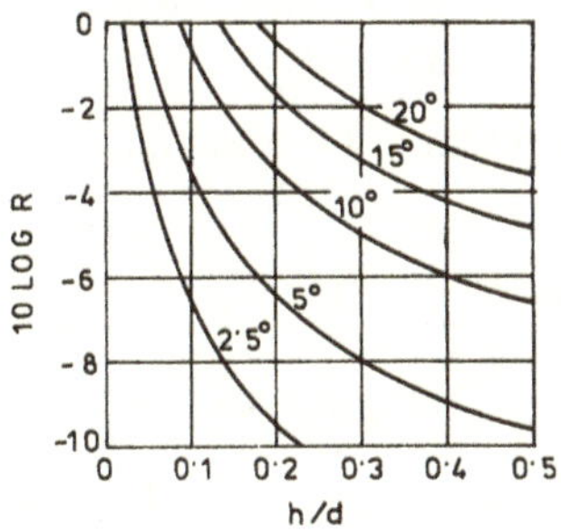

Figure 4.15 Approximate RCS reduction based on mean returns averaged over the indicated interval (from [4,5]).

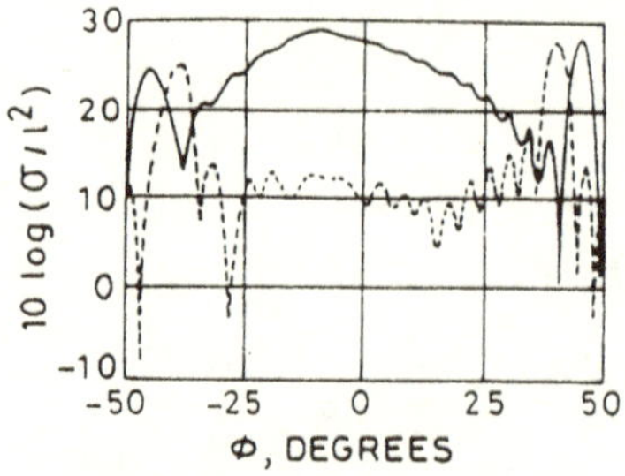

Figure 4.16 Calculated RCS patterns of dihedral corner for $ka = 45$ and $kb = 30$ (solid curve is for conventional 90° and dashed curve is for 100° dihedrals) (from [2]).

4.2.2 Plane Surface Made Cylindrical

Large specular returns from a flat metallic plate can be reduced by deforming the plate into a cylindrical plate; however, this method may not reduce the mean RCS by more than 1 dB. The reduction would be appreciable when the angular extent of the plate is considerably larger than the angular sector in which the RCS reduction is desired. The geometry of an electrically large cylindrical structure is shown in Figure 4.14; its axial length is assumed to be 1. Detailed analyses for this structure are given in [5–7]. The physical optics backscattering RCS for this structure is given by [5, 6]

$$\sigma = \frac{\lambda^2}{\pi} | s |^2 \tag{4.10a}$$

with

$$S = \frac{ik^2}{\pi} \left| \int_A \hat{n} \cdot \hat{i} \exp(i2k\mathbf{r} \cdot \hat{i}) \right| da \tag{4.10b}$$

where

$\hat{i}$ = the unit vector in the direction of incidence,
$\hat{n}$ = the vector normal to the surface at the point of incidence,
r = the position vector for the point of incidence on the surface,
da = an elementary area on the surface A.

For a flat plate of length $2d$ and width l, (4.10a) and (4.10b) yield the familiar result:

$$\sigma_{\text{flat plate}} = \frac{l^2}{\pi} \left[2kd \cos\theta \frac{\sin(2kd \sin\theta)}{2kd \sin\theta} \right]^2 \tag{4.11}$$

For the cylindrical configuration (4.10b) reduces to

$$S = ika \exp(-ika) \int_{-(\gamma+\theta)}^{(\gamma-\theta)} \cos\gamma \exp[i2ka(1 - \cos\tau)]\, d\tau \tag{4.12}$$

which may be evaluated numerically. If τ is small, $\cos \tau \simeq 1 - \tau^2/2$, and then (4.12) assumes the following form:

$$S \simeq \frac{\mathrm{i}ka}{\tau} \exp(-\mathrm{i}ka) \int_{-(\gamma+\theta)}^{(\gamma-\theta)} \exp(\mathrm{i}ka\tau^2)\, \mathrm{d}\tau \tag{4.13}$$

which can be expressed in terms of Fresnel integrals. The RCS for a cylindrical segment can now be determined by using (4.10a) and (4.13):

$$\sigma_{\text{cyl}} = \frac{kal^2}{2} \left| F\left[2\sqrt{a/\lambda}(\gamma - \theta)\right] + F\left[2\sqrt{a/\lambda}\,(\gamma + \theta)\right] \right|^2 \tag{4.14}$$

where

$$F(x) = \int_0^x \exp\left(\frac{\mathrm{i}\pi t^2}{2}\right) \mathrm{d}t \tag{4.15}$$

Figure 4.17 shows the RCS pattern for a cylindrical segment of width $2d = 30\lambda$ for a variety of h. The important conclusions that can be drawn from the results shown

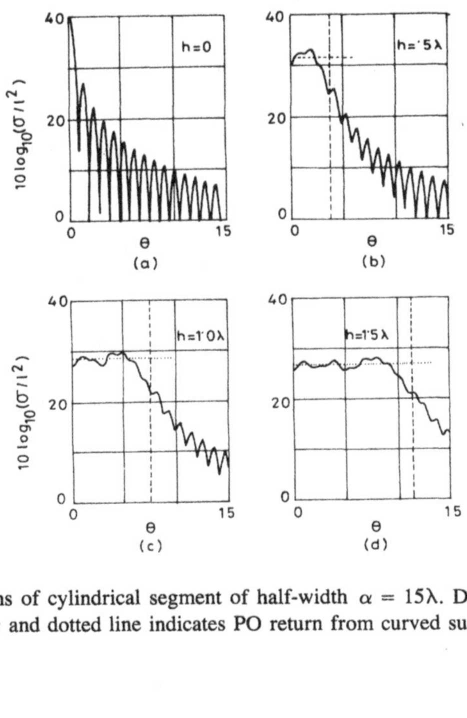

Figure 4.17 RCS patterns of cylindrical segment of half-width $\alpha = 15\lambda$. Dashed lines indicate segment of half-angle γ and dotted line indicates PO return from curved surface (from [5]).

in Figure 4.17 are these: as the plate becomes more and more curved, the main lobe of the return becomes broadened and reduces in intensity, which means a suppression of the specular echo accompanied by an enhancement for angles in the neighborhood of the specular region. To a radar target RCS control engineer, the mean returns rather than the details of returns are of interest. Figure 4.18 (a)–(c) show the mean RCS values for a cylindrical segment obtained by averaging over a given aspect angle interval. Observe from Figure 4.18 that the mean RCS is approximately independent of the increase in h until the base interval is equal to γ, after which the mean cross section decreases rapidly.

The RCS reduction factor as defined in [5] for cylindrical segments is given by (4.9), which is rewritten as

$$R \approx \frac{\Delta\theta}{2}\left(\frac{d}{h} + \frac{h}{d}\right), \quad \theta < \gamma$$

$$\approx 1, \quad \theta > \gamma,$$

with

$$a = \frac{d}{2}\left(\frac{h}{d} + \frac{d}{h}\right), \quad \text{in Figure 4.14}$$

For a small $\Delta\theta$,

$$R = \frac{\lambda}{16h}\,\frac{h^2 + d^2}{d^2} \tag{4.16}$$

A plot of R *versus* h/d for different base segments $\Delta\theta$ is given in Figure 4.15.

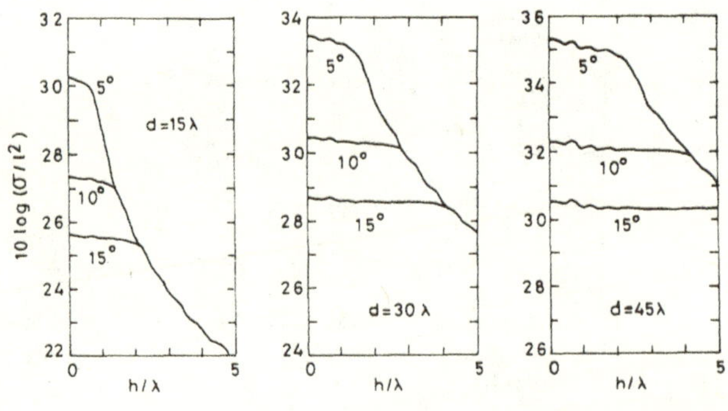

Figure 4.18 Mean RCS obtained by averaging over indicated angular sector for different values of a interval (from [5]).

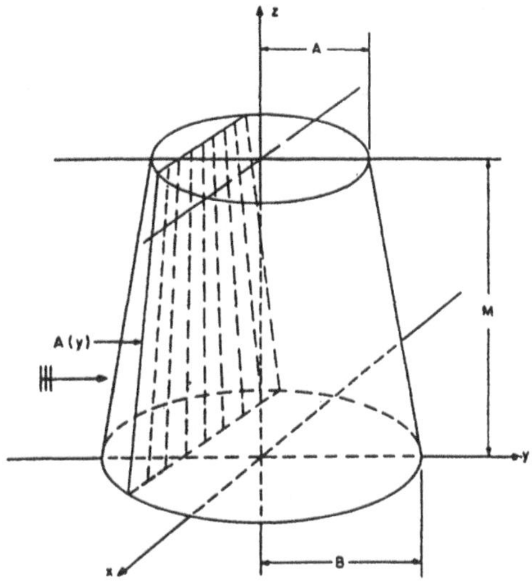

Figure 4.19 A truncated cone.

The RCS reduction is independent of wavelength when the mean RCS is used, but this depends on the aspect angle sector over which the mean is taken. For a small $\Delta\theta$, the reduction becomes wavelength dependent. Maximum reduction is possible by replacing the flat plate by a semicylinder with $h = d$ in which case $R = \Delta\theta$. The essential condition for a cylindrical segment to provide RCS reduction is that it should have an angle larger than the averaging interval.

4.2.3 Shaping of Support Columns

The question of shaping support columns for antennas and scatterers plays an important role because they may cause unwanted reflections. This has been discussed in several references [8–10]. When fabricated from low-density plastic foam, a truncated cone model support has been found to be superior to a cylindrical column [8]. Geometry of a truncated cone is shown in Figure 4.19; experimental backscattering cross sections for the truncated cone are shown in Figure 4.20, which also shows a theoretical backscattering cross section for a circular cylinder for comparison. In applications where dielectric structural support rods must be placed parallel to the electric field vector of an incident electromagnetic wave, the scattering by such supports can be reduced considerably by canceling the capacitance with the help of inductive tunes. Measured scattering patterns for the support dielectric post with and without tuning are shown in Figure 4.21. Average suppression of the scattered power of the support post by a typical tuning

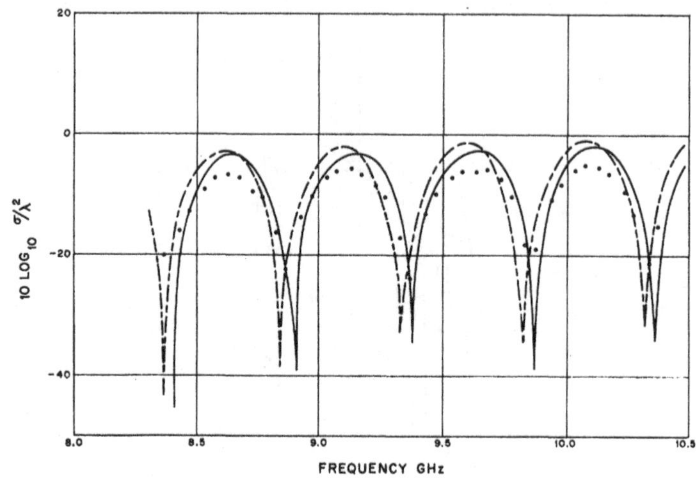

Figure 4.20 Backscattering from a truncated cone (A = 15.16 cm, β = 15.32 cm, M = 39.85 cm): ——— theoretical; - - - - experimental; —— theoretical backscattering from a cylinder (A = B = 15.32 cm; M = 39.85 cm). (From [8].)

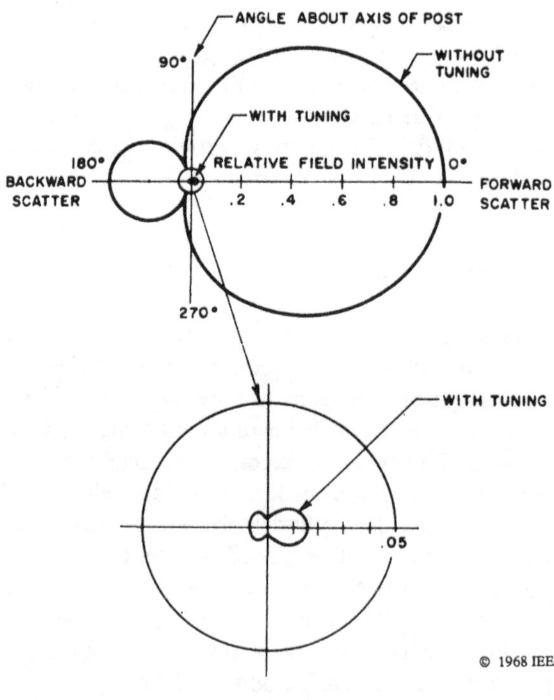

Figure 4.21 Measured scattered field patterns of support post (from [9]).

post is shown in Figure 4.22. A simple approximate formula for the radius a of the metallic wire used for tuning is [9].

$$a = \frac{b}{\sqrt{\varepsilon}} \exp\left[-\frac{2}{k-1}\left(\frac{\lambda}{2\pi b}\right)^2\right] \tag{4.17}$$

where b is the radius of the dielectric post.

The echo widths (RCS per unit length) of dielectric cylinders made of foam material ($\varepsilon_r = 1.035$) and having circular, square, and triangular cross sections have been investigated theoretically and experimentally in [10]. The researchers found, in general, that a cylinder with square cross section has an echo width at least 13 dB below that of a cylinder with circular cross section, although a circular cylinder can have low echo width for a narrow band of frequencies. On the basis of these observations it is recommended that dielectric support structures or pedestals used during antenna or scattering measurements be shaped to have a square cross section. The basis of the recommendation is further confirmed by Figure 4.23 (a) and (b), which gives results for the echo widths for dielectric cylinders having triangular and square cross sections.

4.2.4 The Cone-Sphere

A metallic cone-sphere having a smooth transition between the base of the cone and the spherical section has a low backscattering cross section for some range of aspect angles. For this reason, such configurations have attracted attention for many possible applications, and its scattering properties have been studied in great detail by Senior [11]. The

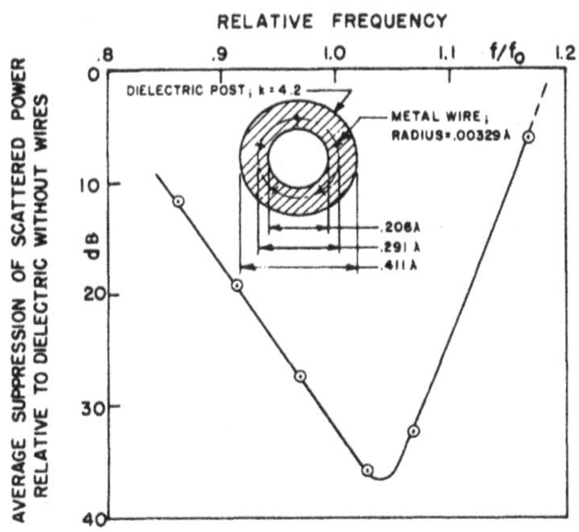

Figure 4.22 Average suppression of scattered power as a function of frequency for the illustrated tuned post (from [9]).

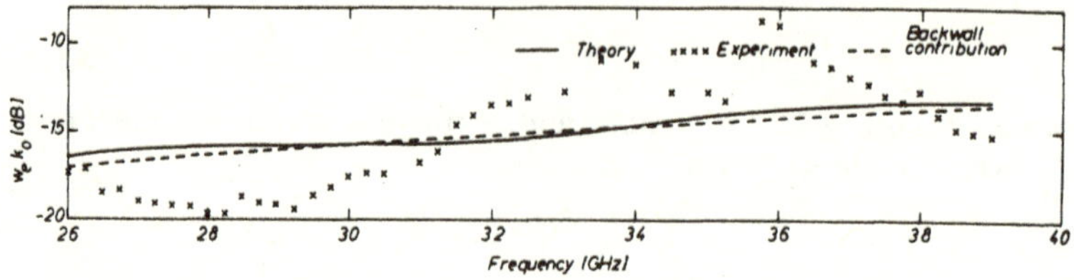

(a)

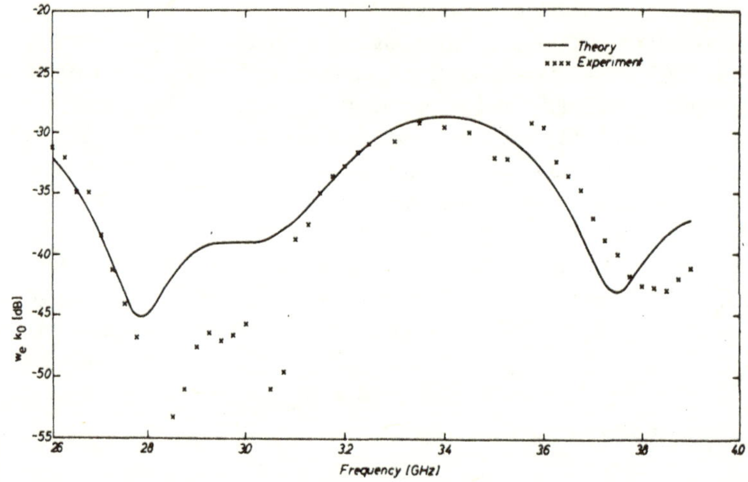

(b)

Figure 4.23 (a) The echo width for a cylinder with triangular cross section multiplied by the free-space propagation constant with frequency (from [10]); (b) The echo width for a cylinder with square cross section multiplied by the free-space propagation constant with frequency (from [10]).

physical optics approximation to the nose-on scattering function for a cone-sphere is [11]

$$S = -\frac{i}{4}\left[\tan^2\alpha \exp(-\mathrm{i}ka \operatorname{cosec} \alpha) - \sec^2\alpha \exp(-\mathrm{i}2ka \sin \alpha) + 1\right] \quad (4.18)$$

where

2α = the total angle of the cone,
a = the radius of the base.

In (4.18) the first, second, and third terms may be identified as contributions from the tip, cone-sphere joint, and the shadow boundary, respectively. Further details are available in [11, 12]. The shadow boundary contribution given by the PO expression in (4.18) has been found to give incorrect results; analytical expression for this based on creeping wave analysis has been obtained by Senior and Goodrich [12]. For details of the creeping wave analysis the reader is referred to [12].

Knott and Senior showed [13] that connecting fins of suitable size makes the cross section depend critically on the roll and aspect angle. It is then possible to significantly reduce the nose-on RCS over small angular sectors. Figure 4.24 shows the normalized nose-on RCS *versus* aspect angle for a cone-sphere ($\theta = 40°$) with and without fins.

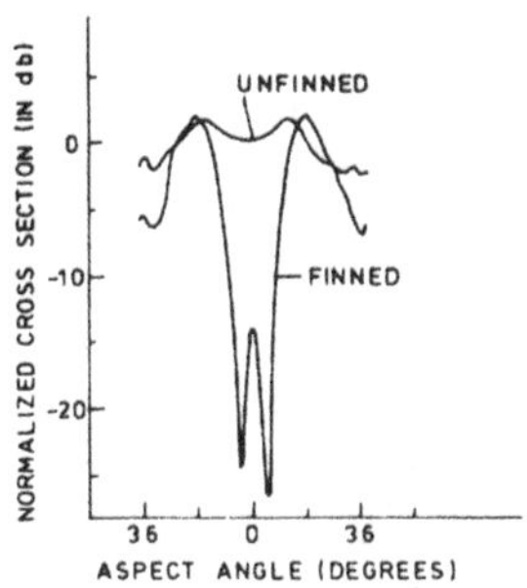

Figure 4.24 The normalized cross sections as a function of aspect angle with $\alpha = 40°$ for a cone sphere with and without fins (from [5]).

The reduction of nose-on RCS by fins is clear from Figure 4.24. Generally, we can obtain some reduction in nose-on RCS for all roll angles with particular fin size and at a particular frequency but such reduction is quite small.

4.2.5 Limitations

RCS reduction of a target by modifying its shape has the following limitations:

1. Normally, reduction is achieved in a limited angular region and at the cost of increased RCS in other regions.
2. The method becomes more effective at higher frequencies; in fact, at low radar frequencies shaping becomes almost ineffective.
3. After the contribution due to a major scattering center is reduced, there often appears a tendency of a previously unimportant scattering area of the body to contribute significantly.
4. Often, the principal radii of curvature of the modified shape may dominate the nature of specular return.

4.3 DISCRETE LOADING

RCS reduction by this method consists in impedance loading at discrete points on the target, which disturbs the induced current thereby leading to reduced scattered fields in a specified range of aspect angles and, hence, to reduced RCS. Being a phase cancellation method, it also can give rise to large scattered fields at angles slightly outside the specified range. Although much work was done on this technique during the 1960s, it has now been discarded as a viable RCS reduction technique. For historical reasons and also for completeness we briefly discuss the method along with some representative results that may find selective applications.

Iams [14] was the first to use impedance loading to reduce backscattering of a conducting body of revolution and Sletten first applied the idea [15] of loading to reduce RCS of objects in space. Various aspects of scattering by impedance loaded conducting cylindrical structures are discussed in [16–21]. Chen and Liepa [18–20] systematically studied the modification of backscattering cross section of a conducting cylinder by discrete impedance loading. They found that, by centrally loading the cylinder, the induced current can be modified to minimize the broadside RCS of the cylinder, and there is an optimum impedance that produces zero broadside scattering. This optimum impedance is real (i.e., passive) for cylinder lengths shorter than 1λ and active for lengths larger than 1λ. Loading at a single point was found ineffective in reducing off-broadside scattering, which required loading at least at two discrete points. Also, with appropriate impedance loading at double points, it was possible to eliminate the broadside scattering completely.

The geometry of a centrally loaded thin-cylindrical dipole illuminated by a plane electromagnetic wave is shown in Figure 4.25. The induced current has been determined in [18] by solving an integral equation, Figure 4.26 gives the theoretical induced current for different values of the load. With no loading ($Z_L = 0$) the induced current is very large and the current distribution takes the standard form of a shifted cosine curve. Infinite loading ($Z_L = \infty$) produces a double-hump distribution with a null at the center. With Z_L capacitive and finite loading, the induced current lies between those with $Z_L = 0$ and $Z_L = \infty$. With Z_L inductive the induced current amplitude is smaller than that with $Z_L = \infty$, and the distribution has three loops of current one of which has its phase reversed; this latter property gives rise to a reduced scattered field in the broadside direction. The optimum impedance for zero broadside scattering from the configuration shown in Figure 4.25 is [18]

$$(Z_L)_0 = \frac{-i60T_{sd}(1 - k_0 h \cot k_0 h)}{2\cos k_0 h - 2 + k_0 h \sin k_0 h} \tag{4.19}$$

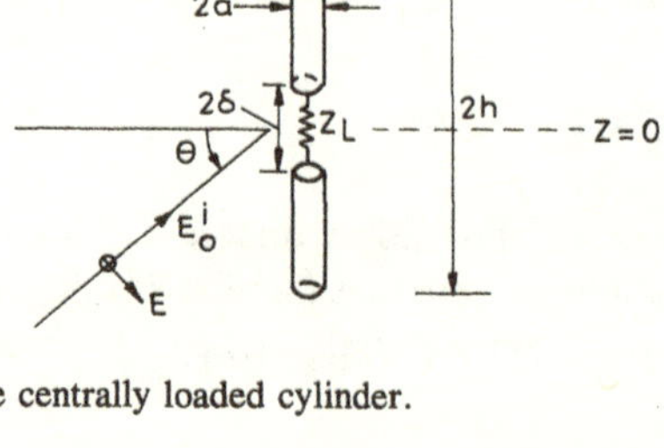

Figure 4.25 The geometry of the centrally loaded cylinder.

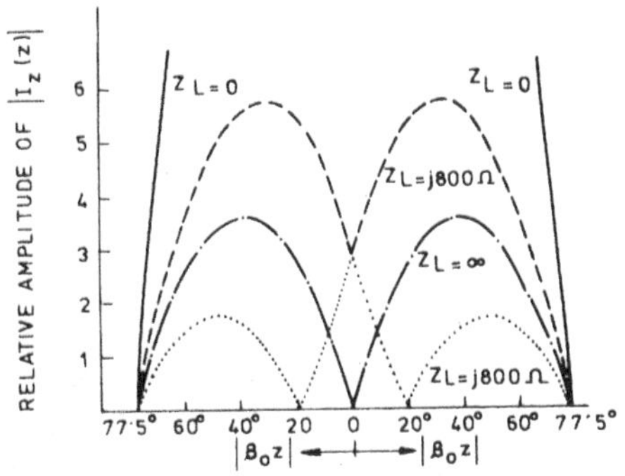

Figure 4.26 Current distribution $I_z(z)$ on cylinder of $h = 0.215\lambda$; $a = 0.0173\lambda$ for different central loads Z_l (theoretical) (from [18]).

where

$$T_{sd} = \int_{-h}^{h} \sin k_0(h - |z'|) K_d(0, z')\, \mathrm{d}z' \tag{4.20}$$

$$K_d(z, z') = K_a(z, z') - K_a(h, z') \tag{4.21}$$

$$K_a(z, z') = \frac{\exp\left[-\mathrm{i}k_0 \sqrt{(z - z')^2 + a^2}\right]}{\sqrt{(z - z')^2 + a^2}} \tag{4.22}$$

For a cylinder having $a = 0.0173\ \lambda$ the optimum impedance *versus* normalized cylinder length as determined from (4.19) is shown in Figure 4.27. Examination of the results in Figure 4.27 leads to the following important observations: (1) the optimum

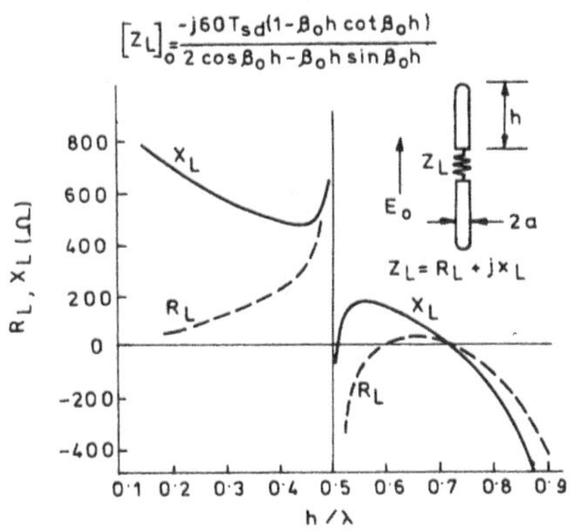

Figure 4.27 Optimum central impedance Z_l *versus* cylinder length for a centrally loaded cylinder ($a = 0.0173\lambda$) (from [18]).

central impedance for zero backscattering has both resistive and reactive components; (2) for $h < 0.5\ \lambda$ the optimum load has resistive and inductive components; and (3) for $h > 0.5 \cdot \lambda$ the optimum load has a negative resistive and either capacitive or inductive reactance.

Impedance loading at a single point is not effective in reducing off-broadside backscattering. In such cases a proper loading at two selected points has been found to be very effective [20]. Here, again, the aim is to find an optimum loading for zero backscattering in the broadside direction, a suitable loading for reduction of off-broad-side direction, a suitable loading for reduction of off-broadside backscattering, and optimum loading combinations for reduction of cross section over a range of aspect angles. A typical geometry is shown in Figure 4.28, where it is assumed that $\lambda/4 < h < 2\ \lambda$ and $k_0^2 a^2 \ll 1$. The optimum load impedance for zero backscattering in the broadside direction for the configuration shown in Figure 4.28 is [20]

$$(Z_L)_0 = -\mathrm{i}15 T_{id} \cos k_0 h\,(\sin k_0 h - k_0 h \cos k_0 h) / \Big\{\sin k_0(h-d) \cdot (\cos k_0 d - \cos k_0 h)^2 - \cos k_0 d \sin^2[k_0(h-d)]\,(\sin k_0 h - k_0 h \cos k_0 h)\Big\} \tag{4.23}$$

where

$$T_{id} = \int_{-h}^{h} [2 \sin k_0 h \cos k_0 h \cos k_0 z - \cos k_0 h\,(\sin k_0 |z' - d| + \sin k_0 |z' + d|)]\,K_d(0, z')\,\mathrm{d}z'$$

and $K_d(z, z')$ is as given by (4.22)

The required impedance for low off-broadside backscattering is given in [20]. Real and imaginary parts of the optimum load determined from (4.23) and applicable for the configuration given in Figure 4.28 are shown in Figures 4.29 and 4.30. The results indicate that $(Z_L)_0$ depends on the position of the loading points. For low backscattering over a wide range of angles, the design of optimum impedances and their

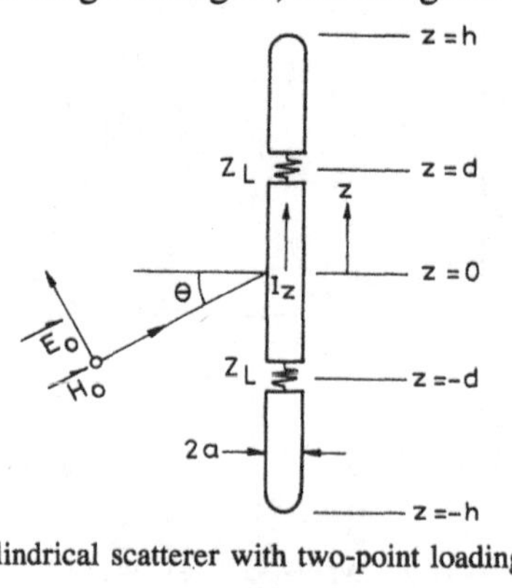

Figure 4.28 Geometry of a cylindrical scatterer with two-point loading.

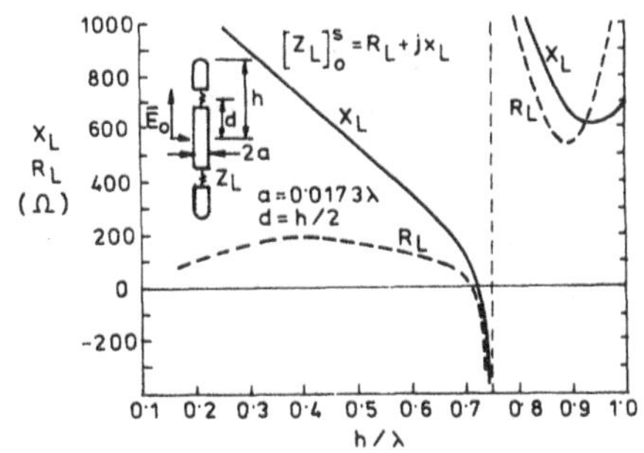

Figure 4.29 Optimum impedance for zero-broadside backscattering *versus* cylinder length with $d = h/2$ (from [20]).

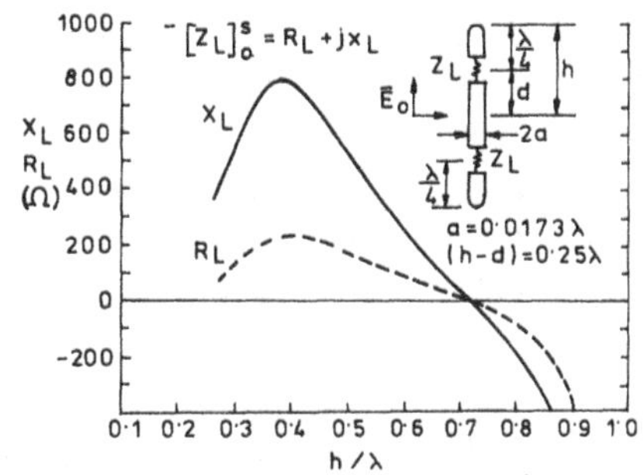

Figure 4.30 Optimum impedance for zero-broadside backscattering *versus* cylinder length with $h - d = \lambda/4$ (from [19]).

locations are more complicated, as discussed in [20]. Figure 4.31 shows the variation of optimum impedance and optimum position for double loading to obtain minimum backscattering over a wide range of aspect angles.

The results presented indicate that we can design an appropriate double-impedance loading for a given conducting cylinder to minimize its backscattering cross

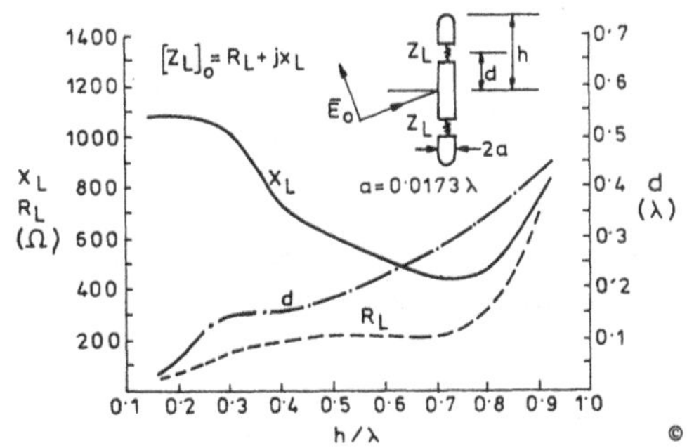

Figure 4.31 Optimum impedance and optimum position for double-loading to minimize backscattering over a wide aspect angle range (from [20]).

section over a wide range of aspect angles. The general theory of *N*-port loading of a scatterer by reactive loading to obtain reduced scattering has been developed in [21] by using the method of characteristic modes. A pattern synthesis procedure is available for extracting the real current whose radiation field distribution is the least mean square approximation to a desired scattered field pattern. A discussion on this topic is beyond the scope of the present book and the reader is referred to [21, 22] for further details.

4.4 SURFACE MODIFICATIONS

The RCS of a body can be directly controlled by physically modifying its surface. Although this method could have been classified under target shaping, discussed in Section 4.2, we have given it a separate name because of its application to RCS reduction associated with a variety of targets. The surface can be altered in a number of ways. A plane conducting plate may be altered by introducing apertures on its surface or by making the surface corrugated. It is well known that the former will cause controllable transmission and reflection; the latter may act as a resonant configuration under favorable condition and thereby control its RCS.

4.4.1 Apertures on a Plane Surface

Although there has been extensive study of the electromagnetic energy leakage through a perforated conducting plate, initially less attention was given to the problem of its RCS. Kieburtz and Ishimaru [23] investigated scattering by a conducting plate having a periodic distribution of apertures on its surface. In general, electromagnetic scattering by a perforated thin conducting plate can be solved by using the induced dipole approximation, if the apertures are electrically small, and by the GTD method, if the apertures are electrically larger. Let us consider a flat conducting plate perforated by a doubly periodic distribution of apertures and illuminated by a plane electromagnetic wave as shown in Figure 4.32. The RCS of the configuration shown in Figure 4.32 can be determined by using either the dipole method or the GTD method.

Dipole Method

With the incident field as shown in Figure 4.32 and assuming parallel polarization, we can show that the overall backscattering cross section in the (θ, φ) direction for the perforated plate is [24]

$$\sigma_{\|} = \left| \left(\cos x - \mathrm{i}\frac{\sin x}{\sin\theta} \right) k^{-1}x + \frac{2k^2a_0^3}{3\sqrt{\pi}} \left(1 + \frac{\sin^2\theta}{2} \right) x \right.$$

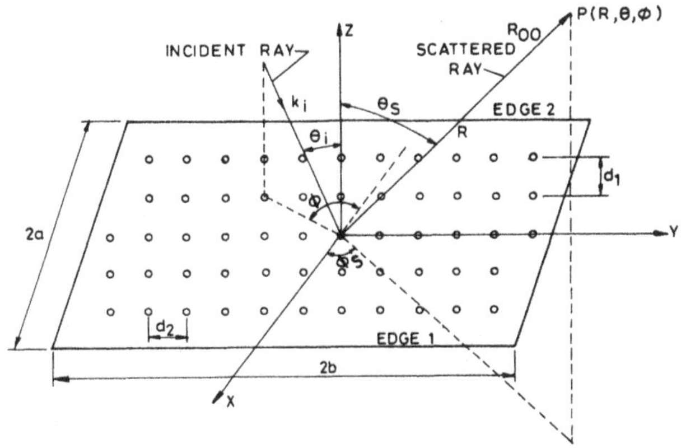

Figure 4.32 Geometry and coordinate system of a plate with perforations.

$$\left.\sum_{n_1}\sum_{n_2}\exp[2k\sin\theta(n_1d_1\cos\varphi+n_2d_2\sin\varphi)]\right|^2 \tag{4.24}$$

$$x = 2ka\sin\theta \tag{4.25}$$

where $n_1 = 0, \pm 1, \pm 2, \pm N_1$; $n_2 = 0, \pm 1, \pm 2, \ldots, \pm N_2$, a_0 is the radius of each aperture and the other parameters are as described in Figure 4.32

Similarly, for perpendicular polarization the backscattering cross section of the perforated plate is [24].

$$\sigma_{\perp} = \left|\left(\cos x + i\frac{\sin x}{\sin\theta}\right)k^{-1}\frac{2a_0^3k\omega\mu_0}{3\sqrt{\pi}}\cos^2\theta\right.$$

$$\left.\sum_{n_1}\sum_{n_2}\exp[-i2k\sin\theta(n_1d_1\cos\varphi+n_2d_2\sin\varphi)]\right|^2 \tag{4.26}$$

with x and other parameters as defined in connection with (4.24).

The GTD Method

The GTD method has also been applied to determine the backscattering cross section of the perforated plate shown in Figure 4.32. The appropriate backscattering cross sections for parallel and perpendicular polarizations are [24]

$$\sigma_{\parallel} = \left| \left(\cos x - i\frac{\sin x}{\sin\theta} \right) k^{-1} + \frac{1 + \csc\theta}{2} \sqrt{\frac{a_0}{2\pi k \cos\theta}} \right.$$

$$\times \left\{ \sum_{n_1} \sum_{n_2} \exp[-i2k \sin\theta(-a_0 + n_1 d_1 \cos\varphi + n_2 d_2 \sin\varphi)] \right. \quad (4.27)$$

$$\left. \left. + \sum_{n_1} \sum_{n_2} \exp[-i2k \sin\theta(a_0 + n_1 d_1 \cos\varphi + n_2 d_2 \sin\varphi)] \right\} \right|^2$$

$$\sigma_{\perp} = \left| \left(\cos x + i\frac{\sin x}{\sin\theta} \right) k^{-1} + \frac{1 - \csc\theta}{2} \sqrt{\frac{a_0}{2\pi k \cos\theta}} \right.$$

$$\times \left\{ \sum_{n_1} \sum_{n_2} \exp[-i2k \sin\theta(-a_0 + n_1 d_1 \cos\varphi + n_2 d_2 \sin\theta)] \right. \quad (4.28)$$

$$\left. \left. + \sum_{n_1} \sum_{n_2} \exp[-i2k \sin\theta(a_0 + n_1 d_1 \cos\varphi + n_2 d_2 \sin\varphi)] \right\} \right|^2$$

with x given by (4.25) and the other notations are as explained earlier.

Figure 4.33(a) and (b) compare monostatic RCS of a square flat plate with and without punched circular apertures, determined by using the two methods just described. For comparison, we have also shown the corresponding RCS of the undisturbed plate obtained by the GTD method (Section 2.9.1). The results presented in Figure 4.33 are for the case when $2a = 2b = 1.2192$ m, $N_1 = N_2 = 50$, and all the apertures are assumed to be circular each with a radius $a_0 = 1.5$ mm ($= \lambda/20$ at $f = 10$ GHz). The area occupied by the total number of apertures is only 0.3505×10^{-3} m^2, which is less than 1% of the physical area of the plate (1.48 m^2). It has been found that $\sigma_{\parallel}$ and $\sigma_{\perp}$ are independent of the number of apertures for the broadside ($\theta = 0$) case, but not for $\theta \neq 0°$.

Table 4.2 gives the calculated reduction of RCS for normal incidence as a function of frequency for a perforated conducting plane; at normal incidence they are found to be independent of the two principal polarizations.

The reduction of RCS at an oblique angle of incidence is a function of the angle of incidence, as is evident from Figure 4.33(a) and (b).

Sometimes, it may be of practical interest to estimate the reduction of the RCS of a metal plate by a slot on its surface. A single resonant slot properly placed on the plate can considerably reduce the RCS of the entire plate. This was originally investigated by Green [25]. Figure 4.34 shows the normal incidence backscattering cross section of a

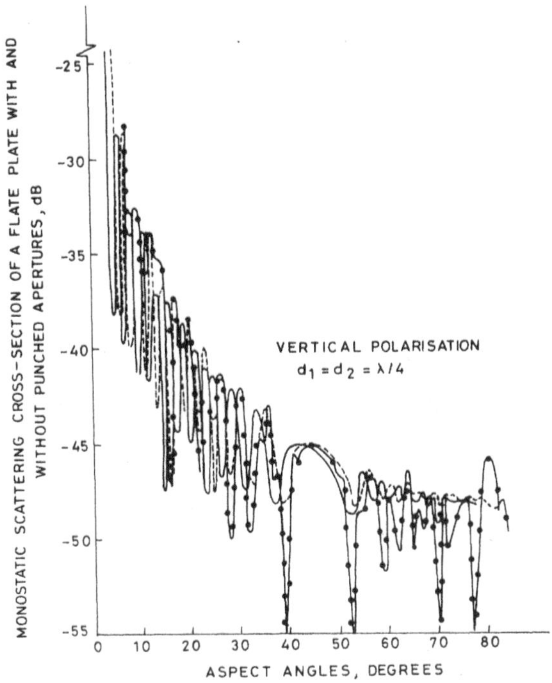

(a)

Figure 4.33 Monostatic RCS of flat plate with and without punched apertures: (a) vertical polarization (b) horizontal polarization; $A = B = 0.6096$ m; $N_1 = N_2 = 50$; $f = 10$ GHz; ——— plate without aperture, - - - - plate with apertures using approach a, ••• plate with apertures using approach b (from [24]).

slot-loaded metal plate as a function of the slot length. The results clearly indicate considerable reduction in the RCS for $L \approx \lambda$; that is, when the slot is resonant. The RCS *versus* aspect angle for a rectangular plate centrally loaded with a $\lambda/2$-long slot is shown in Figure 4.35, which also shows the RCS variation for the same plate without slot. The results show that the slot yields an RCS reduction of about 10 dB over a range of aspect angles $\pm 30°$ from normal incidence.

4.4.2 Corrugations on a Surface

As is well known, corrugated surfaces act as filters. Surface corrugations can be used to reduce RCS of bodies at least over a selected range of aspect angles, frequencies, and polarizations. Such a technique could be used in the design of the roofs of large reflecting structures in the vicinity of an operational radar. Radars operating in crowded

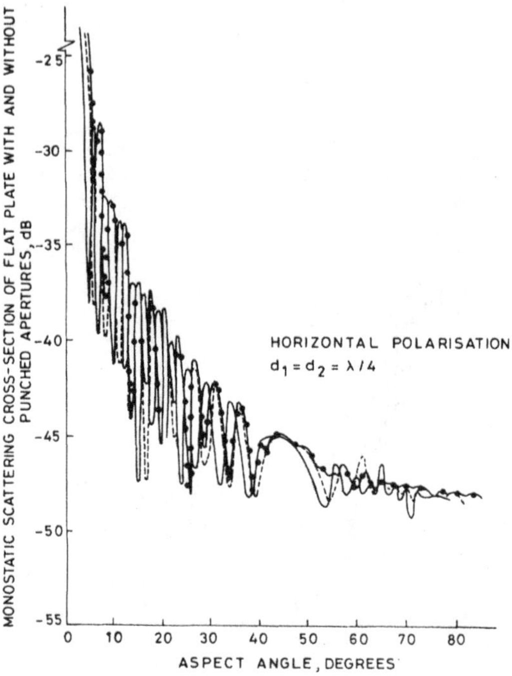

(b)

Figure 4.33 *Continued*

Table 4.2
Reduction of RCS for Normal Incidence Due to Perforations in a Conducting Plate, where $N_1 = N_2 = 20$, $a = b = 0.6096$ m, $d_1 = d_2 = 3\lambda/8$

Frequency (GHz)	*RCS Reduction* (in dB)
8.0	5.74
8.5	5.48
9.0	5.21
9.5	4.98
10.0	4.71
10.5	4.56
11.0	4.36
11.5	4.15
12.0	3.97

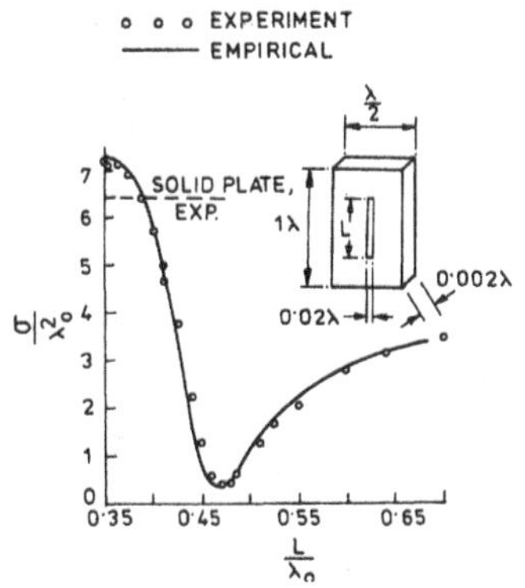

Figure 4.34 Normal incidence backscattering RCS of a slot-loaded rectangular plate ($\lambda_0 \times \lambda_0/2$) *versus* the slot length in wavelengths (from [25]); perpendicular polarization.

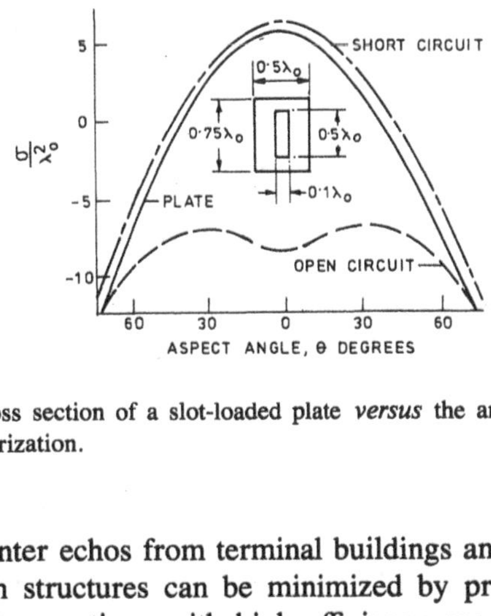

Figure 4.35 Backscattering cross section of a slot-loaded plate *versus* the angle of incidence (from [25]); perpendicular polarization.

airports frequently encounter echos from terminal buildings and large aircraft hangars. The scattering from such structures can be minimized by properly corrugating their surfaces. Blazed diffraction gratings with high efficiency are useful in radio applications, such as frequency multiplexers and scanners.

Electromagnetic scattering by periodic surfaces like corrugated surfaces and gratings has been a subject of intense investigation from the viewpoint of reduction of specular reflection and monostatic RCS. Jull and his colleagues have done extensive work in this area [26–32]. They found that a conducting corrugated surface having a period between $\lambda/2$ and λ scatters only on two modes: specular reflection and backscatter. If the period is chosen for constructive interference in the direction of incidence, the surface is "nonreflecting" in the sense that the specular reflection is reduced to a minimum and the incident power mostly is transferred to the backscatter mode. With proper design the reverse is true for the backscattering mode of interest in this section.

Electromagnetic scattering of plane *H*-polarized waves by a grating of narrow grooves has been investigated in [26], where an appropriate equation was derived and simplified by applying quasistatic approximations. The simplified equation was identified with an integral equation arising from an electrostatic problem that then was solved by conformal transformation. Using scattering matrix and integral transform techniques, Ebbeson [33] investigated the same problem with TM wave illumination. Kalhor [34] analyzed the problem by an approximate method based on the Leontovich boundary conditions. Shafai and Iskander [35] used impedance boundary conditions to determine the scattered fields associated with a corrugated surface.

This present section discusses some theoretical and experimental results for electromagnetic scattering from a uniformly corrugated surface illuminated by a parallel polarized plane wave with the incident plane being normal to the surface. Although most results available in the literature are for specular reflections, the analyses are similar for monostatic RCS. The geometry of the general backscattering problem associated with corrugated conducting surface is shown in Figure 4.36.

4.4.2.1 Grating Dimensions

The uniformly periodic structure shown in Figure 4.36 is assumed to be of infinite dimensions in the x and y directions, and the conducting fins are infinitely thin and of uniform depth. For constructive interference of the contributions from adjacent cells in the scattering or diffraction direction θ_n (θ_n counted positive when measured clockwise from the normal direction in Figure 4.36), the following grating equation must be

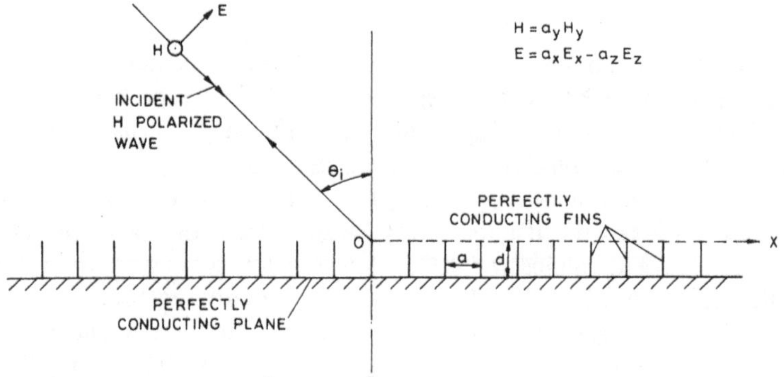

Figure 4.36 Fin-corrugated surface with incident H-polarized wave.

satisfied:

$$\sin \theta_n = \sin \theta_i + \frac{n\lambda}{a} \tag{4.29}$$

where $n = 0, \pm 1, \pm 2, \ldots$ is known as the spectral order.

We can show by using (4.29) that, for $\lambda/2 < a < \lambda$, the allowed real value of θ_n satisfying the grating equation are those for $n = 0$ and -1, which represent the cases of specular reflection and backscattering, respectively. No such simple expression exists for the required corrugation depth d, which depends not only on the angle of incidence but also on the surface profile as well as on the incident wave polarization. With $n = 0$, $\theta_0 = \theta_i$, which is the specular reflection direction in which maximum power is transferred.

For the same corrugated surface, the condition for constructive interference in the backscattering direction is given by Bragg's law:

$$ka \sin \theta_i = \pi, \quad \text{with } k = 2\pi/\lambda$$

which yields the optimum angle of incidence θ_{opt} for enhanced backscattering as:

$$\theta_i = \theta_{\text{opt}} = \sin^{-1}\left(\frac{\lambda}{2a}\right) \tag{4.30}$$

Note that (4.30) can also be obtained from (4.29) with $n = -1$, $\sin \theta_{-1} = -\sin \theta_i$, and $\theta_i = \theta_{\text{opt}}$. For other angles of incidence (i.e., $\theta_i \neq \theta_{\text{opt}}$), we can design a fin corrugated surface having appropriate fin period and depth so that a maximum of energy is transferred to a specularly reflected mode.

The scattering and reflection of waves from the corrugated surface depend on the nature of TM waveguide mode excited in the parallel fin region. Therefore, to know the attenuation of the dominant TM mode propagating in the $+ z$-direction is of interest. By using a well known perturbation technique [36], we can show that the attenuation suffered by the lowest-order TM mode (propagating in the z-direction) is

$$\alpha = \frac{2R_m}{\eta_a\left[1 - (\lambda/2a)^{1/2}\right]}, \quad \lambda < 2a \tag{4.31}$$

where $R_m = (\omega\mu)/2\sigma$ is the surface resistivity of the fins.

4.4.2.2 Design for Minimum Backscattering

In general, the reflection of scattering from a corrugated surface are functions of the fin spacing a, and fin height d, and the angle of incidence θ_i. The parameters a and d

must be optimized to obtain minimum backscattering. One such design for *H*-polarized incident TEM waves is discussed in [37], which also gives expression for monostatic RCS of the corrugated surface. The corrugated configuration design consisted of metallic square plate having each side equal to 53.3 cm and a total of 31 metallic fins placed 1.51 cm apart and parallel to each other. The fin thickness used was 1/16 in. which was about $\lambda/20$ at 10 GHz. The optimum value of d has been found to be 0.5λ. Observe that, at $f = 10$ GHz, $\lambda = 3$ cm, which is slightly smaller than $2 \times 1.51 = 3.02$ cm. The maximum reflection coefficient has been found to be 0.2 for $\theta_i = 85°$. For an angle of incidence $\theta_i = 85°$ the relative powers of $n = 0$ and $n = -1$ modes as functions of fin height are shown in Figure 4.37 for $a = 1.51$ cm and $\theta_i = 85°$. The results clearly indicate that $d = 0.75\lambda$ yields minimum power in the backscattering direction of θ_i. The relative power and phase of the $n = 0$ and $n = -1$ mode returns as functions of the angle of incidence are shown in Figure 4.38, where experimental results for backscattering are given for comparison. Note that the power reflected in the backscattering ($n = -1$) mode is sensitive to the frequency of the incidence wave. In the present case if $f > 10$ GHz power from the specular mode will be transferred to the backscattering mode. For example, at 11.3 GHz the power divides equally between the two modes. It was found experimentally that for incident angles up to 80.6° the fin corrugated surface yields a backscattered return about 35 dB less than that from a conducting plate of the same size. This is a case of narrowband RCS reduction. To obtain wider band performance it may be necessary to modulate the groove depth and the spacing of the fins. This is a subject for further research.

Figure 4.39 gives the corrugation depth h *versus* the angle of incidence θ_i for total TM polarized backscatterer from a rectangular groove surface having $d/\lambda = 1/(2 \sin \theta_i)$ see (4.30). The reduction of the TM polarized specular return from the configuration as a function of θ_i is shown in Figure 4.40. For very thin slots, the

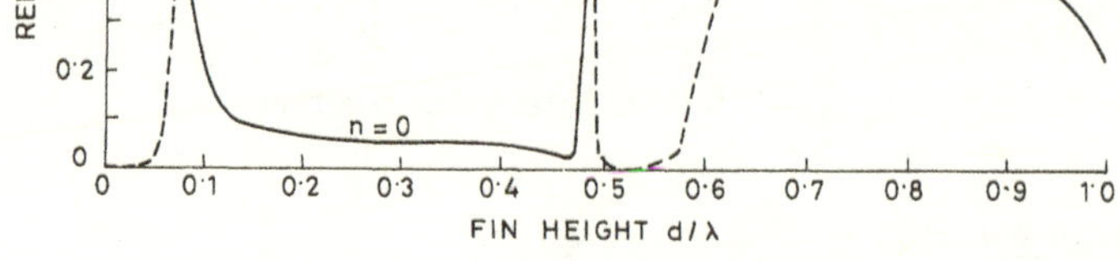

Figure 4.37 Relative power of $n = 0$ and $n = -1$ modes *versus* fin height; $a = 0.5033\lambda$; 100% reduction at $\alpha = 0.5\lambda$; 99% reduction at $d = 0.56\lambda$. For the $n = -1$ mode, the angle of incidence is $\theta_i = 85°$ (from [37]).

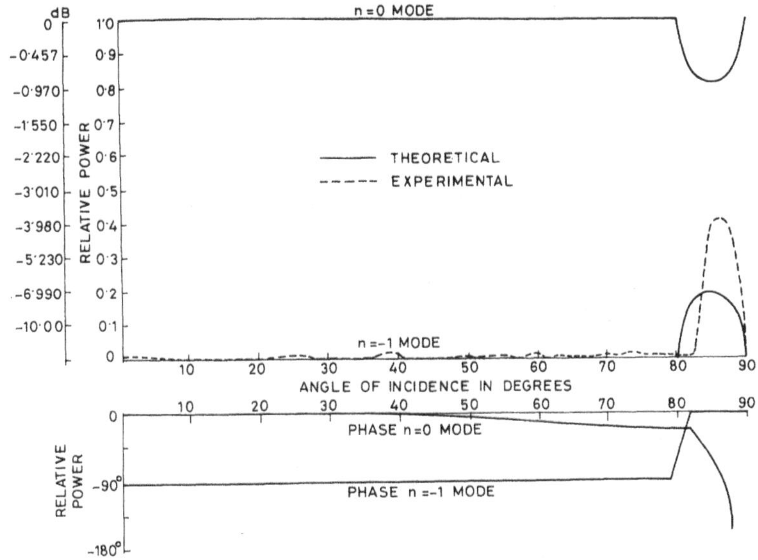

Figure 4.38 Relative power and relative phase of the $n = 0$ and $n = -1$ modes *versus* angle of incidence (no attenuation) with $a = 0.5033\lambda$ and $d = 0.5\lambda$ (from [37]).

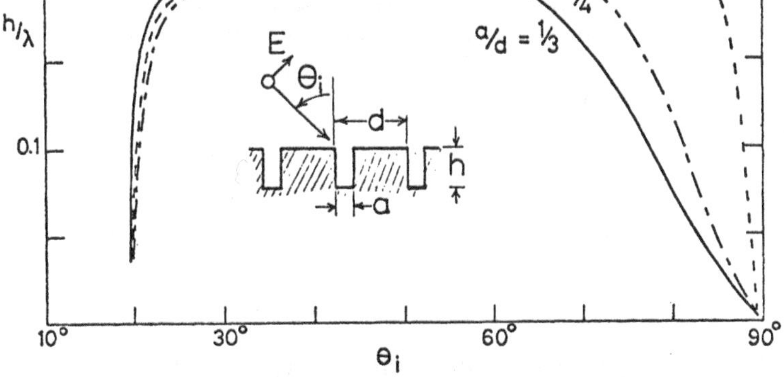

Figure 4.39 Corrugation depth *versus* angle of incidence (from [38]).

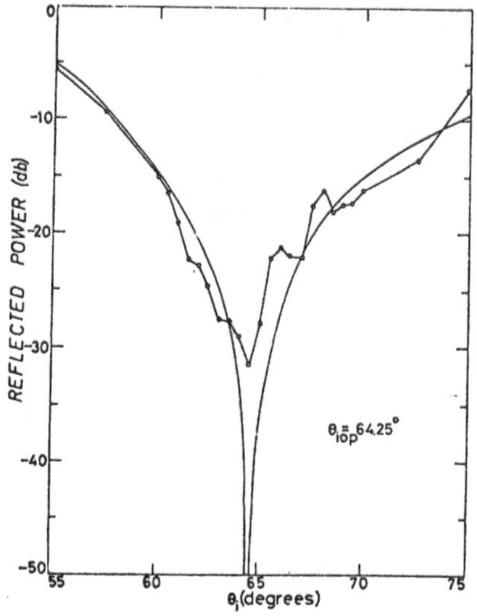

Figure 4.40 Reduction of TM polarized specular return (from [38]).

groove depth for total TM backscatter is approximately $\lambda/4$. The corrugations can reduce TM polarized specular reflection by about 30 dB at the design angle. Similar work has also been extended to triangular groove gratings [38], and strip gratings [39, 40].

4.4.3 Corrugated Ogive Structures

The nose-on echo area of the familiar ogive structure can be reduced by using corrugations that are inversely tapered to excite a cut-off mode so that there is very little or no return from the tip at the far end [41]. The structure is shown in Figure 4.41(a) and its nose-on backscattered RCS or echo area *versus* aspect angle is shown in 4.41(b), where the RCS of the smooth ogive is also shown for comparison. The results indicate that by this technique an RCS reduction on the order of 15 dB is obtained in the vicinity of the nose-on incidence.

4.5 ABSORBER COATING

4.5.1 Introduction

Radar absorbing materials have found important applications in a variety of practical situations; for example, RCS reduction in vehicles, design of reflectionless anechoic

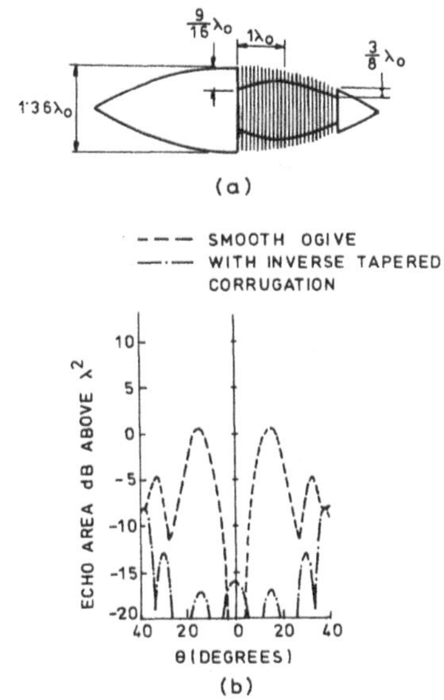

Figure 4.41 (a) The corrugated ogive structures; (b) nose-on echo area *versus* aspect angle.

chamber, low-sidelobe antennas utilizing RAM for rim loading. Generally, the suitability of a RAM for use in RCS reduction is judged on the basis of the following requirements:

1. The absorbing material should have adequate frequency response.
2. It should work for two orthogonal polarizations.
3. It should work with the specified aspect angle characteristics.

In addition, for use with flying objects the absorber should have high tensile and compressional strength, resistance to impact, tolerance to fuels and hydraulic fluids, and be lightweight.

To choose a RAM that simultaneously satisfies all of these requirements is difficult, if not impossible. Frequently we use multilayered coatings of RAM, each having a suitably chosen complex dielectric constant. Also, considerations of weight and environment (e.g., temperature, rain, snow) play an important role in deciding the thickness of any RAM coating.

Absorbers can be grouped under two broad classes: dielectric and magnetic. Magnetic ferrite materials with suitable dielectric constants also are important radar absorbers because they require less thickness for the same amount of absorption; however, they are heavier than the dielectric kind. The composition of magnetic absorber is an elastomer spread with ferrite material and designed to yield certain absorption characteristics. The elastomers are isoprene, neoprene, nitrile, silicon, urethane, or fluoroelastomers.

4.5.2 Theorems for Absorbers

Weston [42] developed two theorems for backscattering from absorber coated conducting bodies illuminated by plane waves. Because of their important applications they are stated here without proof.

Theorem 1

If a target has equal values for relative permittivity and permeability, that is, $\varepsilon/\varepsilon_0 = \mu/\mu_0$, the far-zone backscattered fields are zero if the shape and material of the body remain unchanged for a 90° rotation of the body around the direction of incidence, which is also the direction of observation.

Theorem 2

For plane wave incidence on a target, the backscattered fields will be zero if the shape and constituent of the target is invariant with respect to 90° rotation around the direction of incidence (observation), and if $\eta = \eta_0$, and the following impedance boundary condition is satisfied on the surface of the body:

$$\mathbf{E} = (\hat{n} \cdot \mathbf{E})\hat{n} = \eta(\hat{n} \times \mathbf{H}) \tag{4.32}$$

where $\hat{n}$ is the unit outward normal to the surface:

$$\eta = \sqrt{\frac{\mu}{\varepsilon}} = \eta_0\sqrt{\frac{\mu_r}{\varepsilon_r}}$$

$$\eta_0 = \sqrt{\frac{\mu_0}{\varepsilon_0}}$$

4.5.3 Some Analytical Techniques

This subsection briefly describes selected techniques available for analyses of reflection and scattering of electromagnetic waves by layered absorbing media. The subject is vast: we touch upon only a few relevant topics giving appropriate expressions having application to control of RCS of bodies.

4.5.3.1 Transmission Line Techniques

Transmission line analysis has been found very useful for investigating the reflection and transmission of electromagnetic waves by layered media. Figure 4.42 describes the

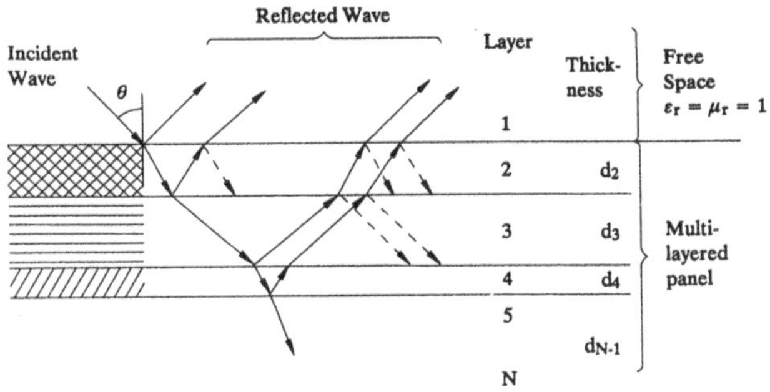

Figure 4.42 Geometry of reflection from a multilayer panel.

geometry of the problem in general, where the media are assumed to be infinitely extending.

The reflection and transmission coefficients for a plane wave obliquely incident on a plane interface between two infinitely extending nonmagnetic dissimilar media are given by Fresnel relations, discussed in many places. Referring to Figure 4.42 and considering only the interface between media 1 and 2, we can show that the reflection coefficients for parallel and perpendicular polarizations are

$$R_{\parallel} = \frac{\left(\dfrac{\varepsilon_2}{\varepsilon_1}\right)\cos\theta - \sqrt{\dfrac{\varepsilon_2}{\varepsilon_1} - \sin^2\theta}}{\left(\dfrac{\varepsilon_2}{\varepsilon_1}\right)\cos\theta + \sqrt{\dfrac{\varepsilon_2}{\varepsilon_1} - \sin^2\theta}} \tag{4.33a}$$

$$R_{\perp} = \frac{\cos\theta - \sqrt{\dfrac{\varepsilon_2}{\varepsilon_1} - \sin^2\theta}}{\cos\theta + \sqrt{\dfrac{\varepsilon_2}{\varepsilon_1} - \sin^2\theta}} \tag{4.33b}$$

where θ is the angle of incidence as shown in Figure 4.42. If the second medium is perfectly conducting, (4.33) indicates

$$R_{\parallel} = R_{\perp} = -1$$

Let us now consider the reflection of plane waves from the multilayered medium shown in Figure 4.42. The complex reflection coefficients for parallel and perpendicular polarizations are given by [43]

$$R_{\parallel} = \frac{Z_{i\parallel}^{(2)}\cos\theta^{(1)} - Z^{(1)}\cos\theta^{(2)}}{Z_{i\parallel}^{(2)}\cos\theta^{(1)} + Z^{(1)}\cos\theta^{(2)}} \tag{4.34}$$

$$R_{\perp} = \frac{Z^{(1)} \cos \theta^{(1)} - Z_{i\perp}^{(2)} \cos \theta^{(2)}}{Z^{(1)} \cos \theta^{(1)} + Z_{i\perp}^{(2)} \cos \theta^{(2)}} \tag{4.35}$$

where $Z_{i\|}^{(2)}$, $Z_{i\perp}^{(2)}$ are determined by the following recurrence relations:

$$Z_{i\|}^{(n)} = Z^{(n)} \frac{Z_{i\|}^{(n+1)} \cos \theta^{(n)} - \mathrm{i} Z^{(n)} \cos \theta^{(n+1)} \tan(c^{(n)} d^{(n)})}{Z^{(n)} \cos \theta^{(n+1)} - \mathrm{i} Z_{i\|}^{(n+1)} \cos \alpha^{(n)} \tan(c^{(n)} d^{(n)})} \tag{4.36}$$

$$Z_{i\perp}^{(n)} = Z^{(n)} \frac{Z_{i\perp}^{(n+1)} \cos \theta^{n+1} - \mathrm{i} Z^{(n)} \cos \theta^{(n)} \tan(c^{(n)} d^{(n)})}{Z^{(n)} \cos \alpha^{(n)} - \mathrm{i} Z_{i\perp}^{(n+1)} \cos \theta^{(n+1)} \tan(c^{(n)} d^{(n)})} \tag{4.37}$$

where $Z_{i\|}^{(n)}$ and $Z_{i\perp}^{(n)}$ are the input impedances of the nth layer for the parallel and perpendicular polarizations, respectively, and $n = 2, 3, \ldots, (N-1)$, with N being the total number of layers. For $n = N$, (4.36) and (4.37) should be replaced by

$$Z_{i\|}^{(N)} = Z_{i\perp}^{(N)} = Z^{(N)}$$

The other quantities needed in (4.36) and (4.37) are defined as follows:

$$Z^{(n)} = \sqrt{\mu_0 / \varepsilon_0} \sqrt{\mu_r^{(n)} / \varepsilon_r^{(n)}}$$

$$\varepsilon_r^{(n)} = \varepsilon_r'^{(n)} - i\varepsilon_r''^{(n)}, \quad \mu_r^{(n)} = \mu_r'^{(n)} - i\mu_r''^{(n)}$$

$$c^{(n)} = k^{(n)} \cos \theta^{(n)}$$

$$k^{(n)} = \omega \sqrt{\varepsilon_0 \varepsilon_r^{(n)} \mu_0 \mu_r^{(n)}}$$

$$d^{(n)} = \text{the thickness of the } n\text{th layer}$$

$$\cos \alpha^{(n)} = \sqrt{1 - \left(\mu_r^{(1)} \varepsilon_r^{(1)}\right) / \left(\mu_r^{(n)} \varepsilon_r^{(n)}\right) \sin^2 \alpha^{(1)}} \tag{4.38}$$

The general equations (4.34) and (4.35) have been found useful for the effects of different types of *radar absorbing structures* (RAS) and will be discussed in a later subsection.

Using the transmission line approach, the reflections by three- and four-layered media have been investigated and the results are shown in Figure 4.43. The period of oscillations was found to be very close to $\lambda/2$ in the material of the first layer; that is, plywood in this case. The oscillation periods in Figure 4.43 approach their respective

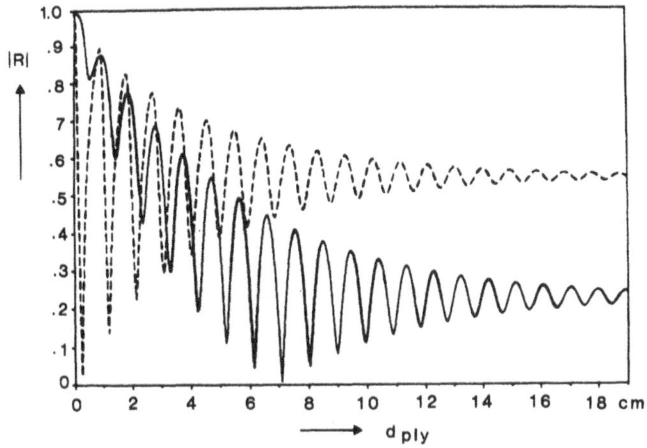

Figure 4.43 Reflection coefficient for vertical incidence as function of layer thickness: $f = 10$ GHz (——); three-layered medium; free space, plywood $\varepsilon_r = 2.5 - j0.16$; iron $\varepsilon_r'' = 1.8 \times 10^7$; $\mu_r' = 470$ (- - - -); four-layered medium: free space, absorber material $\varepsilon_r = 1.5 - j3$; $d_2 = 1$ mm; plywood, iron (from [43]).

values (0.226 for the three-layer and 0.53 for the four-layer) when the thickness of the first medium is made infinite.

4.5.3.2 Conducting Surface Coated with a Single Layer of Absorber

Let a plane electromagnetic wave (in free space) be incident obliquely on a perfectly conducting infinite plate coated with a single layer of absorber having complex permeability and permittivity. Let d be the layer thickness and θ_i be the angle of incidence measured from the normal to the plate.

The tangential component of the total E-field is required to satisfy the impedance boundary condition, (4.32), on the outer surface of the coating. The impedance parameter η in (4.32) depends on θ_i, μ, ε, d, and the frequency and polarization of the incident wave. The impedance η for the two orthogonal polarizations are [44]

$$\eta_{\parallel} = i\eta_0 \frac{\mu_r}{N} \cos\theta_i \tan(Nk_0 d \cos\theta_i) \tag{4.39a}$$

$$\eta_{\perp} = i\eta_0 \frac{\mu_r}{N} \frac{\tan(Nk_0 d \cos\theta_i)}{\cos\theta_i} \tag{4.39b}$$

where

$N = \sqrt{\mu_r \varepsilon_r}$ is the refractive index of the absorber,
$k_0 = \omega \sqrt{\mu_0 \varepsilon_0}$.

The two reflection coefficients for the two orthogonal polarizations are then given by

$$R_{\parallel} = \frac{\eta_{\parallel} - \eta_0 \cos \theta_i}{\eta_{\parallel} + \eta_0 \cos \theta_i} \tag{4.40a}$$

$$R_{\perp} = \frac{\eta_{\perp} \cos \theta_i - \eta_0}{\eta_{\perp} \cos \theta_i + \eta_0} \tag{4.40b}$$

The following special cases are of practical importance. First, for a perfect conductor $\eta_{\perp} = \eta_{\parallel} = \eta = 0$ and $R_{\parallel} = R_{\perp} = -1$. Second, for normal incidence ($\theta = 0°$) it follows from (4.39) that

$$\eta_{\parallel} = \eta_{\perp} = \eta = i\eta_0 \frac{\mu_r}{N} \tan(Nk_0 d) \tag{4.41}$$

Third, for large values of refractive index; that is, $kd \ll |N| \gg 1$, the wave travels in the lossy material almost in the normal direction. In this case,

$$\eta_{\parallel} = \eta_{\perp} \approx i\eta_0 \frac{\mu_r}{N} \tan(Nk_0 d)\left[1 + 0\left(\frac{k_0 d}{N}\right) + 0\left(\frac{1}{N^2}\right)\right] \tag{4.42}$$

Hence, for large N, the impedance parameter is the same as that for normal incidence; therefore the impedance is independent of polarization. However, the reflection coefficient will carry a large value because the index of refraction of free space is unity.

Fourth, when $Nk_0 d$ has a large negative imaginary part (large thickness of coating) then $\tan Nk_0 d \approx -i$. Hence, as expected, the impedance parameter becomes equal to that of the dielectric; that is, from (4.41),

$$\eta_{\parallel} = \eta_{\perp} \approx \eta_0 \frac{\mu_r}{N} = \sqrt{\frac{\mu}{\varepsilon}} = \eta \tag{4.43}$$

Fifth, for a very thin layer of absorber $|Nkd| \ll 1$, and the impedance is given by

$$\eta \approx i\eta_0 \mu_r k_0 d \tag{4.44}$$

Sixth, the condition for zero backscattering is determined by using Theorem 2 given in Subsection 4.5.2. Because the impedance parameter η is to equal η_0, the following relation then must be satisfied:

$$\tan(Nk_0 d) = -i\frac{N}{\mu_r} \tag{4.45}$$

The following cases are of interest. When $d \gg \delta$, δ being the skin depth of the material, $\tan nk_0 \approx -i$, and (4.45) becomes independent of d. This is essentially the case of $\mu' = \varepsilon$ absorber illustrated in the first theorem of Weston.

When the absorber is electrically thin, (4.45) reduces to

$$k_0 d = -i/\mu_r$$

which means that the complex permeability of the material should have a large negative imaginary part and a small real part.

For an intermediate electrical thickness of the layer, (4.45) must be used to obtain the desired solution.

4.5.3.3 Effects of Curvature

The effects of curvature of a target on the efforts to reduce its RCS are important to know. Utilizing the concepts associated with the plane surface absorber theory various mathematical models have been developed for the curved cases.

The specular RCS of a curved surface can be obtained approximately from the plane surface theory, particularly for surfaces with large radii of curvature. The analysis makes the following assumptions:

1. The principal radii of curvature of the surface are large compared to the wavelength in the dielectric.
2. The thickness of the coating should be much smaller than the radii of curvature and $kd \ll N$.
3. The refractive index of the absorber material is large and has a large imaginary part so that the fields vary much more slowly in free space than in the dielectric.

Under these assumptions, the flat plate absorber theory can be extended to the case of a curved surface. Figure 4.44 shows the geometry of a concentric spherical configuration, which is chosen here for simplicity and for which exact results are known. The GO technique can be used here to find the specular RCS [44]. In Figure 4.44 the total field at the observation point is the superposition of the contributions of two reflections, one by the interface of media 1 and 2, and the other by the interface of media 2 and 3. The specularly reflected far field at the observation point due to reflection at the dielectric-dielectric interface is

$$E_1 = R_{12}\frac{r_2}{2l}\,e^{-ikl} \tag{4.46}$$

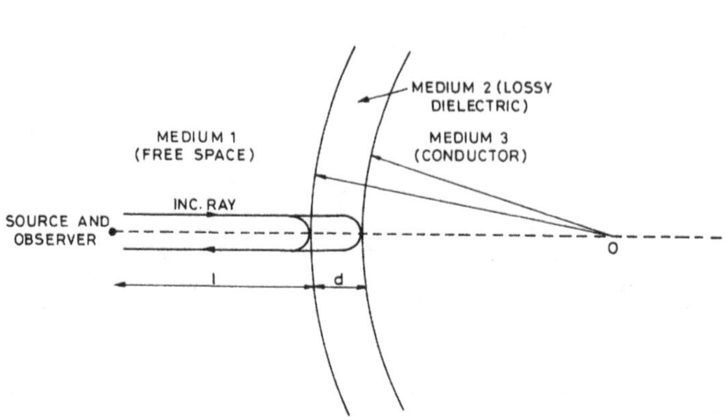

Figure 4.44 Reflection from a concentric spherical geometry.

where

$$R_{12} = \frac{1 - \sqrt{\varepsilon_r}}{1 + \sqrt{\varepsilon_r}} \tag{4.47}$$

r_2 = the radius of curvature of the separating (medium 1 and 2) surface in both principal planes,
ε_r = the real part of the dielectric constant of the lossy medium.

The far field at the observation point due to reflection at the dielectric-conductor interface is

$$E_2 = T_{12} R_{23} T_{21} \sqrt{\varepsilon_r} \frac{r_1}{2l} e^{-ikl+2\gamma d} \tag{4.48}$$

where

T_{ij} = the transmission coefficient when the ray travels from medium i to medium j,
R_{23} = the reflection coefficient at the interface of medium 2 and 3,
r_1 = the radius of curvature of the separating (medium 2 and 3) surface in both principal planes,

γ = the propagation constant in medium 2.

The RCS for the spherical metallic target coated with a concentric spherical shell can now be obtained, given by

$$\frac{\sigma}{\pi r_2^2} = \left| R_{12} + T_{12} R_{23} T_{21} \sqrt{\varepsilon_\gamma} \left(1 - \frac{d}{r_2}\right) e^{i\gamma d} \right|^2 \tag{4.49}$$

For a more general configuration, similar to the preceding but having doubly curved surfaces, the RCS is given by

$$\frac{\sigma}{\pi r_{1a} r_{1b}} = \left| R_{12} + T_{12} R_{23} T_{21} \sqrt{\varepsilon_r} \frac{\sqrt{\gamma_{1a}\gamma_{1b}}}{\gamma_{2a}\gamma_{2b}} e^{-2\gamma d} \right|^2 \tag{4.50}$$

where r_{1a}, r_{1b} and r_{2a}, r_{2b} are the radii of curvatures of the surfaces in the two principal planes.

4.5.3.4 Some Results

Figure 4.45 compares the results of the GO solution with those of the exact solution for a conducting sphere coated with a lossy spherical absorber. As expected, the GO results appear to be the mean of the exact results. The approximation used here may be improved by considering the contributions from the creeping waves around the sphere and the glory rays [44, 45].

The analysis discussed here may be extended to the cases of an arbitrary metallic body coated with absorbing material and multiply coated targets. Some of the results associated with absorber-coated targets will be discussed in a later subsection. The curvature effects on the reflection coefficient have been studied by Bowman and Weston [45] for the two basic cases of scattering from singly and doubly coated metallic spheres and cylinders. The solution obeying the exact boundary condition is asymptotically expanded with a Luneberg-Kline series, the first term of which is the GO term that can be obtained using impedance boundary conditions. The second term can be thought of as a correction term due to the coating characteristics and radii of curvature of the body. The reflection coefficient depends on the curvature. For further details and results, the reader is referred to [46]. Typical RCS results obtained with absorber-coated targets will be discussed in a later subsection.

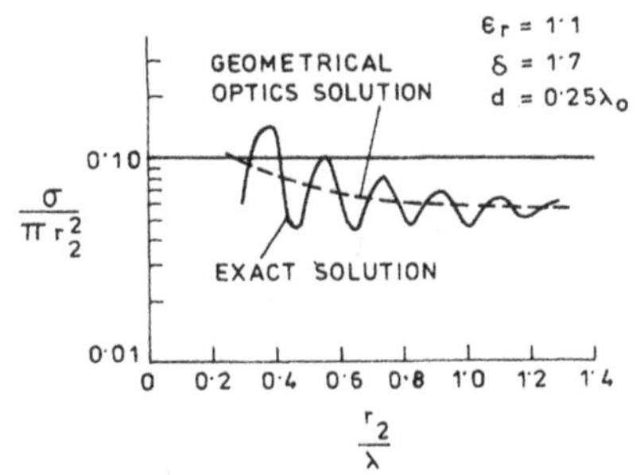

Figure 4.45 The geometrical optics RCS normalized with respect to πr_2^2 *versus* r_2/λ compared with exact solution (from [44]).

4.5.3.5 *High-Frequency Scattering by Impedance-Covered Half Planes*

Electromagnetic scattering by wedges or half planes with different face impedances has many important practical applications. Lossy dielectric loadings of the edges of horns and parabolic reflector antennas reduce their sidelobes; absorber coatings of surfaces are utilized to reduce the RCS of complex targets. Maliuzhinets [47] investigated the scattering by a wedge with two different face impedances; however, the high-frequency asymptotic expansion of his solution becomes singular at the reflection and shadow boundaries. Williams [48] investigated the excitation of surface waves on wedges with different face impedances. Mohsen and Hamid [49] developed an approximate solution for a conducting wedge loaded with a thin dielectric slab on one side. Tiberio and Kouyoumjian [50] utilized Pauli-Clemmow method [51] to investigate the problem of scattering by a half plane with two face impedances; similar methods have been used in [52] for conducting strips having two-face impedances. By using fields scattered by electric and magnetic currents and edge diffraction coefficients involving Maliuzhinets's function, Senior [53] developed a general treatment of electromagnetic scattering by uniform resistive, conductive, and nonmetallic half planes. Bucci and Franceschetti [54] carried out a systematic study of electromagnetic scattering by half planes with two different face impedances. They revised Maliuzhinets's solution for the scalar case and obtained a uniform asymptotic expansion of this solution by using the Pauli-Clemmow method. The original field expressions were modified to free the Maliuzhinets's function from any singularity in the range of arguments of interest. They also examined the condition for surface wave excitation and generalized the results of Williams [55]. Recently, a uniform asymptotic expansion of the Maliuzhinets's solution has been obtained in [56] by using van der Waerden's method of saddle points [57].

Because of the important applications associated with the RCS reduction of variety of bodies, a few significant aspects, selected from among the theories just referenced, of electromagnetic scattering by impedance-coated half planes and strips will be briefly outlined. Because a majority of analyses in this topic assumed a time dependence of $e^{-i\omega t}$ we have modified the relevant relations appropriately to be consistent with our assumed time dependence of $e^{i\omega t}$.

Maliuzhinets's Solution

We confine our attention to the case of scattering by a half plane, although Maliuzhinets's results are valid for a wedge, of an arbitrary wedge angle, with arbitrary face impedances. Let us consider a unit amplitude plane wave incident at an angle φ_0 on a half plane with two face impedances $Z^{\pm}$ as shown in Figure 4.46. Without loss of generality, we assume that $0 \leq \varphi_0 \leq \pi$. The incident polarization could be either TM (or E-) or TE (H-) polarization with respect to the z-axis. The total E_z and H_z for the

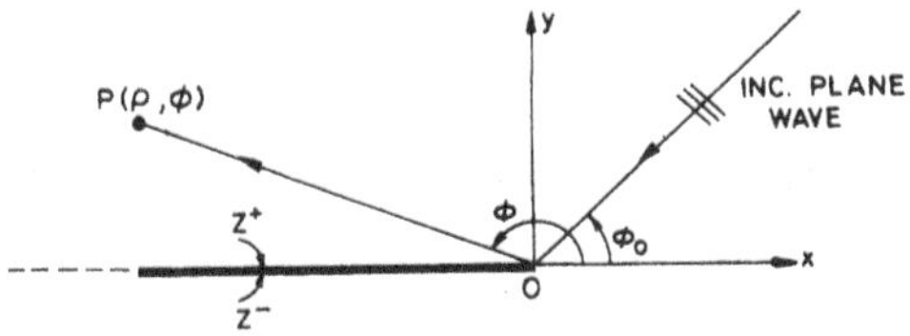

Figure 4.46 Electromagnetic scattering by a half-plane with two face impedances.

two cases can be expressed as [47]

$$E_z = \frac{1}{2\pi i}\int_\Gamma e^{ik\rho\cos\alpha} S_e(\alpha+\varphi)\,d\alpha \tag{4.51}$$

$$H_z = \frac{1}{2\pi i}\int_\Gamma e^{ik\rho\cos\alpha} S_h(\alpha+\varphi)\,d\alpha \tag{4.52}$$

where

Γ = the Sommerfeld contour for integration in the complex α plane,
S_e and S_h = functions associated with E and H polarizations, respectively, to be determined.

The boundary conditions require that the following be satisfied at $\varphi = \pm\pi$:

$$\frac{1}{\rho}\frac{\partial E_z}{\partial e} \pm ik\sin\theta_e^{\pm} \pm E_z = 0, \quad \text{for TM case} \tag{4.53}$$

$$\frac{1}{\rho}\frac{\partial H_z}{\partial e} \pm ik\sin\theta_h^{\pm} + H_z = 0, \quad \text{for TE case} \tag{4.54}$$

where E_Z and H_Z are given by (4.51) and (4.52), respectively;

$$\sin\theta_h^{\pm} = \eta/Z^{\pm}, \quad \sin\theta_h^{\pm} = \frac{Z^{\pm}}{\eta} \tag{4.55}$$

with the upper and lower signs on the left-hand sides corresponding to those on the right-hand sides;

$$k = \omega\sqrt{\mu\varepsilon} = \frac{2\pi}{\lambda}, \quad \eta = \sqrt{\mu/\varepsilon}$$

θ_e and θ_h correspond to E and H polarizations, respectively. (4.51)–(4.54) now give the following integral equations:

$$\int_{\Gamma} \exp(ik\rho\cos\alpha)(\sin\alpha \pm \sin\theta_e^{\pm})S_e(\alpha \pm \pi)\,d\alpha = 0 \qquad (4.56)$$

$$\int_{\Gamma} \exp(ik\rho\cos\alpha)(\sin\alpha \pm \sin\theta_h^{\pm})S_h(\alpha \pm \pi)\,d\alpha = 0 \qquad (4.57)$$

Maliuzhinets showed [47] that the solutions to (4.56) and (4.57) satisfy the following equations (after dropping the subscripts *e* and *h* for convenience):

$$(\sin\alpha \pm \sin\theta^{\pm})S(\alpha \pm \pi) - (-\sin\alpha \pm \sin\theta^{\pm})S(\alpha \pm \pi) = 0 \quad (4.58)$$

Note that the desired functions S_e and S_h may be identified with E and H polarizations, respectively, through the parameters $\theta_e^{\pm}$ and $\theta_h^{\pm}$. The expression for $S(\alpha)$ can be expressed as

$$S(\alpha) = \frac{1/2\cos(\varphi_0/2)}{\sin(\alpha/2) - \sin(\varphi_0/2)}\frac{\psi(\alpha)}{\psi(\varphi_0)} \qquad (4.59)$$

where the auxiliary function $\psi(\alpha)$ should be regular in the strip $-\pi \le \mathrm{Re}[\alpha] \le \pi$; and $\psi(\alpha)$ is given by

$$\begin{aligned}\psi(\alpha) = M\left(\alpha + \frac{\pi}{2} + \theta^{+}\right)M\left(\alpha + \frac{3\pi}{2} - \theta^{+}\right)\\ M\left(\alpha - \frac{\pi}{2} - \theta^{-}\right)M\left(\alpha - \frac{3\pi}{2} + \theta^{-}\right)\end{aligned} \qquad (4.60)$$

with the Maliuzhinets's function $M(u)$ expressed by

$$M(u) = \exp\left\{-\frac{1}{8\pi}\int_0^u \left[\left(\pi\sin t - 2\sqrt{2}\,\pi\sin(t/2) + 2t\right)/\cos t\right]dt\right\} \quad (4.61)$$

Again, the auxiliary functions $\psi_e(\alpha)$, $\psi_h(\alpha)$ are obtained using $\theta_e^{\pm}$, $\theta_h^{\pm}$, respectively, on the right-hand side of (4.60). The properties of the Maliuzhinets's function and its representation in terms of easily computable expressions are discussed in Appendix A.

Use of (4.59)–(4.61) in (4.51) and (4.52) indicates that we now have exact integral expressions for the total E_z and H_z in terms of Maliuzhinets's functions. The diffraction integral involved in Maliuzhinets's solution can be evaluated by the methods of Pauli-Clemmow and Van Der Waerden leading to UTD and UAT results, respectively, for the problem. These are briefly described.

Pauli-Clemmow Expansion (UTD Results)

After properly deforming the integration contours in (4.51) and (4.52) and again omitting the subscripts e and h for E and H polarizations, respectively, we can show [54] that the total field (E_z or H_z) is given by

$$\begin{aligned} &H(\varphi - \varphi_0 + \pi)\exp\left[ik\rho\cos(\varphi - \varphi_0)\right] \\ &\quad - H(\varphi + \varphi_0 - \pi)\frac{\psi(\varphi_0 + 2\pi)}{\psi(\varphi_0)}\exp\left[ik\rho\cos(\varphi + \varphi_0)\right] \\ &\quad + H(\varphi - \xi^+)\frac{-\sqrt{2}\cos(\varphi_0/2)}{\sin(\varphi_0/2) + \sin(\theta^+/2)}\,\frac{\psi(2\pi + \theta^+)}{\psi(\varphi_0)} \\ &\quad \times\frac{M(3\pi/2)}{M(7\pi/2)}\exp\left[ik\rho\cos(\theta^+ - \varphi)\right] \\ &\quad + H(-\varphi - \xi^-)\frac{\sqrt{2}\cos(\varphi_0/2)}{\sin(\varphi_0/2) - \sin(\theta^-/2)}\,\frac{\psi(2\pi + \theta^-)}{\psi(\varphi_0)} \\ &\quad \times\frac{M(3\pi/2)}{M(7\pi/2)}\exp\left[ik\rho\cos(\theta^- + \varphi)\right] \quad + \frac{1}{4\pi i}\int_{\Gamma'} d\alpha\,\exp(ik\rho\cos\alpha) \\ &\quad \times\frac{\cos(\varphi_0/2)}{\sin\left[(\alpha + \varphi)/2\right] - \sin(\varphi_0/2)}\,\frac{\psi(\alpha + \varphi)}{\psi(\varphi_0)} \end{aligned} \tag{4.62}$$

where the $\psi(\cdot)$ and $M(\cdot)$ functions are as defined by (4.60) and (4.61), respectively; Γ' is the steepest descent integration contour [56] in the complex plane $\alpha = \alpha_1 + i\alpha_2$; $H(x)$ is the well-known Heaviside unit step function see Subsection 2.5.1, (2.18).

The parameters α and ξ^+ are given by

$$\alpha_1 = \pm\pi - \cos^{-1}\left(\frac{1}{\cosh\alpha_2}\right)\operatorname{sgn}(\alpha_2) \tag{4.63}$$

$$\xi^{\pm} = \pi + \theta_1^{\pm} + \cos^{-1}\left(\frac{1}{\cosh\theta_2^{\pm}}\right)\operatorname{sgn}(\theta_2) \tag{4.64}$$

with

$$\begin{aligned} \operatorname{sgn}(x) &= +1, \quad x > 0 \\ &= -1, \quad x < 0 \\ \theta^{\pm} &= \theta_1^{\pm} + i\theta_2^{\pm} \end{aligned}$$

Note that a passive surface impedance (which is under consideration) requires that $0 \leq \theta_1^{\pm} \leq \pi/2$.

The five terms composing the total field by (4.62) may be interpreted as follows: the first two terms are the GO incident and reflected fields; the third and fourth terms are the contributions from the surface fields on the two faces of the half plane; asymptotic evaluation of the last term indicates that it represents the cylindrical wave diffracted from the edge of the half plane.

Uniform Asymptotic Theory

Referring to Figure 4.46 Maliuzhinets's exact solution [47] for the total field E_z or H_z represented by u^t can be written in the spectral or Fourier domain as [56]

$$u^t(\rho, \varphi) = \frac{-i}{4\pi} \int_{\Gamma} X(\varphi_0, \omega) \exp(-ik\rho\cos(\omega - |\varphi|)\, ds \tag{4.65}$$

with

$$\begin{aligned} X(\varphi_0, \zeta) = {} & \frac{\cos(\varphi_0/2)}{\psi(\varphi_0)} \\ & \times \left[\frac{\psi(\zeta - \pi)}{\sin(\zeta - \pi/2) - \sin(\varphi_0/2)} \mp \frac{\psi(\zeta + \pi)}{\sin(\zeta + \pi/2) - \sin(\varphi_0/2)} \right] \end{aligned} \tag{4.66}$$

where Γ is the Sommerfeld contour, the upper and lower signs in (4.66) give the values of x, $X(\varphi_0, \zeta)$ in (4.65) stand for E_z and H_z, respectively, $\Psi(x)$ is defined in [54] and is given by (4.60).

Using Van Den Waerden's method [57], the contour Γ in (4.65) can be deformed, and we can show that the uniform asymptotic expansion for the total field is [56]

$$\begin{aligned} u^t(\rho, \varphi) \sim {} & u^i(\rho, \varphi)\left[F(\xi_i) - \hat{F}(\xi_i)\right] + u^r(\rho, \varphi)\left[F(\xi_r) - \hat{F}(\xi_r)\right] \\ & + u^r(\rho, \varphi)\left[F(\xi_r) - \hat{F}(\xi_r)\right] \\ & + u^s(\varphi_0, Z^+) + u^s(\varphi_0, Z^-) + u^d(\rho, \varphi) + 0(k^{-3/2}) \end{aligned} \tag{4.67}$$

where $F(\xi)$ and $\hat{F}(\xi)$ are defined by (2.16) and (2.17), respectively; $\xi_{i,r}$ are the detour parameters given by

$$\xi_{i,r} = \pm\sqrt{2k\rho}\cos\left(\frac{\varphi \mp \varphi_0}{2}\right) \tag{4.68}$$

with the upper and lower signs corresponding to i and r, respectively.

Various terms in (4.67) can be interpreted as follows: the first and second terms are the incident and reflected fields; the third and fourth terms are contributions from the surface waves on the two surfaces; and the last term is the diffracted field. The surface wave terms are described in detail in [54]. The diffracted field may be represented by

$$U^d(\rho, \varphi) \sim \frac{\exp[-(ik\rho + \pi/4)]}{2\sqrt{2k\rho}} X(\varphi_0, \varphi) U^i(\text{at the edge}) \tag{4.69}$$

RCS of Metal Plates Coated with Lossy Dielectric Materials

The preceding techniques outlined will now be used to determine the high-frequency monostatic RCS of a finite metal plate coated with lossy dielectric materials. For simplicity we shall ignore the effects of corner diffraction. The scattering geometry is shown in Figure 4.47, which also depicts the relevant parameters. We assume that $\varphi_0 = \pi$ and $-\pi$ surfaces of the plate (Figure 4.47) have impedances Z^+ and Z^-, respectively. With plane of incidence in the xy-plane, as shown in Figure 4.47, the

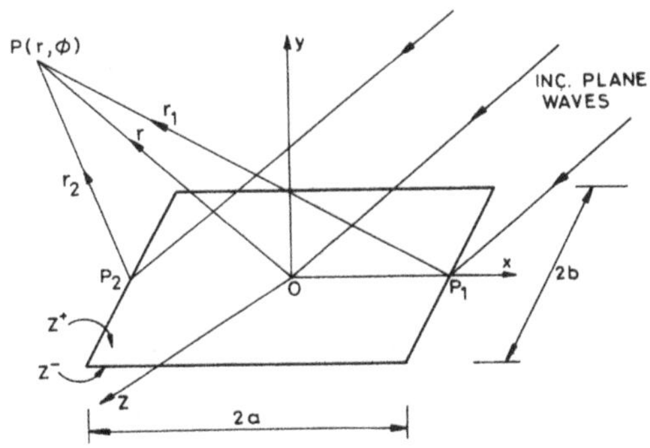

Figure 4.47 Geometry of the finite planar structure and its coordinate system.

diffraction coefficients needed to obtain the field at P are given by [56]

$$D_{s,h} = \frac{e^{-i\pi/4}\cos(\varphi_0/2)}{(2\pi k)^{1/2}} A(\varphi, \varphi_0, \theta_{e,h}^{+}) A(-\varphi, -\varphi_0, \theta_{e,h}^{-}) \times \left[B(\varphi, \varphi_0, \theta_{e,h}^{+}) + B(-\varphi, -\varphi_0, \theta_{e,h}^{-}) \right] T(\varphi, \varphi_0, kr) \tag{4.70}$$

where the subscripts e and h are the right-hand side correspond to E- and H-polarization, respectively, and they correspond to the subscripts s and h on the left-hand side; and

$$\sin\theta_e^{\pm} = \frac{\eta}{Z^{\pm}}, \quad \sin\theta_h^{\pm} = \frac{Z^{\pm}}{\eta}, \quad \eta = \sqrt{\frac{\mu}{\varepsilon}}$$

Here μ and ε are the permeability and permittivity of free space:

$$A(\varphi, \varphi_0, \theta) = \frac{1}{1 - \tan\left(\dfrac{\varphi_0 - \theta}{4}\right)} \frac{M(\varphi + \theta - \pi/2) M(\varphi - \theta + \pi/2)}{M(\varphi_0 + \theta + \pi/2) M(\varphi_0 - \theta + \pi/2)} \tag{4.71}$$

$$B(\varphi, \varphi_0, \theta) = \frac{\tan[(\varphi - \theta)/4]}{\cos(\varphi/2) - \sin(\varphi_0/2)} \frac{1 - \tan\left(\dfrac{\varphi + \theta}{4}\right)}{1 + \tan\left(\dfrac{\varphi - \theta}{4}\right)} \tag{4.72}$$

with Maliuzhinets's function $M(u)$ given by (4.61) and $\theta = \theta_{e,h}^{\pm}$. The last factor in (4.70) is called the *transition function*, given by

$$T(\varphi, \varphi_0, kr) = -\frac{2i\eta_+\eta_-}{\eta_+^2 - \eta_-^2} \times \left[\eta_- e^{i\eta_+^2} \int_{\eta_+}^{\infty} e^{it^2} dt - \eta_+ e^{i\eta_-^2} \int_{\eta_-}^{\infty} e^{-it^2} dt \right] \tag{4.73}$$

where

$$\eta_{\pm} = (2kr)^{1/2} \left| \cos\left(\frac{\varphi \pm \varphi_0}{2} \right) \right| \tag{4.74}$$

Note that in the far zone, $\eta_{\pm} = \infty$ and $T(\varphi, \varphi_0, kr) = 1$.

The total singly diffracted field at an observation point P in Figure 4.47 is the vector sum of the singly diffracted fields of the two edges. Using the appropriate diffraction coefficient from (4.70) and (2.12), we can show that the total field $U_e(P)$ for E-polarization is

$$
\begin{aligned}
U_e(P) = {} & \frac{\exp[-ik(r_1 - a\cos\varphi_0) - i\pi/4]}{(2\pi k r_1)^{1/2}} \\
& \times \cos\left(\frac{\varphi_0}{2}\right) A(\varphi, \varphi_0, \theta_e^+) A(-\varphi, -\varphi_0, \theta_e^-) \\
& \times [B(\varphi, \varphi_0, \theta_e^+) + B(-\varphi, -\varphi_0, \theta_e^-)] T(\varphi, \varphi_0, kr) \\
& + \frac{\exp[ik(r_2 + a\cos\varphi_0) - i\pi/4]}{(2\pi k r_2)^{1/2}} U_i(P) \\
& \times \sin\left(\frac{\varphi_0}{2}\right) A(\pi - \varphi, \pi - \varphi_0, \theta_e^+) A[-(\pi - \varphi), -(\pi - \varphi_0), \theta_e^-] \\
& \times \{B(\pi - \varphi, \pi - \varphi_0, \theta_e^+) + B[-(\pi - \varphi), -(\pi - \varphi_0), \theta_e^-]\} \\
& \times T(\pi - \varphi, \pi - \varphi_0, kr)
\end{aligned}
\tag{4.75}
$$

where $U_i(P)$ is the incident E-field at the edge under consideration. Note that the field given by (4.75) is that due to diffraction by an infinite strip (i.e., $2b \to \infty$ in Figure 4.47) of width $2a$. Therefore, the RCS per unit length is given by

$$
\sigma_{\text{length}} = \underset{r\to\infty}{\text{Limit}}\, 2\pi r \left|\frac{U_e(P)}{U_i(P)}\right|^2 \tag{4.76}
$$

with $U_e(P)$ given by (4.75). Equation (4.76) can be modified to determine the RCS $\sigma = \sigma_{\text{area}}$ for the rectangular plate. On the basis of PO approximation Siegel indicated [58] that one can convert the two-dimensional field given by (4.75) (due to the infinite strip) to the three-dimensional field appropriate for the rectangular plate (Figure 4.47) by simply multiplying the former field by the factor:

$$
\frac{L\, e^{i\pi/4}}{\sqrt{r\lambda}}
$$

where L is the length of the edge, r is the distance of the observation point from the edge, and λ is the wavelength. Using this approximation the RCS of the rectangular plate (of dimensions $2a \times 2b$) is

$$\sigma = \sigma_{\text{area}} = \frac{8b^2}{\lambda} \sigma_{\text{length}} \tag{4.77}$$

which has been found to be a valid approximation for edge dimensions as small as 2λ.

Thus, on the basis of singly diffracted rays, the RCS of the rectangular plate illuminated as shown in Figure 4.47 can be obtained by using (4.75) to (4.77), given by

$$\begin{aligned}\sigma = \sigma_{\text{area}} = \frac{4b^2}{\pi} \Big| \{ & \exp(i2ka\cos\varphi)\cos(\varphi/2) A(\varphi, \theta_e^+) \\ & A(-\varphi, \theta_e^+)[B(\varphi, \theta_e^+) + B(-\varphi, \theta_e^-)] \\ & + \exp(-i2ka\cos\varphi)\sin\left(\frac{\varphi}{2}\right) A(\pi - \varphi, \theta_e^+) \\ & A[-(\pi - \varphi), \theta_e^-][B(\pi - \varphi, \theta_e^+) + B(-(\pi - \varphi), \theta_e^-)]\} \Big|^2\end{aligned} \tag{4.78}$$

If desired, this RCS expression can be modified by taking into account the effects of double diffraction in the total field given by (4.75). However, this is not important for electrically large plates, for which (4.78) is sufficient.

Theoretical and experimental monostatic RCS patterns for a flat metallic plate with one side coated with resin are shown in Figure 4.48. The theoretical values shown were obtained by using (4.78), and within the range of experimental results shown, the agreement between the two may be considered satisfactory. Theoretical TE polarized backscattering patterns for a flat plate with and without resin coating are shown in Figure 4.49, the results for the uncoated case were obtained by using the GTD formula of Ross [59]. The results of Figure 4.49 indicate that, with the resin coating used on one face, the RCS reduction is not significant near edge-on incidence but it increases to a maximum of about 5dB for normal incidence. Similar calculations for other coating materials used as RAM (listed in the Appendix B) can be carried out by using the theory given.

Discussion

Although nonuniform diffraction coefficients associated with respective half planes are used to obtain the total field, (4.75), the contributions from the two edges combine to cancel the singularities in the directions of grazing and normal incidence. As a result, the total field given by (4.75) is bounded. In general, for the plate problem under

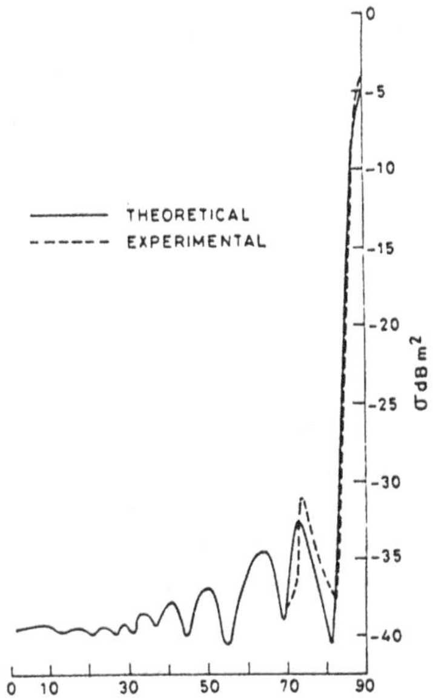

Figure 4.48 Backscattered RCS of a flat tin plate having one side coated with resin. Thickness of coating = $\lambda/20$; $\varepsilon_r = 2.53 - j0.165$; Plate size 12 in. × 16 in. (from [66]).

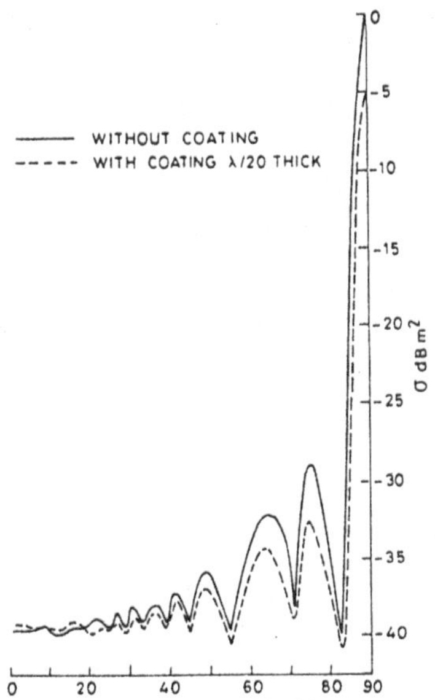

Figure 4.49 Backscattered RCS of a flat tin plate of size 12 in. × 16 in.: f = 10.958 GHz (with and without resin coating) (from [66]).

consideration we can employ the ray solution by properly dividing the plate into strips [60] for which the two-dimensional ray solution is bounded even for backscattering at normal incidence. The surface waves that have been neglected in the theory may be of importance in this class of problems. Even if the surface waves are tightly bound, they would undergo reflections and diffraction at the edges, thereby contributing to the RCS. However, if the dielectric coating is sufficiently lossy the surface wave effects would be relatively small.

4.5.3.6 Scattering by Resistive Strips

Backscattering from a strip decreases in directions away from the normal incidence; at edge-on incidence, the strip becomes invisible for H-polarized (i.e., H-field parallel to the edge) incident waves but it has a nonzero monostatic RCS for E-polarization. Generally, backscattering cross sections of strips made of resistive material are less than those of conducting strips. Hence, it would be advantageous if a resistive sheet could be used instead of a coated conducting surface to reduce the RCS. For this reason the backscattering properties of resistive strips have been investigated extensively both theoretically and experimentally. For resistive strips one can define a surface resistance

$$R_s = \frac{1}{\sigma\tau} \tag{4.79}$$

when σ is the conductivity of the strip material and τ is the thickness of the strip. For thin strips as $\tau \to 0$, a large value of σ could be used so that R_s in (4.7a) stays finite which then is used in the impedance boundary condition (see Subsection 1.2.4) required for solving the scattering problem. Senior [61, 62] used integral equation formulation to investigate scattering by resistive strips. Let us consider a strip of width w and resistance R_s occupying the portion $0 \le x \le w$, $-\infty < z < \infty$ of the $y = 0$ plane of a Cartesian coordinate system and illuminated by an E-polarized plane electromagnetic wave in free space:

$$\mathbf{E}^i = \hat{z}\exp\left[ik\left(x\cos\varphi_0 + y\sin\varphi_0\right)\right] \tag{4.80}$$

where $0 \le \varphi_0 \le \pi$ (Figure 4.50). The total z-directed current induced in the strip is

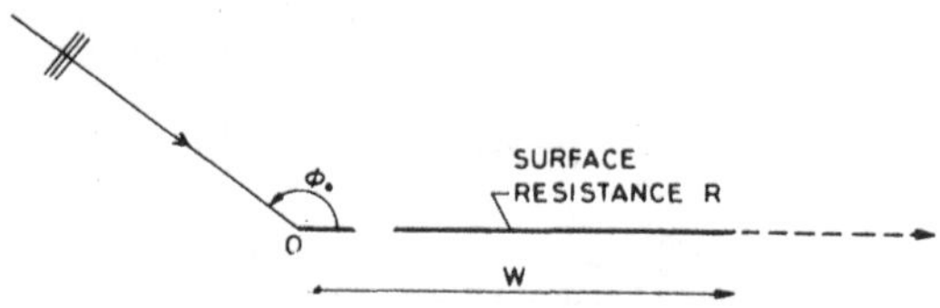

Figure 4.50 Geometry of strip scattering. For edge-on incidence, $\phi_0 = \pi$.

given by

$$J(x) = -\left[H_x(x,0^+) - H_x(x,0^-)\right] \tag{4.81}$$

and the required boundary condition on the surface is

$$E_z(x,0^+) = E_z(x,0^-) = R_s J_z \tag{4.82}$$

where E_z and H_z are the total of E- and H-fields and plus and minus correspond to the upper and lower surfaces of the strip. Note that when $R_s = 0$ the strip is perfectly conducting, and it no longer exists when $R_s = \infty$.

Application of the boundary conditions (4.81) and (4.82) on the strip yields the following integral equation for the unknown induced current [61]:

$$R_s J(x) = \exp(\mathrm{i}kx\cos\varphi_0) - \frac{k_0\eta_0}{4}\int_0^w J(x')H_0^{(2)}(k\,|\,x-x'\,|)\,\mathrm{d}x,$$

$$\text{for } 0 < x < w \tag{4.83}$$

where $H_0^{(2)}$ is the zeroth-order Hankel function of the second kind.

In the far zone the scattered electric field is

$$E_z^s \sim \sqrt{\frac{2}{\pi k_0 \rho}}\exp[-\mathrm{i}(k_0\rho - \pi/4)]\,P(\varphi,\varphi_0) \tag{4.84}$$

where (ρ, φ) are cylindrical polar coordinates such that $x = \rho\cos\varphi$ and $y = \rho\sin\varphi$, and the complex scattering amplitude $P(\varphi, \varphi_0)$ is

$$P(\varphi,\varphi_0) = -\frac{k_0\eta_0}{4}\int_0^w J(x')\exp(-\mathrm{i}k_0x'\cos\varphi)\,\mathrm{d}x' \tag{4.85}$$

The two-dimensional scattering cross sections (per unit length) are then obtained from

$$\sigma(\varphi,\varphi_0) = \frac{2\lambda}{\pi}\left|P(\varphi,\varphi_0)\right|^2 \tag{4.86}$$

Note that for edge-on incidence ($\varphi_0 = \pi$) the front edge of the strip is the one coincident with the z-axis. Analytical solutions for the current distribution and scattered fields under variety of conditions are discussed in [61, 62].

In [61] moment method was used to solve Eq. (4.83) for the induced current from which backscattering cross sections were determined for various values normalized surface resistance $\eta = 2R_s/\eta_0$ and strip width w. The edge-on backscattering cross sections of strips with $\eta = 1$, 4, and 20 are shown as functions of the strip width w in Figure 4.51, where the rather regular oscillation with minima $\lambda/2$ apart indicates an interference between the front- and rear-edge contributions. The backscattering cross sections of four different strips of widths 1.25λ are shown as functions of φ in Figure 4.52, which illustrates the cross-section reductions achieved. When $\eta = 4$, for example, the reduction (compared to the case with $\eta = 0$) exceeds 14 dB at all aspects compared with the 15.6 dB edge-on. Some other observations gleaned from [61, 62] that may find application in cross-section reduction using the resistive strip concepts are these:

1. The choice for R_s would be $R_s/\eta_0 \to \infty$ for low backscattering at edge-on incidence. Under this condition the induced current is zero and the strip does not exist. In practice, however, R_s can be large.

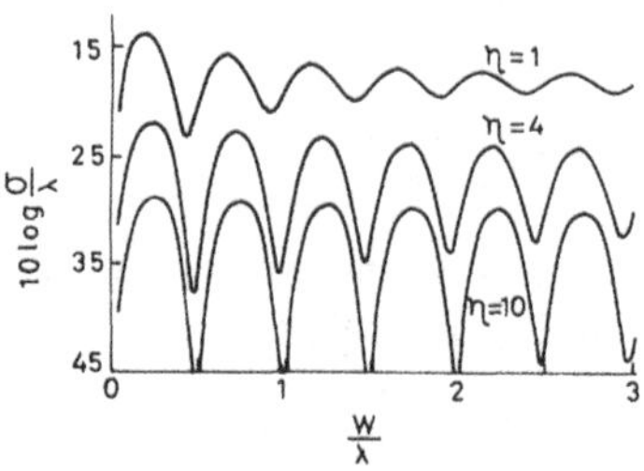

Figure 4.51 Computed edge-on backscattering cross sections for three resistive strips (from [61]).

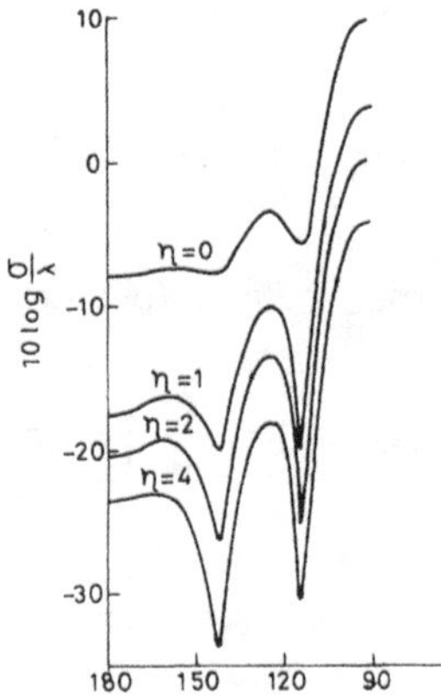

Figure 4.52 Backscattering cross section as a function of aspect angle from edge-on ($\phi = 180°$) to broadside ($\phi = 90°$) for far strips of width 1.25λ (from [61]).

2. For $\eta = 10$, ($R_s = 1885\,\Omega$), the far-field edge-on contribution from the front-edge is 26.6 dB below its value for a perfectly conducting strip.
3. For very narrow strips, maximizing η also minimizes the rear-edge contribution, but for any given η there is a strip width beyond which the rear-edge contribution is least if it is perfectly conducting. For $\eta = 10$, the transition occurs at $w \approx 0.6\lambda$.
4. For wider strips the edge-on scattering would be a minimum if the resistance value were kept maximum at the front edge and tapered down to zero at the rear-edge. With uniform strips, the best that can be done is to choose the maximum resistance available.

An excellent summary of scattering from uniform resistive strips, tapered resistive strips, and synthesis of resistive tapers to control electromagnetic scattering patterns of strips is available in [63]. Resistive loading of the rim or edge of a conducting surface has been used to reduce RCS so that the scattering characteristics are maintained within a certain tolerance. For E-polarization a reduction of induced edge current has been achieved by Ray and Mittra [64] at the expense of generating current spikes at the junction of the metal and resistive materials. However, this could have been eliminated by a smooth tapering between the metal and highly resistive edges. Such a taper on a resistive half-plane is discussed in [65]. Haupt [63] developed and experimentally verified the synthesis techniques for achieving specified levels of monostatic and bistatic scattering patterns for a strip over a desired angle of aspect angles; in addition, an error analysis also had been carried out. Figure 4.53(a)–(d) shows the normalized resistively distribution, induced current, bistatic, and monostatic RCS patterns for a 4λ strip illuminated by E-polarized waves. The resistivity distribution shown in Figure 4.53(b) was synthesized from a Taylor current taper so that nulls at 20° and 50° occur in the bistatic RCS pattern, see Figure 4.53(c). Similar results are shown in Figure 4.54(a)–(d), where the resistive taper was synthesized to have a null at 0° in the backscattering RCS pattern see Figure 4.54(d).

4.5.4 Examples of RCS Reduction

Table 4.3 shows the values of monostatic RCS reduction calculated at X-band frequencies for a conducting rectangular plate whose lit side is coated with lossy dielectrics. The values shown are for normal incidence and TM polarization, and the RCS reduction shown in each case is compared to the RCS of a perfectly conducting plate of the same size. The dielectric used is epoxy resin having $\varepsilon_r = 2.53 - i0.145$. The quantities Z^+ and Z^- refer to the surface impedances on the lit and dark sides of the plate. The results shown in Part (a) were obtained by methods found in [66]; those in Parts (b) and (c) were obtained by using UAT [67] and spectral Galerkin methods [68], respectively.

Figure 4.55 shows the monostatic RCS pattern of a model engine inlet (inset in

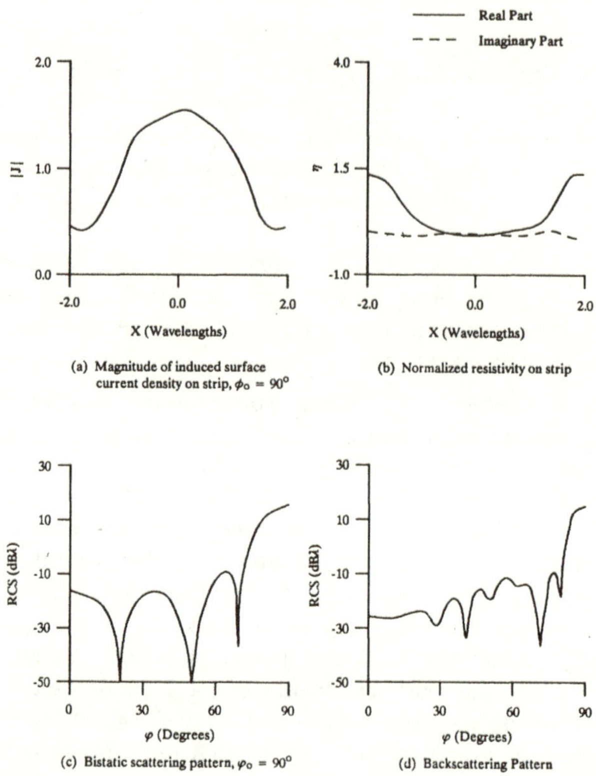

Figure 4.53 E-polarized scattering results from integral equation formulation for a 4λ strip with η synthesized from a Taylor current taper (nulls of the bistatic scattering pattern are at 20° and 60°): (a) magnitude of induced surface current density on strip ($\phi_s = 90°$); (b) normalized resistivity of the strip; (c) bistatic scattering pattern, ($\phi_s = 90°$); (d) backscattering pattern (from [63]).

the figure) obtained under the following conditions of the inlet walls: without RAM coating, with RAM coating applied only on the inside walls, and applied on the inside and outside walls. The RAM used is the Emmersion Cumming AN-73 X-band three-layer RAM; Figure 4.56 sketches the case when the AN-73 three-layer RAM is applied on both sides of the walls. The results of Figure 4.55 indicate considerable reduction of backscattering RCS in directions away from the normal ($\theta > 90°$). The maximum reduction is found to be as large as 28 dB. Theoretical results shown were obtained by using the finite-difference time-domain (FD-TD)method [69].

4.5.5 Cross-Polarized RCS

So far we have discussed the reduction of copolarized RCS of a surface by using coatings of absorbing materials. Although the principal aim has been maximum reduction of the copolarized RCS, it is also important to estimate the cross-polarized

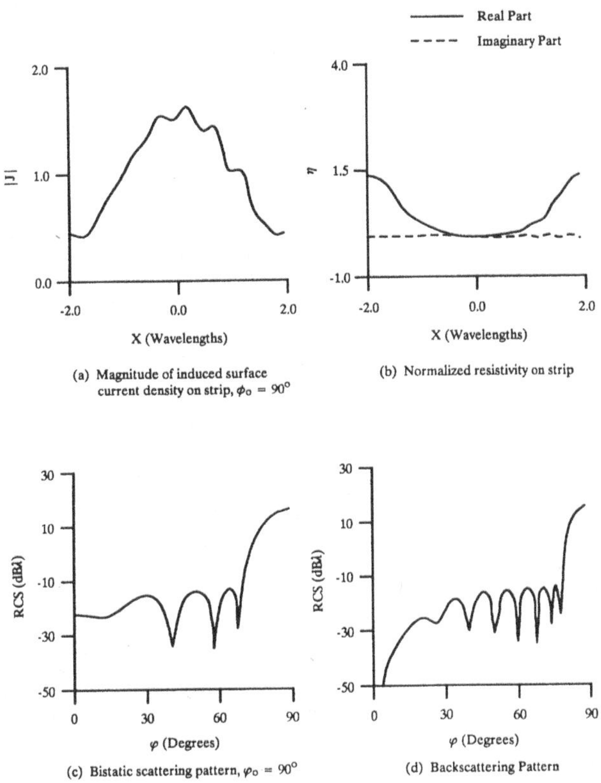

Figure 4.54 E-polarized scattering results from integral equation formulation for a 4λ strip with η synthesized from a Taylor current taper (with a null in the backscattering pattern at 0°): (a) magnitude of induced surface current density on strip ($\phi_s = 90°$); (b) normalized resistivity on strip; (c) bistatic scattering pattern ($\phi_s = 90°$); (d) backscattering pattern (from [63]).

component, which generally become appreciable at oblique incidence. Chu and Semplak reported [70] that when the copolarization response requirements are severe, depolarization suffered by microwave radiation obliquely incident on a painted surface may be significant; the cross-polarized response generally depends on the uniformity of the painted surface. For such a case, the reflection coefficients for parallel and perpendicular polarizations have unity amplitude except near grazing incidence for a slightly lossy and thin layer of paint ($\sigma/\omega\varepsilon < 0.01$, $d < \lambda_0\sqrt{\varepsilon_r}$, d being the thickness of the coating). The parallel and perpendicular reflection coefficients at oblique incidence on a painted surface of thickness l is [70]

$$\rho_{\|} = \frac{\mathrm{i}kl\cos^2\theta_p - \cos\theta_i}{\mathrm{i}kl\cos^2\theta_p + \cos\theta_i} \tag{4.87}$$

$$\rho_{\perp} = \frac{\mathrm{i}kl - \sec\theta_i}{\mathrm{i}kl + \sec\theta_i} \tag{4.88}$$

Table 4.3
Reduction of RCS of a Coated Conducting Plate*, Normal Incidence: TM polarization

(a) Epoxy resin on the lit side: $\varepsilon_r = 2.53 - i0.145$; $Z^- = (0.,0.)\ \Omega$

Frequency (GHz)	*Z^+ in Ω*	*Plate Size (in. × in.)*	*RCS Reduction in dB*
8.542		4 × 4	6.64
8.542		5 × 5	6.81
8.542		6.5 × 6.5	6.92
10.958	496 + i242	12 × 12	9.89

(b) Plate size 5″ × 5″; $f = 10.958$ GHz; $Z^- = (0.,0)\ \Omega$

Z^+ in Ω	*RCS Reduction in dB*
50 + i0	3.21
100 + i0	4.32
200 + i0	5.07
496 + i242	6.81

(c) Plate size = $(3.125 \times 3.125)\lambda^2$; $Z^-(0.,0)\ \Omega$; $F = 10.958$ GHz

Z^+ in Ω	*UAT [67]*	*RCS Reduction in dB Spectral Galerkin [68]*
50 + i0	2.38	3.00
100 + i0	4.21	3.75
200 + i0	7.01	6.50

*Reduction with respect to RCS of a perfectly conducting plate of the same size.

where

θ_p = the angle between surface normal and the propagation direction inside the layer of paint,
θ_i = the angle of incidence,
k = the propagation constant.

The two reflection coefficients have unit amplitude but differential phase shift $\Delta\varphi$ given by

$$\Delta\phi = 2kd\left[\frac{\cos^2\theta_p}{\cos\theta_i} - \cos\theta_i\right] \tag{4.89}$$

If the dielectric constant of the paint is large, for nongrazing incidence, $\theta_p \approx 0$ and the differential phase shift [70], from (4.87), is

$$\Delta\varphi \approx 2kd\sin\theta_i\tan\theta_i \tag{4.90}$$

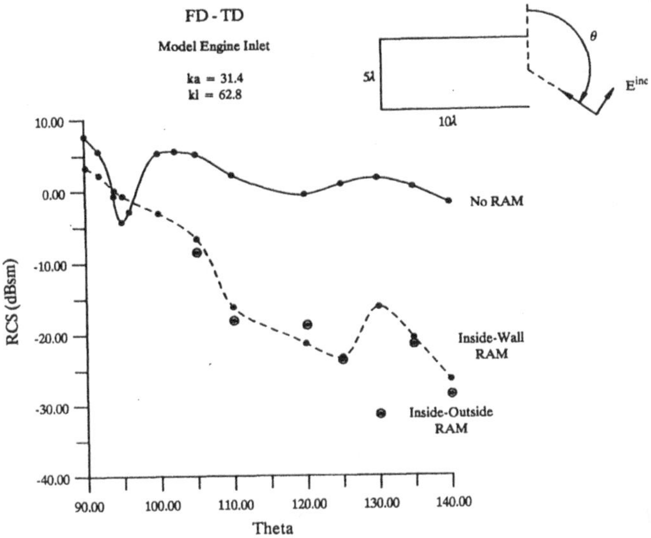

Figure 4.55 Model of an engine inlet and its radar cross section in dB/mW *versus* aspect angle.

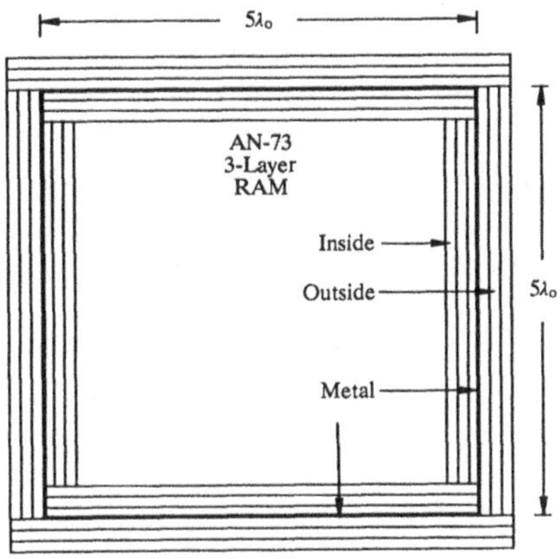

Figure 4.56 X-band RAM (Emerson & Cuming AN-73) complex dielectric constants for layers 1, 2, and 3 are $\varepsilon_{r1} = (1.4 - i0.35)$, $\varepsilon_{r2} = 1.6 - i0.9$ and $\varepsilon_{r3} = 3.4 - i10.0$ (*Source*: Emerson & Cuming, Bulletin No. 150).

and independent of the dielectric constant of the paint. Because the depolarization suffered by an incident wave is dependent on $\Delta\varphi$, (4.88) now indicates that it is insensitive to the dielectric constant of the paint. For example, let us estimate the cross-polarized component for a case with $d = 0.04$ mm and $f = 19$ GHz ($\lambda =$ 1.578927 cm), $\theta_i = 45°$ from the plane of incidence. From (4.88) $\Delta\varphi \approx 0.023$, which leads to a cross-polarized component of −39 dB less than the desired polarization. Hence, we may conclude that the cross-polarized component is not small for such a case if the desired polarization requirement is extremely stringent or a large number of reflections occur. On the contrary, depolarization is expected to be very small near normal incidence or if incident polarization is parallel or perpendicular to the plane of incidence.

4.5.6 Narrowband Radar Absorbing Structures

Different types of dielectric and magnetic structures known as *radar absorbing structures* (RAS) are used to reduce RCS of surfaces. This subsection describes the narrowband structures commonly used.

The Salisbury screen is a classical resonant absorber described in many places, for example in [71, 72]. We shall mention only the key features of the screen and give the relevant expressions leading to its design. Salisbury screens can be electric or magnetic. A typical screen with four layers is shown in Figure 4.57, which shows that the screen consists of a thin resistive sheet or a screen in front of a conducting plane separated by either a dielectric or magnetic slab called a *spacer*. The thin screen used is electric or magnetic with an electric or magnetic slab, respectively. The permeability and permittivity of the thin screen may vary arbitrarily in the z-direction but be homogeneous in the xy-plane. If the fourth layer is metal, as in Figure 4.57, the

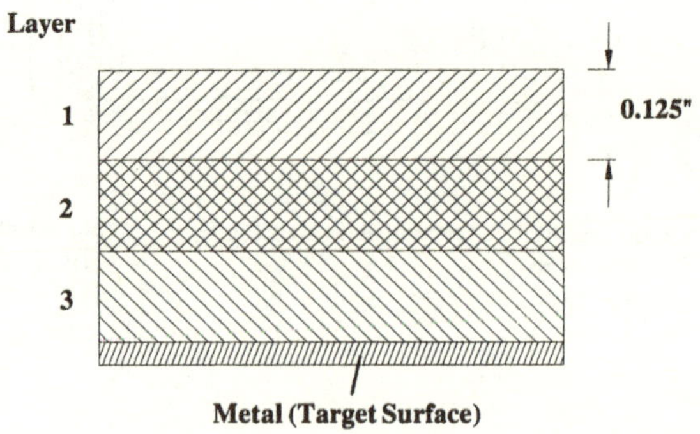

Figure 4.57 Sketch of a Salisbury screen.

following conditions must be satisfied by Salisbury screen parameters:

1. The effective input impedances for medium 2 must be large compared to the intrinsic impedance of medium 2;
2. $\text{Im}(\varepsilon_r) = \varepsilon_r'' > \text{Re}(\varepsilon_r) = \varepsilon_r'$ for the electric screen;
   $\text{Im}(\mu_r) = \mu_r'' > \text{Re}(\mu_r) = \mu_r'$ for the magnetic screen;
3. $1/d\omega\varepsilon'' = \mu_0/\varepsilon_0$ for the electric screen;
   $\omega\mu'' d = \mu_0/\varepsilon_0$ for the magnetic screen.

Amplitudes of reflection coefficients for parallel and perpendicular polarizations as functions of the angle of incidence are shown in Figure 4.58 for an electric Salisbury screen. Figure 4.59 shows amplitudes of reflection coefficients *versus* the angle of incidence for an electric Salisbury screen designed for zero reflection of $\theta = 15°$ for perpendicular polarization.

Recently, Fante and McCormack [71] investigated the design of multiple electric and magnetic Salisbury screens. They developed analytical relations for the maximally flat design of two- or three-screen planar stack-ups and gave appropriate values for the spacer materials used to separate the screens. The main conclusions of this study are as follows:

1. Using relatively few Salisbury screens we can achieve, a relatively large reduction of RCS in the specular direction.

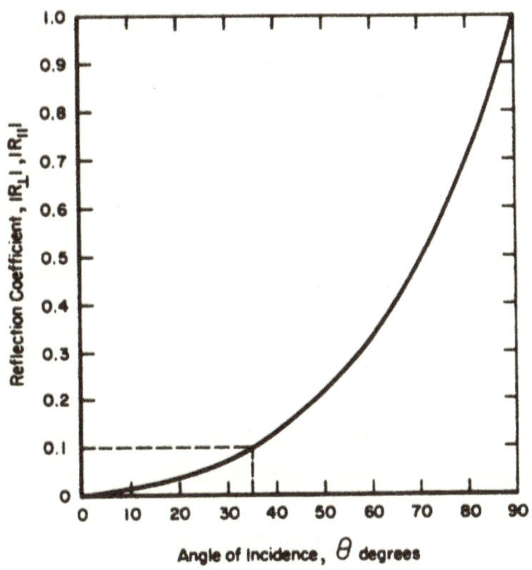

Figure 4.58 Reflection coefficient of an electric Salisbury screen *versus* the angle of incidence (from [73]).

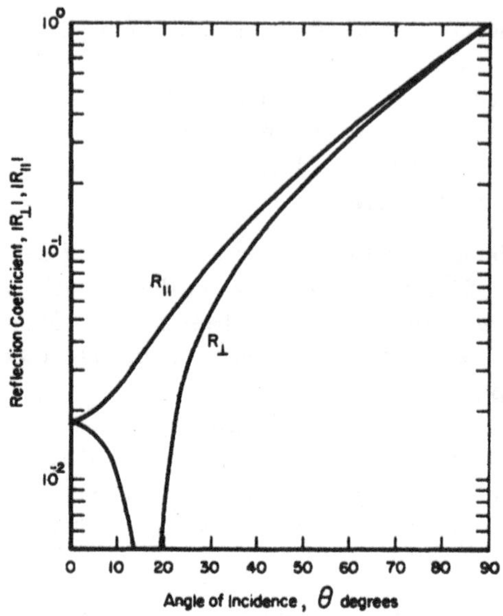

Figure 4.59 Reflection coefficient *versus* the angle of incidence for an electric Salisbury screen designed for zero reflection at $\theta = 15°$ for perpendicular polarization (from [73]).

2. The RCS reduction achieved does not deteriorate too much in directions away from the normal or if the surface is curved or contains fabrication errors.
3. The ideal absorber is a magnetic screen placed directly on the conducting surface, having its normalized complex magnetic impedance (with respect to the free space impedance) equal to unity. Alternatively,

$$\omega\mu''(\omega)\,\Delta = \eta_0$$

for all frequencies where ω is the angular frequency, $\mu''(\omega)$ is the imaginary part of the complex permeability, and the Δ is the thickness of the screen (see Figure 4.57).

Figure 4.60 plots the magnitude of the reflection coefficient at normal incidence *versus* normalized frequency $F(= f/f_0 - 1$, with f_0 as the center frequency) for a configuration consisting of two electric screens in front of a perfectly conducting ground plane with the spacer relative admittance y as a parameter. The parameters involved are explained in the inset of Figure 4.60 where d_1 and d_2 are the thicknesses of the two screens. Figure 4.61 shows similar plots of reflection coefficient for different values of the relative admittance of the spacer; this case corresponds to $d_1 = d_2 = \lambda/4$ of the

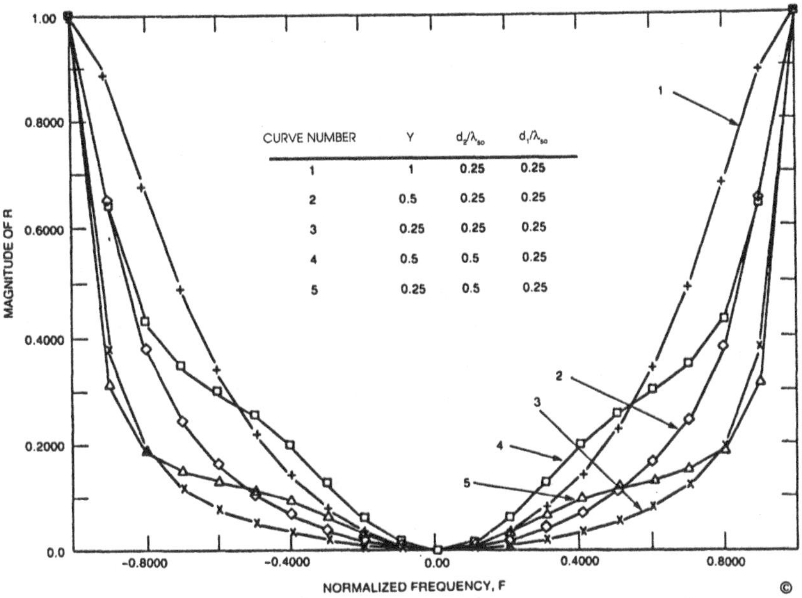

Figure 4.60 Magnitude of the reflection coefficient (normal incidence) *versus* normalized frequency for two electric screens in front of a ground plane (λ_{so} = midband electrical wavelength in the spacer) (from [71]).

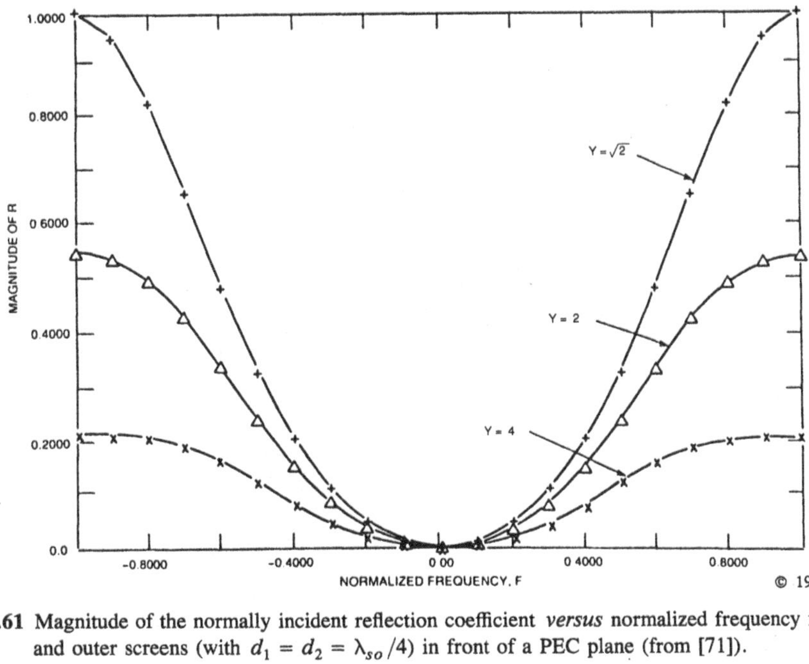

Figure 4.61 Magnitude of the normally incident reflection coefficient *versus* normalized frequency for inner and outer screens (with $d_1 = d_2 = \lambda_{so}/4$) in front of a PEC plane (from [71]).

midband frequency and its inner and outer screens are electric and magnetic, respectively. In contrast to Figure 4.60, the results of Figure 4.61 indicate that the bandwidth of the minimum reflection increases with increase of y, the spacer relative admittance; however, unlike the case with two electric screens, the electric and magnetic screen combination does not necessarily give unity reflection at $F = \pm 1$.

The reflection characteristics of a configuration consisting of three screens in front of a perfectly conducting ground plane are shown in Figure 4.62(a) and (b). The thickness of each screen is equal to $\lambda/4$ in the spacer at the midband frequency. Normal incidence reflection coefficient *versus* normalized frequency $F(= f/f_0 - 1)$ is shown in Figure 4.62(a), which can be used to judge the bandwidth properties; the results shown in Figure 4.62(b) give the specular reflection coefficient *versus* incidence angle and apply when the spacer relative admittance is unity. Figures 4.63 and 4.64 show the reductions in backscattered RCS obtained for perfectly conducting infinite cylinders in the presence of magnetic and electric Salisbury screens, respectively. Results indicate a reduction in RCS as large as 50 dB over a narrow band of frequencies.

4.5.7 Dahlenbach Absorbers

A Dahlenbach absorber layer, shown in Figure 4.65, consists of a thick homogeneous lossy layer backed by a metallic plate. Transmission line techniques discussed in an earlier section can be used to show that no reflection occurs when

$$Z_0 = \mathrm{i} Z_1 \tan \alpha_1 \tag{4.91}$$

where Z_0 and Z_1 are the normalized input impedance at $z = 0$ and normalized characteristic impedance of medium 1, respectively, and $\alpha_1 = k_1 \cos\theta$. Use of (4.91) leads to the following design relations: for parallel polarization,

$$\begin{aligned} 1 + \mathrm{i}\sqrt{\mu_r/\varepsilon_r}\, \frac{\left[1 - (\sin^2\theta_i)/(\mu_r\varepsilon_r)\right]^{1/2}}{\cos\theta_0} \\ \tan\left[\frac{2\pi l}{\lambda_0}\sqrt{\mu_r\varepsilon_r}\left(1 - \frac{\sin^2\theta_i}{\mu_r\varepsilon_r}\right)^{1/2}\right] = 0 \end{aligned} \tag{4.92}$$

and, for perpendicular polarization,

$$\begin{aligned} 1 + \mathrm{i}\sqrt{\mu_r/\varepsilon_r}\, \frac{\cos\theta_i}{\left[1 - (\sin^2\theta_i)/(\mu_r\varepsilon_r)\right]^{1/2}} \\ \tan\left[\frac{2\pi l}{\lambda_0}\sqrt{\mu_r\varepsilon_r}\left(1 - \frac{\sin^2\theta_i}{\mu_r\varepsilon_r}\right)^{1/2}\right] = 0 \end{aligned} \tag{4.93}$$

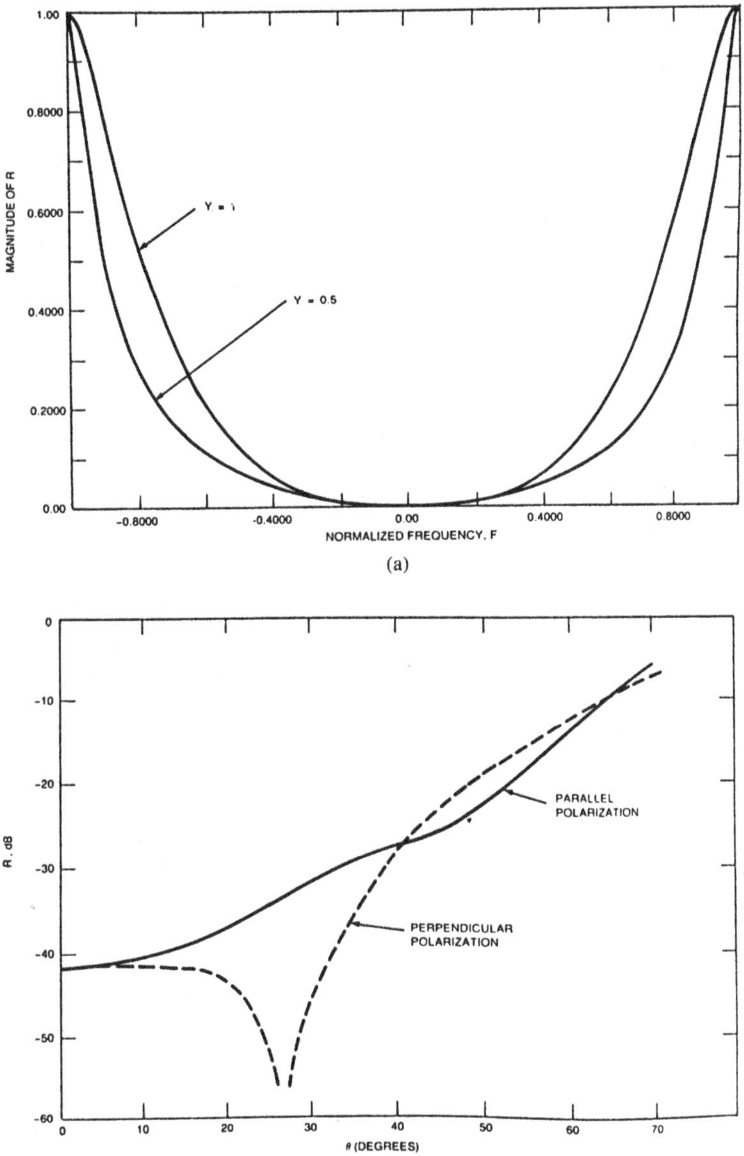

(a)

(b)

Figure 4.62 (a) Magnitude of the normal-incidence reflection coefficient *versus* normalized frequency for the case of three electric screens in front of a PEC ground plane: $d_1 = d_2 = d_3 = \lambda_{so}/4$, where λ_{so} is the midband electrical wavelength in the spacer (from [71]). (b) Magnitude of the specular-reflection coefficient *versus* incidence angle for the case of three electric screens in front of a PEC plate: $y = 1$, $d_1 = d_2 = d_3 = \lambda_{so}/4$ (from [71]).

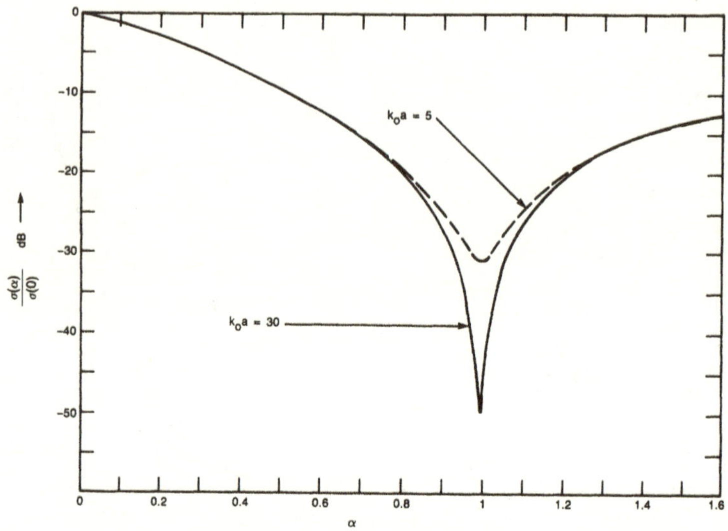

Figure 4.63 Reduction in RCS of infinitely conducting cylinder produced by magnetic Salisbury screen located on cylinder surface (for incident electric field parallel to cylinder axis) (from [71]).

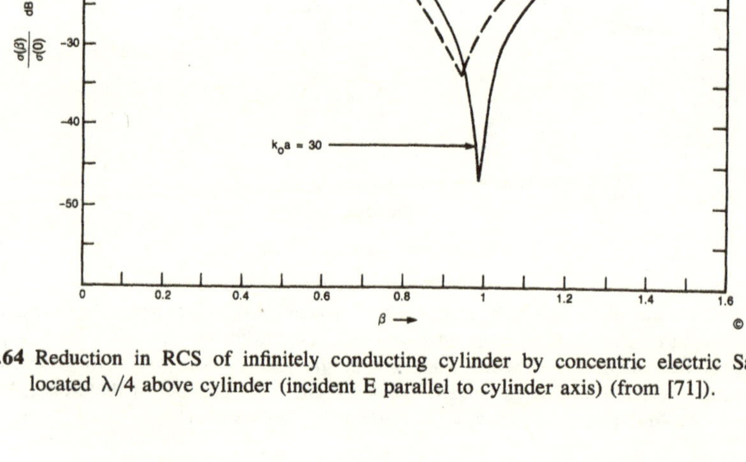

Figure 4.64 Reduction in RCS of infinitely conducting cylinder by concentric electric Salisbury screen located $\lambda/4$ above cylinder (incident E parallel to cylinder axis) (from [71]).

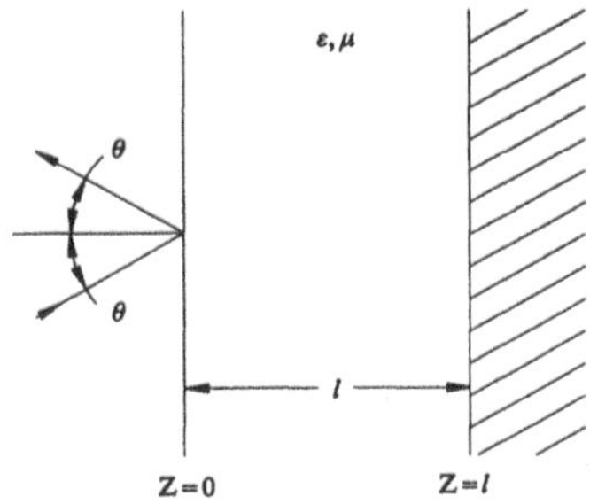

Figure 4.65 Dahlenbach absorber.

where

μ_r, ε_r = the relative permeability and permittivity of the lossy layer,
l = the thickness of the layer,
θ_i = the angle of incidence,
λ_0 = the free space wavelength.

There are certain difficulties in achieving low (zero) reflection from a single-layer absorber backed by a metal plate, as just considered. This is because the layer material must be such that low reflection takes place on the front face of the layer and the layer sufficiently attenuates the wave so that little or no energy reaches the plate to be reflected back. As mentioned earlier, if a material can be found whose impedance relative to free space equals unity, then there will be no reflection at the surface and the attenuation will depend on the loss properties of the material and the electrical thickness of the layer. Unfortunately, such a material is not readily available. Normally, materials causing low reflection will have low loss, thereby causing strong reflection from the metal plate; use of highly lossy material is precluded because such material generally causes strong reflection. Figure 4.66 depicts ε_r', ε_r'' as functions of the layer thickness required for zero reflection at normal incidence for a Dahlenbach layer consisting of a lossy dielectric having $\mu_r = 1$. Figure 4.67 shows μ_r', μ_r'' as functions of the layer thickness required for zero reflection at normal incidence for a Dahlenbach layer consisting of lossy magnetic material having $\varepsilon_r = \varepsilon_r' =$ constant.

It is of practical interest to know the practical bandwidth that can be achieved with the Dahlenbach layer. Such analysis for an ideal case is given in [73]. Approximate expressions for bandwidths (defined as maximum RCS reduction for a given reflection level) Dahlenbach layers are, for a nonmagnetic layer,

$$B = 2\left(\frac{f - f_0}{f_0}\right) = \frac{0.2}{|\pi(l/\lambda_m) - (\pi/16)(\lambda_m/l) - i|} \tag{4.94}$$

$$B = 2\left(\frac{f - f_0}{f_0}\right) \approx \frac{0.2}{|-(2/3\pi)(\varepsilon' l/\lambda_m) + i/2\pi|} \tag{4.95}$$

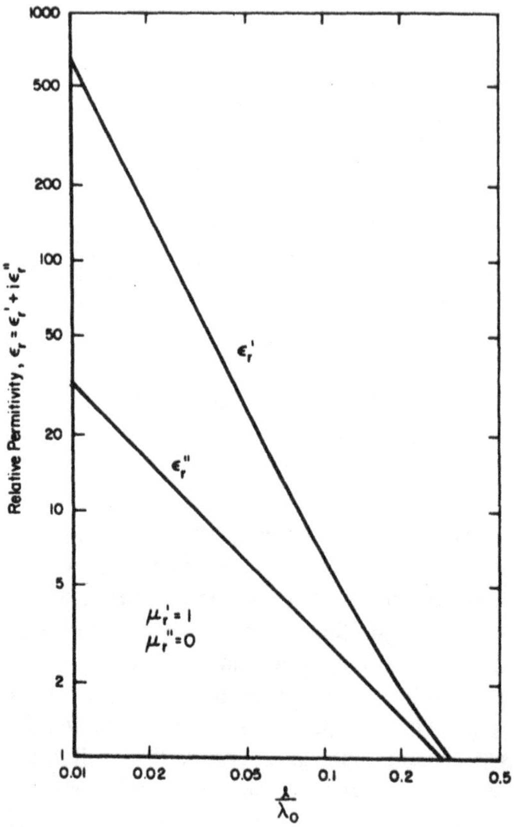

Figure 4.66 The values of ε_r' and ε_r'' *versus* the layer thickness giving a zero normal-incidence reflection coefficient for a homogeneous layer with $\mu_r = 1$ (from [73]).

and, for a layer having magnetic loss only, where we assume that the reflection level is 0.1, and λ_m is the wavelength in the layer medium. These bandwidths are low because both permittivity and permeability of the layer medium are functions of frequency. For larger bandwidths we must use multiple layers.

4.5.8 Wideband Radar Absorbing Structures

So far we have discussed the RAS used in resonant configurations for RCS reduction. Many applications need control of RCS over a wide band of frequencies; consequently if RAM or RAS is used for this purpose it must be broadband. The bandwidth of RCS

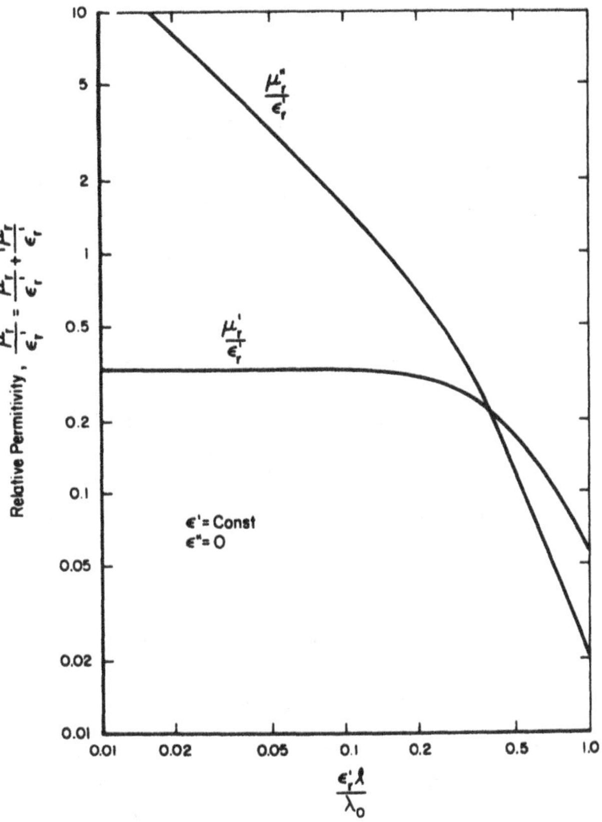

Figure 4.67 The values of μ_r' and μ_r'' *versus* the layer thickness giving a zero normal-incidence reflection coefficient for a homogeneous layer with $\epsilon_r = \varepsilon_r'$ – constant (from [73]).

reduction may be increased by using a slow transition of impedance through the absorber layer to the target or by using a multilayered absorber operating as low impedance resonator and an impedance transformer. Different types of absorber designs may be used for reduction of RCS over a wide band of frequencies:

1. $\mu_r = \varepsilon_r$ absorber (see Subsection 4.5.2);
2. Jaumann absorbers;
3. Stepped absorbers;
4. Circuit analog absorbers;
5. Magnetic absorbers;
6. Pyramidal absorbers.

4.5.8.1 $\mu_r = \varepsilon_r$ Absorbers

If a material is available whose relative permeability and permittivity are equal, then the intrinsic impedance of the medium would be equal to that of free space. Let us consider a layer of such medium placed in front of a conducting plate, the other side of the layer forming a plane interface with free space, and illuminated by normally incident plane waves. Thus, the layer medium being matched to free space, there will be no reflection from the first interface. Unfortunately, this is possible to achieve only in a limited way. In practice the material always will have some loss ($\varepsilon_r'' \neq 0$) so that the continuity of the intrinsic impedance at the first interface will not be satisfied, and the inequality of the impedance will be a function of frequency; moreover, there may be some reflection at the conducting plate, thereby giving rise to waves reflected back to the free space side. It appears that matching is not possible with a lossy thin layer. Desired matching also cannot be achieved for oblique incidence cases. On a limited scale, however, it is possible to fabricate a ferrite material with $\varepsilon_r = \mu_r$, and use a layer of such material to reduce RCS of bodies over a considerable bandwidth.

At a plane interface between free space and an absorbing medium, the amplitudes of the obliquely incident plane wave reflection coefficients for parallel and perpendicular polarizations are

$$|R_{\parallel}| = |R_{\perp}| = \left| \frac{\cos\theta_i - \sqrt{1 - \left(\sin^2\theta_i/\mu_r\varepsilon_r\right)}}{\cos\theta_i + \sqrt{1 - \left(\sin^2\theta_i/\mu_r\varepsilon_r\right)}} \right| \tag{4.96}$$

where μ_r and ε_r are the relative permeability and permittivity of the absorbing medium, and θ_i is the angle of incidence measured from the normal to the interface. Based on (4.96), design relations for a $\mu_r = \varepsilon_r$ absorbing layer have been discussed in [73], given as follows:

1. If $\varepsilon_r' = \mu_r'$ and $\mu_r', \varepsilon_r' \gg \mu_r'', \varepsilon_r''$ and $\mu_r'' \neq \varepsilon_r''$, the magnitude reflection coefficient for normal incidence, for both polarizations, is

$$|R| \approx \frac{1}{4}\left| \frac{\mu_r}{\mu_r} - \frac{\varepsilon_r}{\varepsilon_r} \right| \tag{4.97}$$

2. If the absorbing layer of electrical thickness α is backed up by a perfectly conducting plate, the reflection coefficient for normal incidence is

$$R = \frac{-\mathrm{i}\tan\alpha - 1}{-\mathrm{i}\tan\alpha + 1} = -\mathrm{e}^{\mathrm{i}2\alpha} \tag{4.98}$$

The magnitude of reflection coefficients at the plane interface between free space and a $\mu = \varepsilon$ absorbing medium as functions of the angle of incidence with $|\mu\varepsilon|$ as the parameter are shown in Figure 4.68 where the angle of incidence θ is measured from

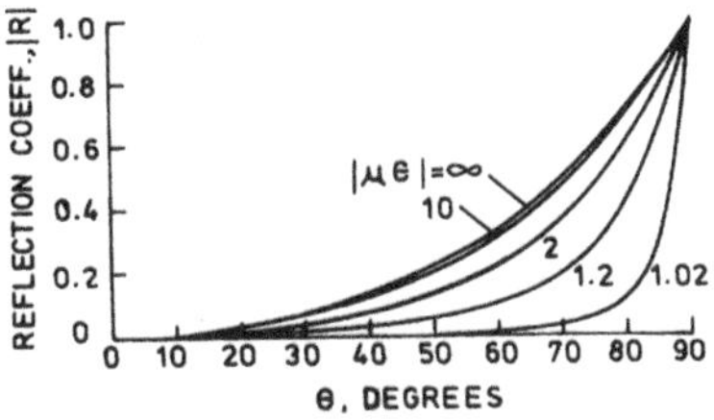

Figure 4.68 The angular performance of a $\mu = \varepsilon$ absorber for different values of $|\mu\varepsilon|$.

the normal to the interface. The results show that for low values $|\mu\varepsilon|$ the reflection coefficient is below 0.2 (i.e., a return loss better than 13.98 dB) even near grazing incidence. But low values of $|\mu\varepsilon|$ imply that any layer of practical importance must be very thick. On the other hand, if $|\mu\varepsilon|$ is large the magnitude of reflection coefficient is ≤ 0.2 up to about 50°; such absorbing layers could be relatively thin and broadband. The exact frequency bandwidth, of course, will depend on the frequency variations of μ and ε.

4.5.8.2 Jaumann Absorbers

Jaumann absorbers are made from alternative layers of thin lossy film and relatively thick layers of low-loss materials. A Jaumann absorber may be considered a cascade of single-sheet Salisbury screens; the cascade process considerably improves the bandwidth of performance. Theoretically the bandwidth may be three times that of single Salisbury screen [74]. The transmission line equivalent circuit of a Jaumann absorber is a parallel combination of resistances [74]. Figure 4.69 shows the calculated reflected power *versus* frequency for absorbers containing variable number of resistive sheets. The results clearly indicate the improvement of theoretical frequency bandwidth with increase of the number of sheets; in fact, the results for four sheets in Figure 4.69 indicate a fractional bandwidth of 1.16. Further information about Jaumann absorbers may be found in [72, Section 9.3].

4.5.8.3 Circuit Analog Absorbers

Circuit analog (CA) absorbers are discussed in [72, 74].These absorbers essentially are sheets of low-loss material on which some specific conducting patterns have been deposited. The patterns constitute resistance, inductance, and capacitance so that the deposited film can be represented by an equivalent *RLC* circuit. The *R*, *L*, *C* parameters can be controlled by the geometric pattern, conductivity of the deposition on the film, and the film thickness. As a result, such an absorber can be tuned as in the case of an *RLC* circuit.

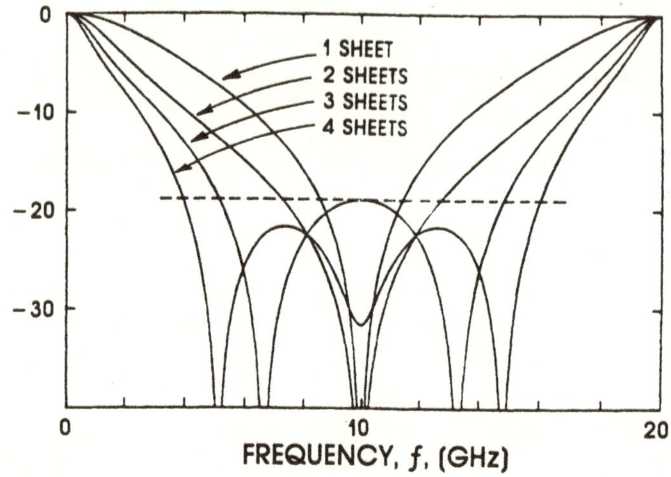

Figure 4.69 Increased bandwidth of Jaumann absorbers with increased number of layers (from [74]).

The equivalent circuit of a circuit analog absorber with a dipole printed on it is shown in Figure 4.70. It is a parallel combination of a lossless parallel resonant circuit and a series resonant circuit. Stacks of such structures can be used to improve the frequency response, and multilayers of such structures are synthesized to achieve wideband frequency response. The major difficulty is to find a proper material whose μ, ε, and α vary with frequency in the desired fashion.

The surfaces used in the CA absorbers are also known as *frequency selective*

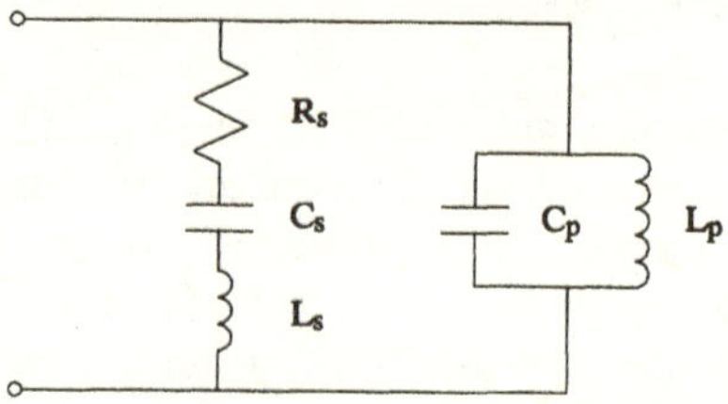

Figure 4.70 Equivalent circuit of a dipole absorber.

surfaces (FSS), which are discussed separately in Section 4.6. Although frequency selective surfaces act as different types of filters, they have the advantage of providing flexibility in the filter design. The Salisbury screens, described earlier, also can be used as circuit analog as discussed in [72, Section 9.5].

4.5.8.4 Magnetic Absorbers

Magnetic absorbers, like ferrite slabs, have distinct advantages over dielectric absorbers in the sense that typically the former require one-tenth of the thickness of dielectric absorbers to cause the same RCS reduction. The frequency variation of the magnetic properties of a ferrite is utilized to keep the electrical thickness of a ferrite slab constant with increases in frequency. The variations of relative permittivity ε_r and permeability μ_r of a typical ferrite material with frequency are shown in Figure 4.71, which indicates that the permittivity stays reasonably constant over the range of frequencies. However, over a range of frequencies, μ_r decreases with increasing frequency, so that it can be used to maintain a constant electrical thickness for a slab of given thickness over a wide band of frequencies and thereby achieve a constant amount of reflection.

Thickness criterion for single-layer radar absorbing material is discussed in [75] and may be useful here. Consider an absorbing material, with $\varepsilon_r = \varepsilon_r' - i\varepsilon_r'' = |\varepsilon_r| e^{i\delta_\varepsilon}$, $\mu_r = \mu_r' - i\mu_r'' = |\mu_r| e^{-i\delta_\mu}$, where δ_ε and δ_μ are the electric and magnetic loss angles for the material. The intrinsic impedance of the material relative to that of free space is $\eta = \mu_r/\varepsilon_r$. An instructive exercise is to examine the families of fixed electrical thickness or fixed levels of reflectance ($20 \log_{10} R$, R being the reflection coefficient) as functions of loss angles δ_μ and δ_ε for fixed values of $|\varepsilon_r|$ and $|\mu_r|$ for a layer of absorbing material. Contours of constant electrical thickness of the material ω wavelengths (as measured in the material) for $|\mu_r| = 1$, $|\varepsilon_r| = 16$ (i.e., $\eta_r < 1$) are

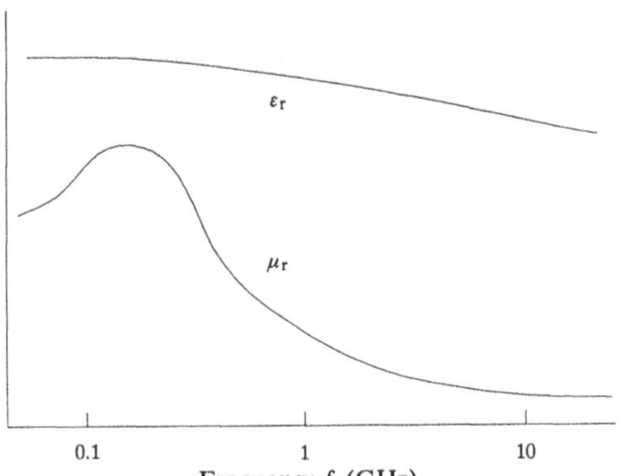

Figure 4.71 Plot of ε_r and μ_r of a magnetic medium with frequency.

shown in Figure 4.72(a), and for $|\mu_r| = 16$, $|\varepsilon_r| = 1$ (i.e., $\eta_r > 1$) in Figure 4.72(b). The numerical labels on the contours of Figure 4.72(a) and (b) denote the electrical thickness in wavelengths. Contours of constant minimum reflectance for the same two cases are shown in Figure 4.73(a) and (b), where the labels on the curves refer to reflectance in dB. From the results shown, we may conclude that, for magnetic absorbers, minimum reflectance of the material may not be realized for an electrical thickness of a quarter of wavelength in the material; and in fact, the thickness may be even more than a half-wavelength [75].

A two-layer broadband absorber is described in [76]. The absorber was constructed from a ferrite-resin mixture incorporated with short metal fibers. With a rather thin configuration (typical thickness is 4.6 mm), it provided nearly 50 percent relative bandwidth in the frequency range 8 to 13 GHz, and over aspect angle variations of ±45° from the normal. The characteristics of the absorber are shown in Figure 4.74.

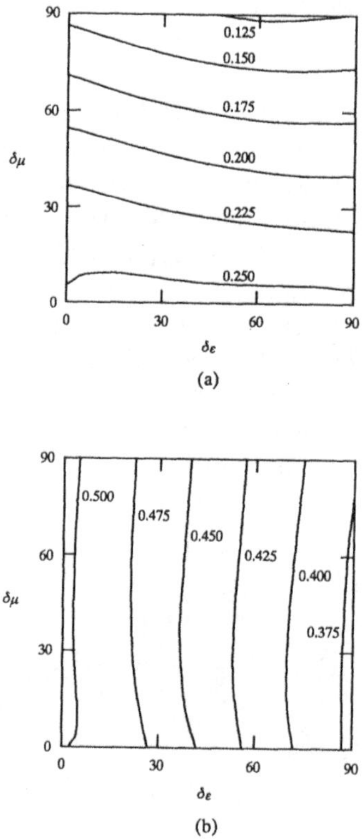

Figure 4.72 Contour of constant electrical thickness: (a) $|\varepsilon_r| = 16$ and $|\mu_r| = 1$; (b) $|\varepsilon_r| = 1$ and $|\mu_r| = 16$ (from [75]).

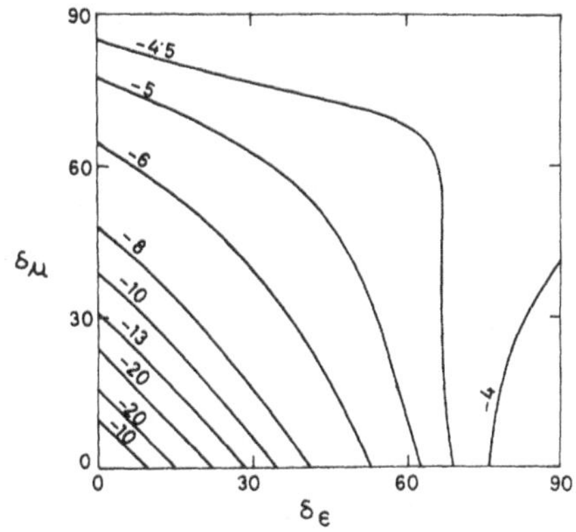

(a)

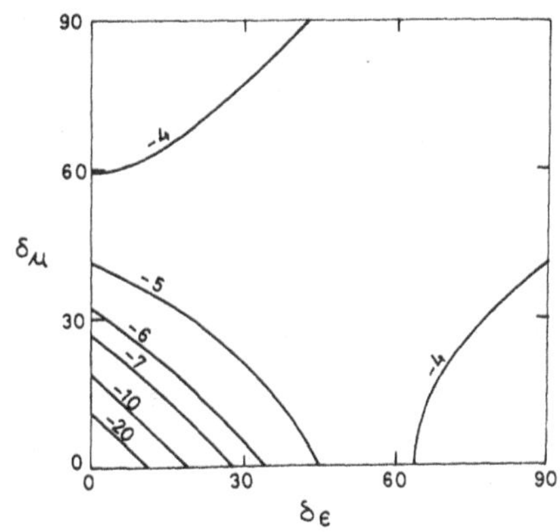

(b)

Figure 4.73 Contours of constant minimum reflectance: (a) $|\varepsilon_r| = 16$ and $|\mu_r| = 1$; (b) $|\varepsilon_r| = 1$ and $|\mu_r| = 16$.

4.5.8.5 Pyramidal Absorbers

In the quest for higher and higher bandwidths of absorbers for applications to wideband measurements, we naturally think of using tapered structures, like cones or pyramids. A planar array of pyramidal absorbers, each with an adequate length, as shown in Figure 4.75(a) constitutes a low reflection structure and commonly is used in an anechoic chamber to reduce reflections from the walls. Pyramidal absorbers having bandwidths from 100 MHz to 100 GHz are now available.

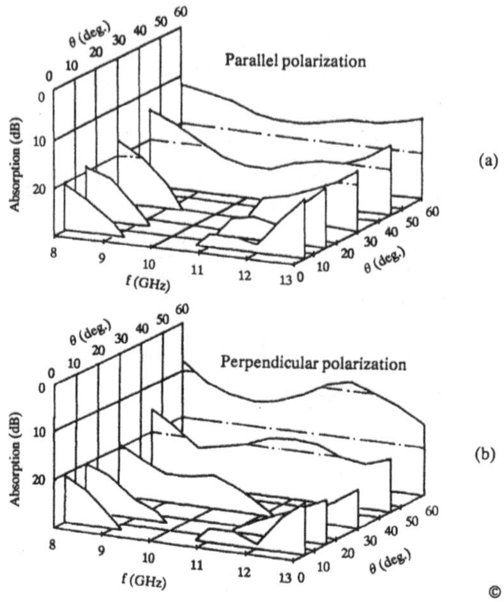

Figure 4.74 Absorption characteristics of a two-layer absorber (from [76]).

To the best of our knowledge no theoretical model for pyramidal absorber panels is available in the open literature apart from [77], which presents an analysis of normal incidence scattering of arbitrarily polarized plane waves by pyramidal absorbers and compares theoretical results with an experiment. The geometry of the scattering problem is shown in Figure 4.75(b), where 2β is the interior wedge angle. Figure 4.76 compares the theoretical (solid lines) and experimental (discrete symbols) reflection coefficients as functions of electrical length (L/λ, L is the length in Figure 4.75(b) and λ_1 is the wavelength in the absorber medium) for the two cases: the electric field parallel to the edge of the wedge (EP) and the magnetic field parallel to the edge of the wedge (HP).

According to their intended use pyramidal absorbers are available in different forms. Table 4.4 shows the performance characteristics of a pyramidal absorber (model no. ECCOORB VHP-NRL) over a wide range of angles. The reflection coefficient *versus* the electrical thickness (at the base of the absorber) for the Emerson Cuming high-performance pyramidal absorber is given in Figure 4.77..

4.6 RCS REDUCTION BY ACTIVE DIODE LOADING

Scattering by a strip dipole loaded with active diode has been discussed in Section 3.4. RCS reduction of bodies by active loading has not been treated in the open literature.

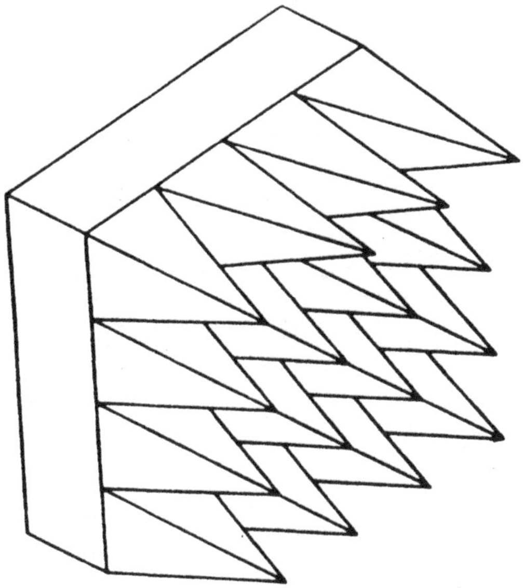

(a)

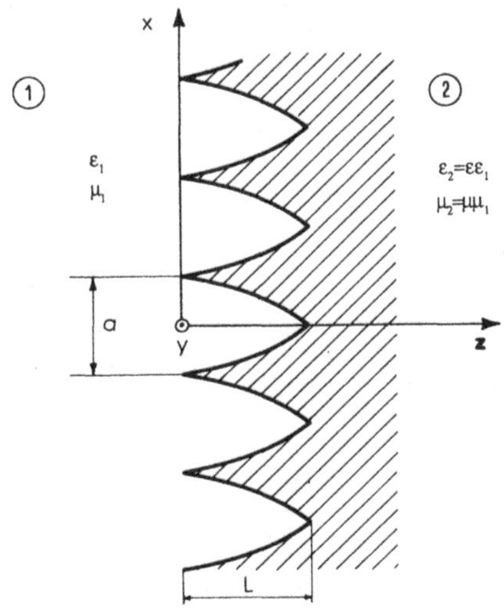

(b)

Figure 4.75 (a) Pyramidal absorbers; (b) model geometry for scattering by pyramidal absorbers.

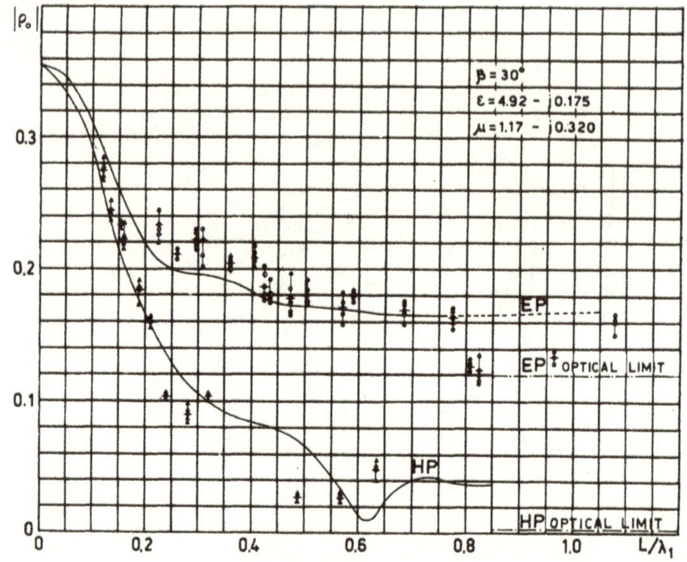

(a)

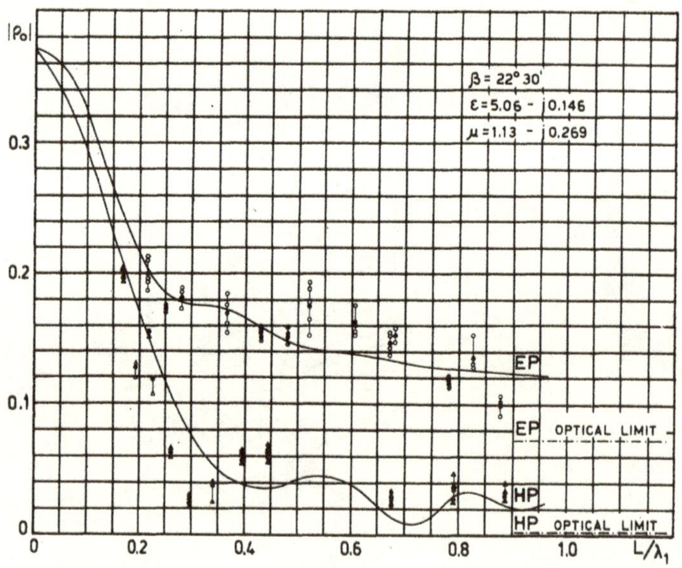

(b)

Figure 4.76 The reflection coefficient of pyramidal absorber panel as a function of electrical length for electrical field parallel to the edge of the wedge and the magnetic field is parallel to the edge of the wedge: (a) $\beta = 30°$; $\varepsilon_r = 4.92 - j0.175$; $\mu_r = 1.17 - j0.32$; (b) $\beta = 22.30°$; $\varepsilon_r = 5.06 - j0.146$; $\mu_r = 1.13 - j0.269$.

Table 4.4
Performance Characteristics of a Pyramidal Absorber (reprinted with permission from Emerson & Cuming)

(a) Guaranteed Minimum Reflection Performance of ECCOSORB VHP-NRL in dB for Angles Near Normal Incidence

	Nominal Thickness (D)		*Nominal Weight per place* $0.372\ m^2$ $4\ ft^2$											
									L	*S*	*C*	*X*	*Ku*	*K*
					VHF	*VHF*	*UHF*	*UHF*	*UHF*	*SHF*	*SHF*	*SHF*	*SHF*	*SHF*
					120	*200*	*300*	*500*	*1*	*3*	*5*	*10*	*15*	*24*
VHP Type	*Meters*	*Inches*	*kg*	*lb*	*MHz*	*MHz*	*MHz*	*MHz*	*kHz*	*kHz*	*kHz*	*kHz*	*kHz*	*kHz*
VHP-150-NRL	3.607	142	42.64	94	40	45	50	50	5.0	5.0	5.0	5.0	5.0	5.0
VHP-110-NRL	2.743	108	32.72	72	35	40	50	50	5.0	5.0	5.0	5.0	5.0	5.0
VHP-70-NRL	1.778	70	20.55	45	30	35	40	45	5.0	5.0	5.0	5.0	5.0	5.0
VHP-45-NRL	1.143	45	14.07	31		30	35	40	5.0	5.0	5.0	5.0	5.0	5.0
VHP-26-NRL	0.660	26	7.72	17			30	35	4.0	5.0	5.0	5.0	5.0	5.0
VHP-18-NRL	0.457	18	5.45	12				30	4.0	4.5	5.0	5.0	5.0	5.0
VHP-12-NRL	0.305	32	3.63	8					3.5	4.0	3.0	5.0	5.0	5.0
VHP-8-NRL	0.203	8	2.63	5.8					3.0	4.0	5.0	5.0	5.0	5.0
VHP-4-NRL	0.127	5	1.50	3.3						3.0	4.0	4.5	5.0	5.0
VHP-2-NRL	0.0889	3.5	1.00	2.2							3.0	4.0	4.5	5.0

(b) Reflection Coefficient in dB *versus* Thickness in Wavelengths, Wide-Angle Performance (⊥ = perpendicular polarization, ‖ = parallel polarization)

$D/\lambda =$	1		1.3		2.2		2.6		3		3.8		5.1		5.7		
Incidence angle	⊥	‖	⊥	‖	⊥	‖	⊥	‖	⊥	‖	⊥	‖	⊥	‖	⊥	‖	⊥
50°	22	20	24	27	29	33	36	30	41	38	43	40	46				55
60°	15	17	17	25	23	31	22	27	34	28	36	38	34	33	42	37	43
70°	12	11	11	19	16	21	15	20	23	21	24	27	25	30	33	28	38

Note: VHP-36-NRL is available for special design, providing performance intermediate between VHP-45-NRL and VHP-26-NRL. Additional VHP-NRL intermediate absorber thicknesses also are available on special order.

Although the theory and approach to the RCS reduction are similar to those discussed for the strip dipole, the objective of this section is to explore the possibility of RCS reduction by utilizing loading with PIN diodes and negative resistance diodes such as tunnel and Gunn diodes. Figure 4.78(a)–(e) gives the broadside backscattering radar cross sections *versus* frequency for a strip dipole located with a diode under a variety of conditions. The results indicate that, depending on the locations of the load, significant cross-section reductions can be achieved at certain frequencies. For example, results in Figure 4.78(d) and (e) show a possible reduction of about 45 dB (compared to

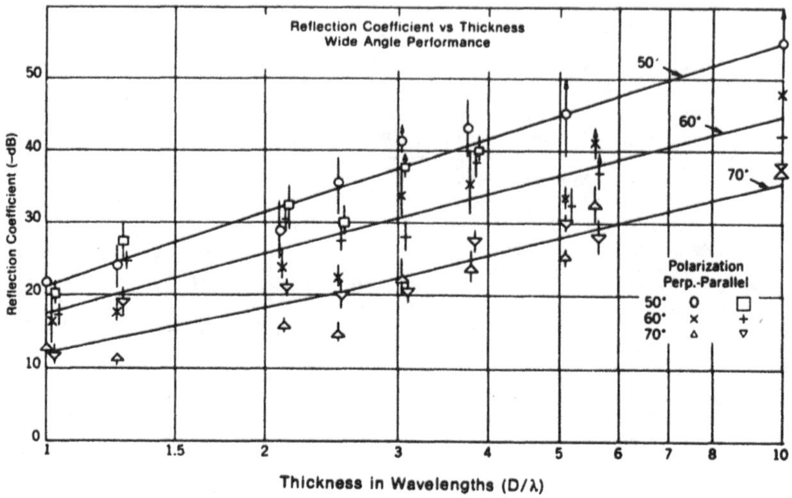

Figure 4.77 Measured reflection coefficient *versus* electrical thickness for Emerson & Cuming high-performance pyramidal absorber [reprinted with permission of Emerson & Cuming].

the RCS of the unloaded case). Results also indicate that RCS reduction obtained for center loading is more than that for asymmetric loading. Further details are discussed in [78].

4.7 FREQUENCY SELECTIVE SURFACES

As indicated earlier (Subsection 4.5.8.3 on CA absorbers), the frequency selective surfaces assume importance where a surface must produce specified forward and backward scattering characteristics. Various patterns used as FSS are shown in Figure 4.79, and they find many practical applications; for example, in wave polarizers, antenna reflections, RCS control, and artificial dielectrics. For a given angle of incidence and polarization we can obtain desired reflection and transmission properties from FSS by proper choice of the size of an individual element in the array and the spacing between the elements. Analyses of various FSS are given in [79–81]. The reflection and transmission characteristics obtained by spectral domain analysis [79] for free-standing circular aperture FSS are shown in Figure 4.80(a) and (b). Figure 4.80(b) compares the spectral domain reflection results with those obtained in Chen [81], who used modal analysis. The results suggest that by proper design it may be possible to design a frequency selective surface yielding low RCS within a reasonable band of frequencies. Figure 4.80(c) compares the transmission characteristics of surfaces using

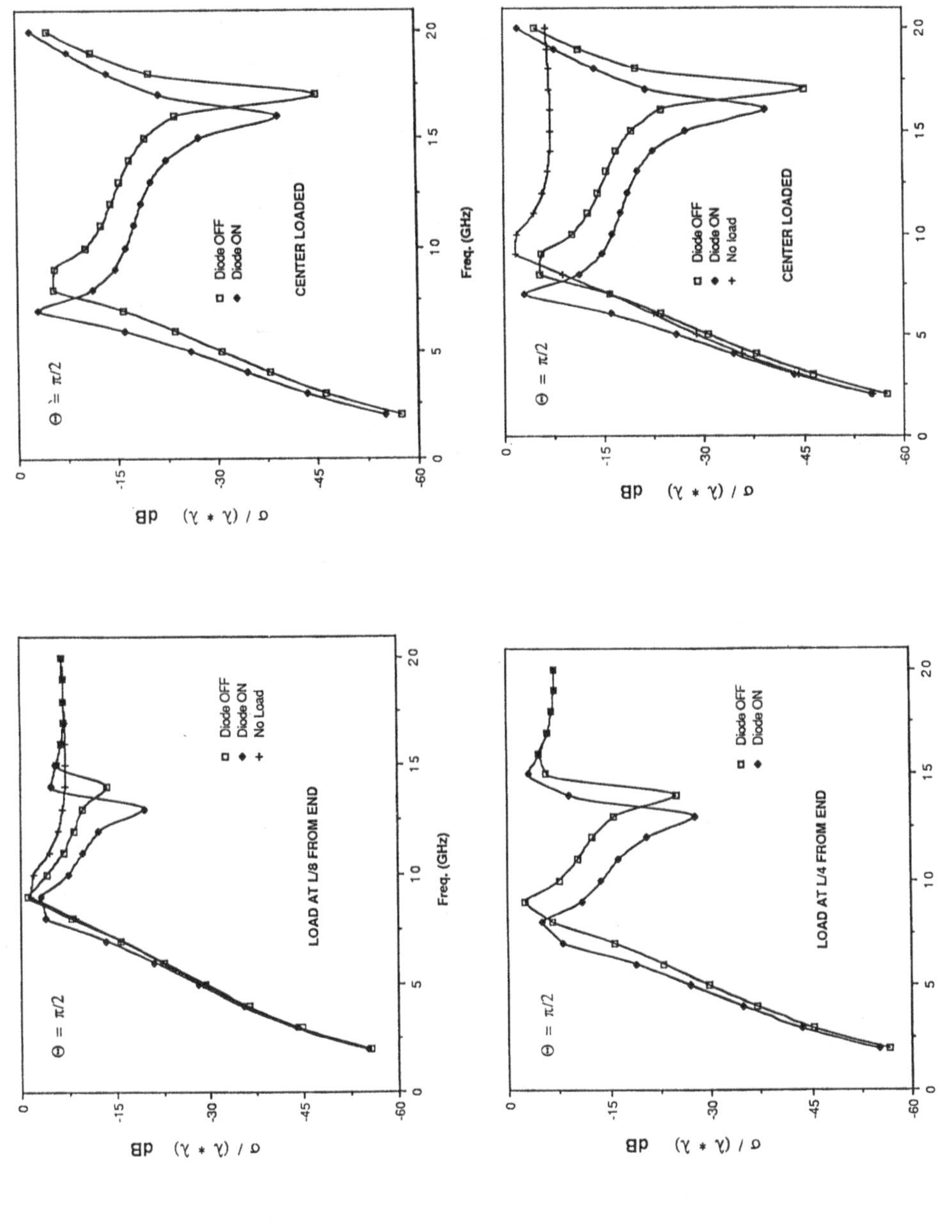

Figure 4.78 Backscattering RCS of a diode-loaded strip dipole with frequency for loading at different point or points and with diode off, on, and no load (from Ref. [11] of Ch. 3).

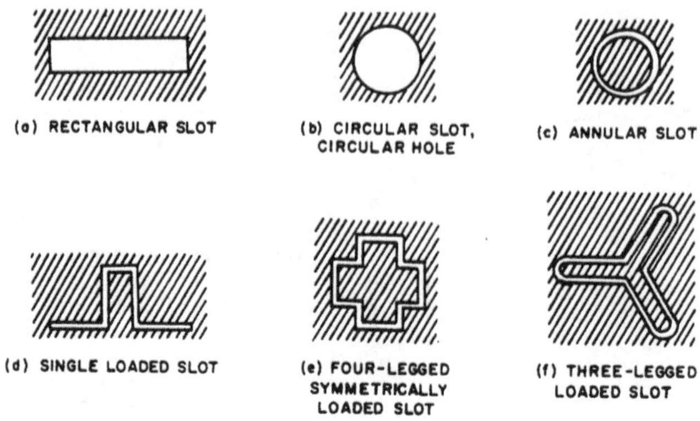

Figure 4.79 Some configurations used in FSS.

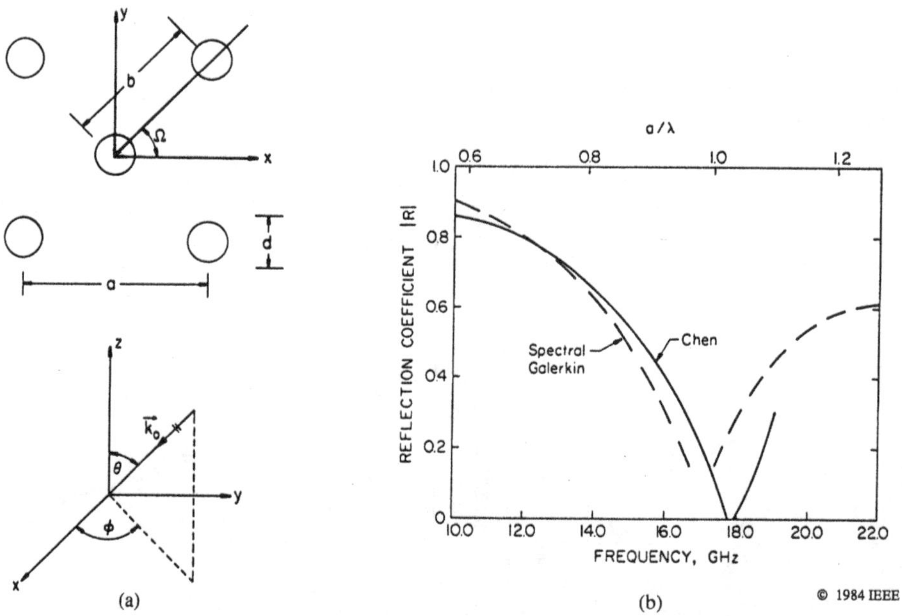

Figure 4.80 Free-standing circular aperture FSS (a) Geometry and coordinate system; (b) computed reflection coefficient $|R|$ *versus* frequency and comparison with Chen's [81] results: $a = b =$ 1.73 cm; $d - 1.2$ cm; $\Omega = 60°$; normal incidence; E parallel to y-axis; (c) transmitted power $|T|^2$ *versus* frequency compared with rectangular aperture: $a = b = 1.0$ cm; $d = 0.79$ cm; $\Omega = 90°$; normal incidence E field parallel to x-axis; (d) transmitted coefficient $|T|$ *versus* frequency. $\theta = 15°, 30°$($a = b = 1.73$ cm; $d = 1.2$ cm; $\Omega = 60°$; E-field in x-z plane) (from [79]).

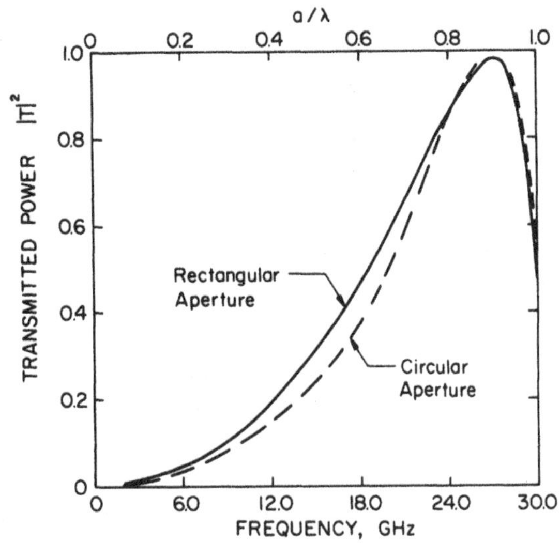

(c)

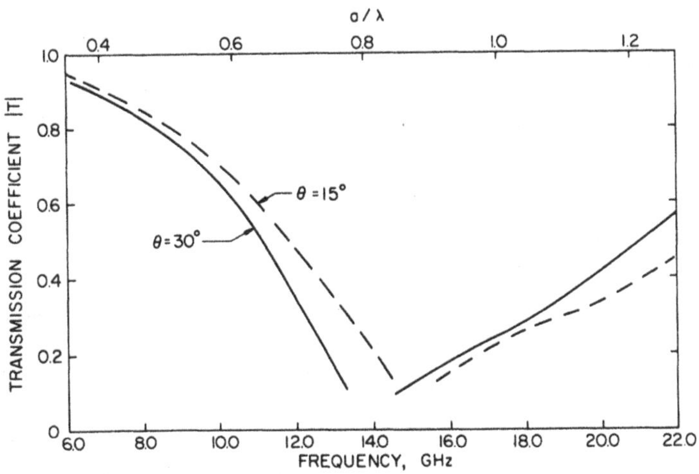

(d)

Figure 4.80 *Continued*

rectangular and circular apertures, and the results indicate that the exact shape of the aperture is unimportant. Transmission coefficients obtained in the normal direction for two oblique angles of incidence are shown in Figure 4.80(d) for the surface with rectangular apertures. The results indicate that the center frequency and the frequency bandwidth for reduction of forward scattering will change with aspect angle. Similar results are obtained with free-standing FSS consisting of crossed dipoles and Jerusalem crosses that have been investigated in [80].

4.8 SCATTERING FROM ANTENNAS

We sometimes want a radar antenna to be invisible to a hostile radar. Even in a friendly environment, a large amount of scattering by an unwanted antenna may cause *electromagnetic interference* (EMI) to our own system. This necessitates that within a specified frequency band, aspect angles and polarization of the RCS of the antenna be minimized as far as possible without significantly affecting its performance. Reducing the RCS of an antenna within the frequency band of its operation is a very difficult, if not impossible, problem and has not been clearly addressed in the open literature. Reduction of out-of-band RCS appears to be an easier task.

Historically, the wire antenna was first treated from the viewpoint of scattering in [16, 17]. Minimum scattering antennas [82] and antenna arrays [83] have been investigated by several authors. Discussions of scattering by antennas are available in [73, 84, 85].

It is convenient to describe the scattering properties of an antenna by using its network model. Generally, the field $\mathbf{E}^S(Z_L)$ scattered by an antenna as a function of its load Z_L is given by [73, Section 8.4]

$$E^S(Z_L) = \mathbf{E}^S(Z_a^*) - \left[I(Z_a^*)\mathbf{E}_r\right]\Gamma_m \tag{4.99}$$

where

$$\Gamma_m = \frac{Z_L - Z_a^*}{Z_L + Z_a} \tag{4.100}$$

is the modified voltage reflection coefficient, $Z_a = R_a + \mathrm{i}x_a$ is the antenna impedance, $\mathbf{E}^r$ is the far field radiated by the antenna with unit current excitation, $I(Z_a^*)$ is the antenna current with conjugate antenna impedance given by

$$I(Z_a^*) = \frac{\mathbf{h}_a^r \cdot \mathbf{E}^i}{2R_a} \tag{4.101}$$

with $\hat{h}_a^r$ is the effective height of the antenna on reception, and $\mathbf{E}^i$ is the incident electric field at the antenna.

Introducing the expression for $\mathbf{E}^r$ in (4.99), the following is obtained for the scattered field:

$$\mathbf{E}^s(Z_L) = \mathbf{E}^s(Z_a^*) - \left[\frac{\mathrm{i}\eta_0}{4\lambda R_a}\hat{\mathbf{h}}_a^t(\hat{\mathbf{h}}_a^i \cdot \mathbf{E}^i)\frac{e^{\mathrm{i}\beta r}}{r}\right]\Gamma_m \tag{4.102}$$

where η_0 is the intrinsic impedance of free space, λ is the wavelength in free space, $\hat{\mathbf{h}}_a^t$ is effective height of the antenna $\beta = 2\pi/\lambda$ is the propagation constant in free space, and r is the distance of the field point from the antenna.

Note that in the case of backscattering $\hat{\mathbf{h}}_a^r = \hat{\mathbf{h}}_a^t$ in (4.102). The first and second terms in (4.102) generally are referred to as the *structural* and *antenna mode scattered fields*, respectively. The aim of an engineer designing a minimum scattering antenna will be to reduce the structural scattered field to as small a value as possible without altering the antenna mode scattered field; that is, the antenna characteristics. A canonical minimum scattering antenna has been defined in [73] as "one which becomes 'invisible' when the accessible waveguide terminals or ports are open-circuits." The complete scattering matrix of such an antenna is unique, but its impedance and admittance matrices are not defined.

The overall scattering cross section σ of an antenna stays within the following bounds [73]:

$$\left|\sqrt{\sigma_s} - \sqrt{\sigma_a}\right|^2 \leq \sigma \leq \left|\sqrt{\sigma_s} + \sqrt{\sigma_a}\right|^2 \tag{4.103}$$

where σ_s, σ_a are the structural and antenna mode scattering cross sections, respectively, of the antenna.

The scattering cross section of an open-circuited antenna is expressed by [73]

$$\sigma = \left|\sqrt{\sigma_s} + \sqrt{\Delta\sigma_e}\, e^{i\psi}\right|^2 \tag{4.104}$$

with

$$\Delta\sigma_e = \frac{\lambda^2}{4\pi} G_r G_t P_r P_t \tag{4.105}$$

where G_t and G_r are the transmitting and receiving mode gains of the antenna, respectively, P_t and P_r are the factors accounting for possible polarization mismatch (P_t is associated with the polarization mismatch between scattering and transmitting mode cases, and P_r is associated with polarization mismatch between scattering and receiving mode cases).

When the transmitting, receiving, and scattering antenna modes are polarization matched

$$\Delta\sigma_e = \frac{\lambda^2}{4\pi} G^2 \tag{4.106a}$$

with G as the gain of the antenna.

Limited results for scattering by antenna are available in the literature. The antenna mode, structural mode, and total cross sections (in square meters) of a dipole antenna as functions of frequency are shown in Figure 4.81(a)–(c). Similar results for the antenna as functions of the normalized length $1/\lambda$ of the dipole are shown in Figure 4.82(a)–(c). Observe from the results that the antenna mode RCS tends to a minimum

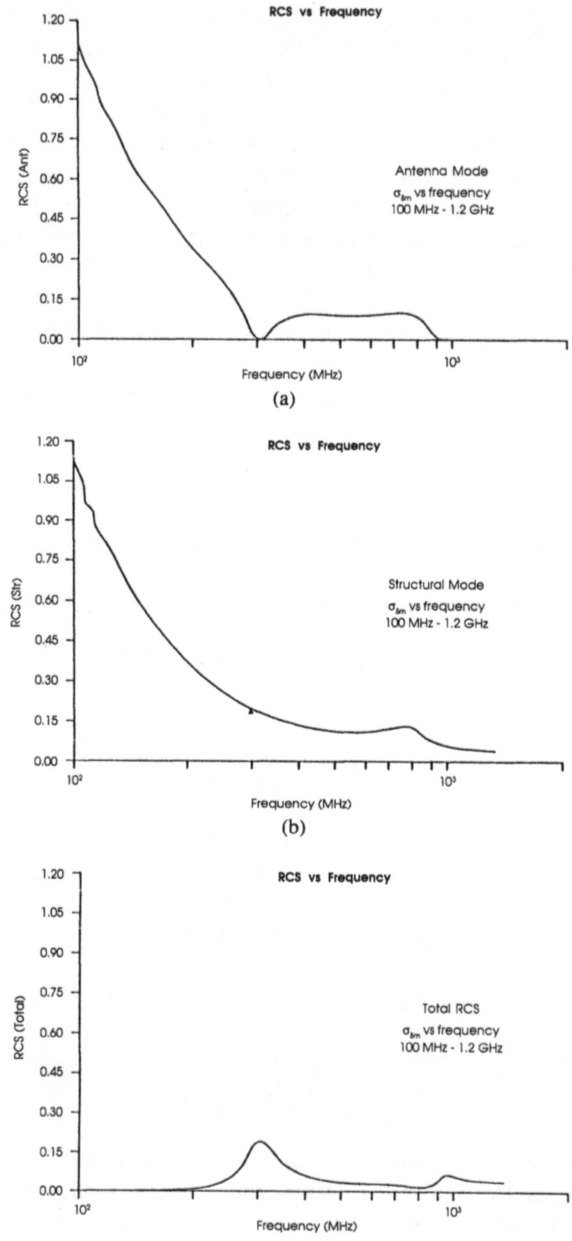

(c)

Figure 4.81 RCS *versus* frequency of a dipole: (a) antenna mode contributions; (b) structural mode contributions; (c) total contributions frequency—100 MHz – 1.2 GHz.

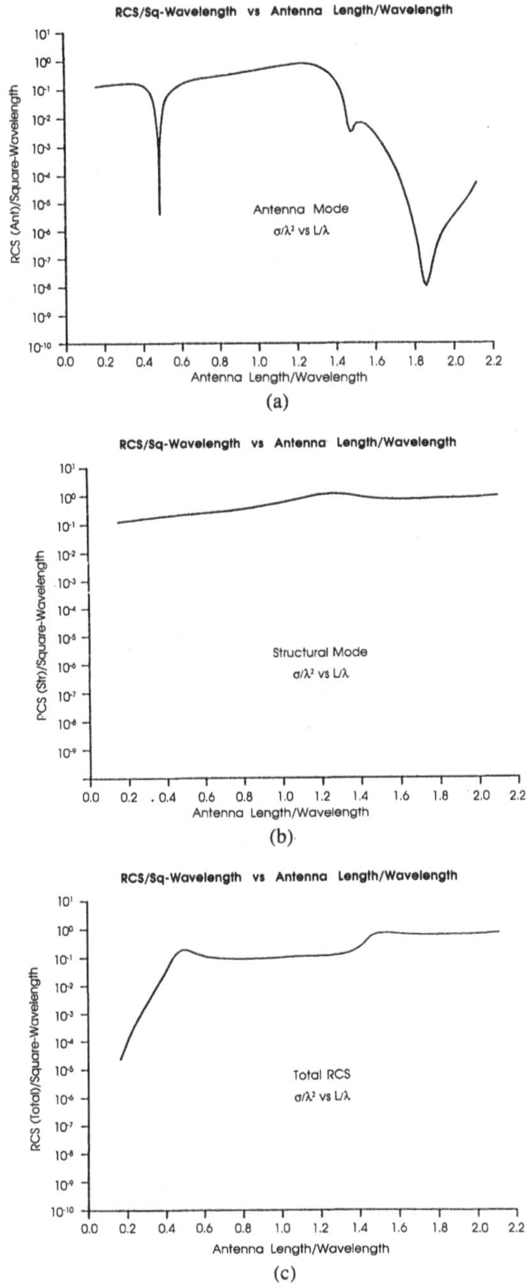

Figure 4.82 RCS *versus* antenna length wavelength: (a) antenna mode contributions; (b) structural mode contributions; (c) total contributions.

for a half wavelength long dipole. Figure 4.83(a) and (b) show the structural RCS (in dB over λ^2, λ being the wavelength) patterns of an X-band horn and a 14-in. parabolic dish antenna. The results shown in Figure 4.84 are for the cases when all the antennas were polarization matched. Under this condition, the structural cross section of an antenna is given by [83, 86]

$$\sigma_s = \frac{\lambda^2}{4\pi} G^2 A^2 \tag{4.106b}$$

where G is the reradiating gain of the antenna, and A is its effective receiving area. Equation (4.106b) indicates that the structural RCS of an antenna is proportional to the square of the product of the receiving area and the reradiating gain of the antenna.

4.9 CHIRAL MEDIUM

Although the RCS of a body is determined significantly by its shape, what the medium is composed of also plays an important role. For example, the RCS of a conducting

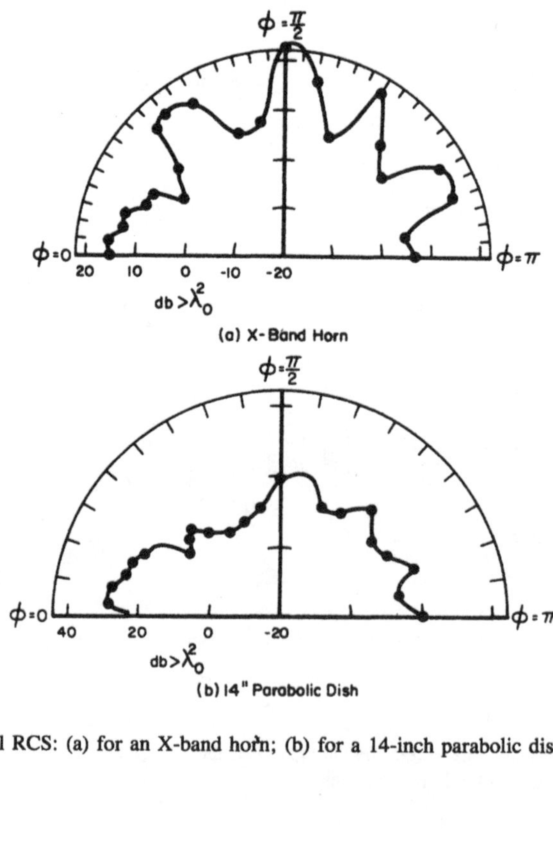

Figure 4.83 Structural RCS: (a) for an X-band horn; (b) for a 14-inch parabolic dish (from [73]).

Figure 4.84 Photograph of a typical low RCS aircraft (courtesy of United States Air Force).

body would be significantly different than that if the entire body of parts of it were made of nonconducting material. Composite materials [87] and their applications to the aircraft, space, and automotive industries have been a subject of great interest in recent years. From the RCS point of view, the most important property of composite materials is that they reflect electromagnetic waves less than metals. Though the exact constituents and primary constants (μ_r, ε_r, σ) of the composite materials are either classified or trade secrets, essentially they are graphic fabric epoxy resins or thermoplastic matrix-graphic composites; these types of materials can be used for RCS reduction of bodies.

Recently, a novel material has been introduced, called Chirosorb™, which is invisible to radar [88]. To achieve zero reflectivity (for electromagnetic waves), the constituent parameters of a material must satisfy certain conditions not ordinarily satisfied by the conventional RAMs. This low-reflectivity property of chiral materials makes them a logical choice for consideration of use to reduce RCS.

Electromagnetic chiral materials combine into themselves the optical activity and circular dichroism. Optical activity and circular dichroism in a medium are related, respectively, to the rotation of the plane of polarization and the change in the polarization ellipticity in the medium. Such an electromagnetic chiral material medium is described by the following chiral relations:

$$\mathbf{D} = \varepsilon\mathbf{E} - \mathrm{i}\xi_c\mathbf{B} \tag{4.107}$$

$$\mathbf{H} = -\mathrm{i}\xi_c\mathbf{E} + (1/\mu)\mathbf{B} \tag{4.108}$$

where **E**, **D**, **B**, and **H** are the standard electromagnetic field quantities, $\varepsilon = \varepsilon_0(\varepsilon'_r - \mathrm{i}\varepsilon''_r)$, $\mu = \mu_0(\mu'_r - \mathrm{i}\mu''_r)$ and the chiral admittance ξ_c is defined by

$$\xi_c = \sqrt{\varepsilon_0/\mu_0}\,(\xi'_c - \mathrm{i}\xi''_c) \tag{4.109}$$

The quantity ξ_c is a measure of the chirality of the medium. Table 4.5 compares the basic electromagnetic relationships and characteristics of simple isotropic and chiral media.

The condition for zero reflectance at the chiral surface is

$$\eta_c = \eta_0 \tag{4.110}$$

which can be written as

$$\begin{bmatrix} \mu' - \mu'' \\ \mu'' - \mu' \end{bmatrix}\begin{bmatrix} \xi'^2_c - \xi''^2_c \\ 2\xi'_c\xi''_c \end{bmatrix} = \begin{bmatrix} \mu' - \varepsilon' \\ \mu'' - \varepsilon'' \end{bmatrix} \tag{4.111}$$

where the parameters are as defined earlier. Because ξ_c is an extra parameter in a chiral medium, it provides extra flexibility in RCS reduction so that a chiral medium can be designed to be an ''invisible'' medium to electromagnetic waves.

One way of fabricating the Chirosorb™ is by embedding randomly oriented identical chiral microstructures such as microhelices, in an isotropic host medium. The

Table 4.5
Comparison of Characteristics of Simple Isotropic and Chiral Media

Characteristics	*Simple Medium*	*Chiral Medium*
Constitutive Relations	$D = \varepsilon\mathbf{E}$ $H = \mathbf{B}/\mu$	$D = \varepsilon\mathbf{E} - \mathrm{i}\xi B$ $H = \mathrm{i}\xi_c\mathbf{E} + B/\mu$
Wave Equation	$\nabla\times\nabla\times\mathbf{E} - k^2\mathbf{E} = 0$	$\nabla\times\nabla\times\mathbf{E} - k^2\mathbf{E} - 2\omega\xi_c\nabla\times\mathbf{E} = 0$
Wave Number	$k = \omega\sqrt{\mu\varepsilon}$	$k_\pm = \pm\omega\mu\xi_c + \sqrt{k^2 + (\omega\mu\xi_c)^2}$
Eigenmodes	Linear polarization allowed	Right and left circular polarizations only
Intrinsic Impedance	$\eta = \sqrt{\dfrac{\mu}{\varepsilon}}$	$\eta_c = \dfrac{\eta}{\sqrt{1 + \eta^2\xi_c^2}}$

parameters ξ_c, ε, and ξ of the chiral medium are determined by the randomness of the embedded microstructures in the host medium. Electromagnetic wave propagation in chiral medium has been discussed in [88–92] and is a topic of on-going research.

4.10 STEALTH VEHICLES

In this section, we provide a qualitative discussion on the reduction of RCS of complex bodies such as cars and aircraft. Such low-RCS vehicles generally are called *stealth vehicles*.

Scattering cross sections of stealth cars are discussed in [73, Section 8.4; 93]. Intrinsically, some cars are designed to reflect less radar energy than others. If desired, the use of composite material in the body instead of metal and properly designed body shaping may be used to reduce the RCS of a car to a certain extent. It has been found [73, Section 8.4] that the detection range of a Corvette is 600 ft compared to 9000 and 1200 ft for a moving van and a Honda Civic, respectively. The lower detection range and, hence, the lower RCS for a Corvette are attributed mainly to its shape. The portion of a car mostly responsible for a large front-on RCS is the radiator (which can be modeled as vertically oriented plates for analysis). It is conceivable therefore that suitable tilting of the radiator plane away from the vertical would reduce the front-on RCS. It is noted [93] that there is an effective reduction of detection distance of a typical car at a cost of $300.

The low RCS aspects of stealth aircraft, discussed in [94–96], will be mentioned briefly. Such factors as shape, use of absorbing materials and structures, and on-board avionics together effectively reduce the RCS of a stealth aircraft. Figure 4.84 shows a photograph of a low RCS aircraft. Generally, several steps are taken to reduce the RCS of an aircraft: (1) the wings are made of mixtures of structures and proper absorbing materials; (2) the waves, incident head-on at the leading edge and tail-on at the trailing edge, are absorbed in an optimized fashion by judicious use of radar absorbing structures consisting of laminated layers of glass fiber and carbon coated plastics; (3) honey-comb absorbers having maximum allowable lengths are used so that they are effective at lower frequencies; (4) resistive cards are used to minimize diffraction from the edges; and (5) as far as practicable from the aerodynamic point of view, metallic portions are replaced by resistive and composite structures.

Typical radar scattering cross section of a stealth aircraft at microwave frequencies is less than 0.1 m^2 whereas that of a large conventional fighter plane is 6 m^2.

REFERENCES

1. W. E. Blore, "The Radar Cross Section of Ogives, Double-Backed Cones, Double-Rounded Cones and Cone-Spheres," *IEEE Trans. on Antennas and Propagation*, Vol. AP-12, Sept 1964, pp. 582–590.
2. E. F. Knott, "RCS Reduction of Dihedral Corners," *IEEE Trans. on Antennas and Propagation*, Vol. AP-25, No. 3, May 1977, pp. 406–409.

3. W. B. Gordan, "Far-Field Approximations to the Kirchoff-Helmholtz Representations of Scattered Fields," *IEEE Trans. on Antennas and Propagation*, AP-23, No. 4, July 1975, pp. 590–592.
4. E. F. Knott, "RCSR Guidelines Handbook," Georgia Institute of Technology, Atlanta, prepared for US Navy, Contract No. N00039-73-0676.
5. E. F. Knott, "Radar Cross Section Reduction Using Cylindrical Segments," *IEEE Trans. on Antennas and Propagation*, Vol. AP-24, No. 6, November 1976, pp. 882–884.
6. S. Sanyal and A. K. Bhattacharyya, "Electromagnetic Scattering by a Curved Plate—Solution Uniform Asymptotic Theory of Diffraction," *IEEE Trans. on Antennas and Propagation*, Vol. AP-32, No. 2, February 1984, pp. 187–189.
7. J. W. Crispin and D. L. Maffett, "Radar Cross Section Estimation for Simple Shapes," *Proc. IEEE*, Vol. 53, No. 8, August 1965, pp. 833–848.
8. M. A. Plonus, E. F. Knott, and N. H. Kuo, "A Truncated Cone Model Support for Anechoic Chambers," *IEEE Trans. on Antennas and Propagation*, Vol. AP-16, March 1968, pp. 256–258.
9. D. F. Hastings, "Dielectric Rod Support Tuned by Included Metal Wires for Reduction of Scattering," *IEEE Trans. on Antennas and Propagation*, Vol. AP-16, March 1968, pp. 350–351.
10. A. H. Jorgen and V. V. Solodukhov, "Echo Width of Foam-Supports Used in Scattering Measurements," *IEEE Trans. on Antennas and Propagation*, Vol. AP-27, No. 2, March 1979, pp. 191–193.
11. T. B. A. Senior, "The Backscattering Cross Section of a Cone-Sphere," *IEEE Trans. on Antennas and Propagation*, Vol. AP-13, March 1965, pp. 271–277.
12. T. B. A. Senior and R. F. Goodrich, "Scattering by a Sphere," *Proc. IEE* (London), Vol. 111, 1964, pp. 907–916.
13. E. F. Knott and T. B. A. Senior, "The Effect of Fins on the Nose-on Cross Section of Cone Sphere," *IEEE Trans. on Antennas and Propagation*, Vol. AP-11, No. 4., July 1963, pp. 504–505.
14. H. A. Iams, Radio Wave Conducting Device, U.S. Patent 2528367, October 31, 1950.
15. Sletten, C. J., ed., *Reflector and Lens Antennas*, Artech House, Norwood, MA, 1988.
16. Y. Y. Hu, "Backscattering Cross Section of a Center-Loaded Cylindrical Antenna," *IEEE Trans. on Antennas and Propagation*, Vol. AP-6, No. 1, January 1958, pp. 140–148.
17. L. Weinberg, "New Techniques for Modifying the Monostatic and Multi-static Radar Cross Section," *IEEE Trans. on Antennas and Propagation*, Vol. AP-11 (Commn.), No. 6, November 1963, pp. 717–719.
18. K. M. Chen and V. V. Liepa, "The Minimization of the Backscattering of a Cylinder by Central Loading," *IEEE Trans. on Antennas and Propagation*, Vol. AP-12, No. 5, September 1964, pp. 576–582.
19. K. M. Chen and V. V. Liepa, "Minimization of the RCS of a Cylinder by Central Loading," Rep. 5548-1-T Radiation Laboratory, University of Michigan, Ann Arbor, April 1964.
20. K. M. Chen, "Minimization of Backscattering of a Cylinder by Double-Loading," *IEEE Trans. on Antennas and Propagation*, Vol. AP-13, March 1965, pp. 262–270.
21. R. F. Harrington and J. R. Mautz, "Control of Radar Scattering by Reactive Loading," *IEEE Trans. on Antennas and Propagation*, Vol. AP-20, No. 4, July 1977, pp. 446–454.
22. R. F. Harrington and J. R. Mautz, "Theory of Characteristic Modes for Conducting Bodies," *IEEE Trans. on Antennas and Propagation*, Vol. AP-19, No. 5, September 1972, pp. 622–628.
23 R. B. Kieburtz, and A. Ishimaru, "Scattering by a Periodically Apertured Conducting Screen," *IEEE Trans. on Antennas and Propagation*, Vol. AP-9, No. 6, November 1961, pp. 506–514.
24. A. K. Bhattacharyya and T. K. Sen, "Electromagnetic Scattering by a Thin Conducting Flat Plate with Doubly Periodic Punched Circular Apertures," *IEEE Trans. on Antennas and Propagation*, Vol. 126, No. 11, November 1979, pp. 1149–1154.
25. R. B. Green, "The General Theory of Antenna Scattering," Antenna Laboratory, Ohio State University, Rep. No. 223-17, November 1963.
26. R. A. Hurd and E. V. Jull, "Theory of a Reflection Grating with Narrow Grooves," *Radio Science*, Vol. 16, No. 3, May-June 1981, pp. 271–277.

27. J. W. Heath and E. V. Jull, "Perfectly Blazed Reflection Gratings with Rectangular Grooves," *J. Optics Society of America*, Vol. 68, No. 9, September 1978, pp. 1211–1217.
28. E. V. Jull, J. W. Heath, and G. R. Ebbesen, "Gratings that Diffract All Incident Energy," *J. Optics Society of America*,, Vol. 67, No. 4, April 1977, pp. 557–560.
29. E. V. Jull and G. R. Ebbesen, "The Reduction of Interference from Large Reflecting Surface," *IEEE Trans. on Antennas and Propagation*, Vol. AP-15, No. 4, July 1977, pp. 565–570.
30. J. W. Heath, and E. V. Jull, "Total Backscatter from Conducting Rectangular Corrugations," *IEEE Trans. on Antennas and Propagation*, Vol. AP-27, No. 1, January 1977, pp. 95–97.
31. E. V. Jull, "Reflection Circular Polarizers," *Electronics Letters*, 15, No. 14, 5 July 1979, pp. 423–424.
32. E. V. Jull, and J. W. Heath, "Reflection Grating Polarizers," *IEEE Trans. on Antennas and Propagation*, Vol. AP-28, No. 4, July 1980, pp. 586–588.
33. G. R. Ebbesen, "TM Polarized Electromagnetic Scattering from Fin-Corrugated Periodic Surface," *J. Optics Society of America*, Vol. 66, No. 12, December 1976, pp. 1363–1367.
34. H. A. Kalhor, "Approximate Analysis of Electromagnetic Scattering from Corrugated Conducting Surface by Surface Impedance Modeling," *IEEE Trans. on Antennas and Propagation*, Vol. AP-25, No. 5, September 1977, pp. 721–722.
35. L. Shafai and K. Iskander, "Investigation of Scattering by Corrugated Surfaces Using Impedance Boundary Conditions," *Electronics Letters*, Vol. 16, No. 5, 2 March 1978, pp. 144–146.
36. G. R. Ebbesen, "The Use of Fin-Corrugated Periodic Surfaces for Reduction of Interference from Large Reflecting Surfaces," MS dissertation, Univ. of British Columbia, July 1974.
37. A. K. Bhattacharyya and S. K. Tandon, "Corrugated Surface with a Low Backscatter," *IEEE Trans. on Antennas and Propagation*, Vol. AP-32, August 1984, pp. 870–872.
38. J. W. Heath and E. V. Jull, "Reflection Polarizers," ISAP, Sendai, Japan 1978.
39. E. V. Jull, N. C. Beaulieu, and D. C. W. Hui, "Perfectly Blazed Triangular Groove Reflection Gratings," *J. Optics Society of America*, Vol. 1, No. 2, February 1984, pp. 180–182.
40. K. A. Jose, C. K. Anandan, and K. G. Nair, "Low Backscattered TM-Polarized Strip Gratings," *Electronics Letters*, 13 August 1987, Vol. 23, No. 7, pp. 905–906.
41. R. E. Lawrie and L. Peters, Jr., "The Control of the Echo Area of Ogives by Cut-off Corrugated Surfaces," *IEEE Trans. on Antennas and Propagation*, Vol. AP-14, November 1966, pp. 605–611.
42. V. H. Weston, "Theory of Absorbers in Scattering," *IEEE Trans. on Antennas and Propagation*, Vol. AP-11, No. 5, September 1963, pp. 578–584.
43. D. Klement, J. Preissner, and V. Stein, "Special Problems in Applying the Physical Optics Method for Backscatter Computations of Complicated Objects," *Trans. on Antennas and Propagation*, Vol. AP-36, No. 2, February 1988, pp. 228–237.
44. J. Mooreland and L. Peters, Jr., "The Specular RCS of Absorber-Coated Bodies," *IEEE Trans. on Antennas and Propagation*, Vol. AP-14, November 1966, pp. 799–800.
45. V. H. Weston and J. J. Bowman, "Some Remarks Concerning the Design of Absorbers in the Resonance Region," *IEEE Trans. on Antennas and Propagation*, Vol. AP-13, pp. 467–468.
46. J. W. Crispin, Jr., and K. M. Siegel, *Methods of Radar Cross Section Analysis*, Academic Press, New York, 1968, p. 288.
47. G. D. Maliuzhinets, "Excitation, Reflection and Emission of Surface Waves from a Wedge with Given Face Impedance," *Soviet Phys. Dokl.*, Vol. 3, No. 3, 1958, pp. 752–755.
48. W. E. Williams, "Propagation of Electromagnetic Surface Waves along Wedge Surfaces," *Quant. J. Mechanics Applied Math.*, Vol. 13, pp. 278–284.
49. A. Mohsen and M. A. K. Hamid, "Diffraction by a Dielectric Loaded Wedge," *Radio Science*, Vol. 8, 1971, pp. 71–80.
50. O. M. Bucci and G. Francischetti, "Electromagnetic Scattering by a Half Plane with Two Face Impedance," *Radio Science*, Vol. 17, September–October 1982, pp. 1199–1210.

51. R. Tieberio and R. G. Kouyoumjian, "A Uniform GTD Solution for the Diffraction by Strips Illuminated at Grazing Incidence," *Radio Science*, Vol. 14, November–December 1959, pp. 933–941.
52. R. Tiberio, F. Bessi, G. Manara, and G. Pelosi, "Scattering by a Strip with Two Face Impedances at Edge-on Incidence," *Radio Science*, Vol. 17, September–October 1982, pp. 1199–1210.
53. T. B. A. Senior, "Half-Plane Edge Diffraction," *Radio Science*, Vol. 10, No. 6, June 1975, pp. 645–650.
54. O.M. Bucci and G. Franceschetti, "Electromagnetic Scattering by Half Plane with Two Face Impedance," *Radio Science*, Vol. 11, No. 1, January 1976, pp. 49–59.
55. W. E. Williams, "Diffraction of an Electromagnetic Plane Wave by a Metallic Sheet," *Proc. Royal Society* (London), A257, pp. 413–419.
56. S. Sanyal and A. K. Bhattacharyya, "Arbitrary Line Source Diffraction by a Half-Plane with Two Face Impedance—A Uniform Asymptotic Expansion," IEEE Int. Symp. on Antennas and Propagation Society held at Boston, June 25–29, 1984.
57. B. L. van der Waerden, "On the Method of Saddle Points," Applied *Science Research*, B2, 1952, pp. 33–45.
58. K. M. Siegel, "Far Field Scattering from Bodies of Revolution," *Applied Science Research*, Sec. B, Vol. 7, 1958, pp. 293–328.
59. R. A. Ross, "RCS of Rectangular Flat Plates as a Function of Aspect Angle," *IEEE Trans. on Antennas and Propagation*, Vol. AP-14, May 1966, pp. 329–335.
60. F. A. Sikta, W. D. Burnside, T. T. Chu, and L. Peters, Jr., "First Order Equivalent Current and Corner Diffraction Scattering from Flat Plate Structures," *IEEE Trans. on Antennas and Propagation*, Vol. AP-31, No. 4, July 1983, pp. 584–589.
61. T. B. A. Senior, "Backscattering by Resistive Strips," *IEEE Trans. on Antennas and Propagation*, Vol. AP-27, No. 6, November 1979, pp. 808–813.
62. T. B. A. Senior, "Backscattering from Resistive Strips," *Radio Science*, Vol. 14, No. 5, September–October 1979, pp. 911–924.
63. R. L. Haupt, "Synthesis of Resistive Tapers to Control Scattering Patterns of Strips," Ph.D. dissertation, University of Michigan, 1987.
64. S. Ray and R. Mittra, "Scattering from Resistive Strips and from Metallic Strips with Resistive Edge Loading," *IEEE 1983 AP-S Int. Symp. Dig.*, pp. 603–606.
65. T. B. A. Senior and V. V. Liepa, "Backscattering from Tapered Resistive Strips," *IEEE Trans. on Antennas and Propagation*, Vol. AP-32, No. 7, July 1984, pp. 747–751.
66. A. K. Bhattacharyya and S. K. Tandon, "Radar Cross Section of a Finite Planar Structures Coated with a Lossy Dielectric," *IEEE Trans. on Antennas and Propagation*, Vol. AP-32, September 1984, pp. 1003–1007.
67. A. K. Bhattacharrya, "Radar Cross Section of a Flat Plate with Two Face Impedance, *Microwave and Optical Tech. Letters*, Vol. 3, No. 9, September 1990, pp. 324–327.
68. C. H. Chan and R. Mittra, "Comparative Study of Iterative Technique, Moments and Spectral Galerkin's Approach for Solving the Problem of EM Scattering by a Rectangular Plate," IEEE AP-S Conf., Blacksburg, VA, June 15–19, 1987.
69. A. Taflove and K. R. Umashankar, "The Finite Difference Time-Domain (FD-TD) Method for Numerical Modeling of Electromagnetic Scattering," *IEEE Trans. on Magnetics*, July 1989.
70. T. S. Chu and R. A. Semplak, "A Note on Painted Reflecting Surface," *IEEE Trans. on Antennas and Propagation*, Vol. AP-24, No. 1, January 1976, pp. 99–101.
71. R. L. Fante and M. T. McCormack, "Reflection Properties of the Salisbury Screen," *IEEE Trans. on Antennas and Propagation*, Vol. AP-36, No. 10, October 1988, pp. 1443–1454.
72. E. F. Knott, J. F. Shaeffer, and M. T. Tuley, *Radar Cross Section*, Artech House, Dedham, MA, 1985.
73. G. T. Ruck, ed., *Radar Cross Section Handbook*, Vol. 2, Plenum Press, New York, 1970.
74. E. Brookner, ed., *Radar Technology*, Artech House, Norwood, MA, 1988.

75. E. F. Knott, "The Thickness Criterion for Single-Layer Radar Absorbing Materials," *IEEE Trans. on Antennas and Propagation*, Vol. AP-27, September 1979, pp. 698–701.
76. K. Hatakeyama and T. Inui, "Electromagnetic Wave Absorber Using Ferrite Absorbing Materials Dispersed with Short Metal Fibers," *IEEE Trans. on Magnetics*, Vol. MAG-20, No. 5, September 1984, pp. 1261–1263.
77. O. M. Bucci and G. Franceschetti, "Scattering from Wedge-Tapered Absorbers," *IEEE Trans. on Antennas and Propagation*, Vol. AP-19, No. 1, January 1971, pp. 96–104.
78. G. T. Ruck, ed., *Radar Cross Section Handbook*, Vol. 1, Plenum Press, New York, 1970, Section 4.1.
79. R. Mittra, R. C. Hall, and G. H. Tsao, "Spectral-Domain Analysis of Circular Patch Frequency Selective Surfaces," *IEEE Trans. on Antennas and Propagation*, Vol. AP-32, No. 5, May 1984, pp. 533–536.
80. C. H. Tsao and R. Mittra, "Spectral Domain Analysis of Frequency Selective Surfaces Comprised of Periodic Array of Cross Dipoles and Jerusalem Crosses," *IEEE Trans. on Antennas and Propagation*, Vol. AP-32, No. 5, May 1984, pp. 478–486.
81. C. C. Chen, "Diffraction of Electromagnetic Waves by a Conducting Screen Perforated Periodically with Circular Holes," *IEEE Trans. on Microwave Theory Tech.*, Vol. MTT-19, May 1971, pp. 475–481.
82. W. K. Khan and H. Kurss, "Minimum Scattering Antennas," *IEEE Trans. on Antennas and Propagation*, Vol. AP-3, September 1965, pp. 671–675.
83. W. Wasylkiwskyj and W. K. Khan, "Theory of Mutual Coupling among Minimum Scattering Antennas," *IEEE Trans. on Antennas and Propagation*, Vol. AP-18, No. 2, March 1970, pp. 206–216.
84. R. C. Hansen, "Relationships between Antennas as Scatterers and as Radiators," *Proc. IEEE*, Vol. 77, No. 5, May 1989, pp. 659–662.
85. R. C. Hansen, "Antenna Mode and Structural Mode RCS: Dipole," *Microwave and Optical Technology Letters*, Vol. 3, No. 1, January 1990, pp. 6–10.
86. D. F. Yaw, "Antenna Radar Cross Section," *Microwave J.*, Vol. 27, No. 9, September 1984.
87. J. Delmonte, *Technology of Carbon and Graphite Fiber Composites*, Van Nortrand, New York, 1988.
88. D. L. Jaggard and N. Engheta, "CHIROSORB^{TM} as an Invisible Medium" *Electronics Letters*, Vol. 25, No. 3, 2 February 1989, pp. 172–173.
89. E. J. Post, *Formal Structure of Electromagnetics*, North-Holland, Amsterdam, 1962.
90. J. A. Kong, *Theory of Electromagnetic Waves*, John Wiley Interscience, New York, 1975.
91. D. L. Jaggard, A. R. Mickelson, and C. H. Papas, "On Electromagnetic Waves in Chiral Media," *Applied Physics*, 1979, Vol. 18, pp. 211–216.
92. S. Bassiri, C. H. Papas, and N. Engheta, "Electromagnetic Wave Propagation through a Dielectric Chiral Interface and through a Chiral Slab," *J. Optics Society of America*, Vol. A, 1988, No. 5.
93. P. Beddard, "Stealth Cars," *Car and Driver*, Vol. 26, No. 5, December 1980, p. 140.
94. B. Sweetman, "Stealth Aircraft," *Motorbook International*, 1986.
95. J. A. Adam, "How to Design an 'Invisible' Aircraft," *IEEE Antennas and Propagation Society Newsletter*, April 1988, pp. 26–31.
96. A. E. Fuhs, *The No-See-Um Book RCS Lectures*, American Institute of Aeromantics and Astronautics, New York, 1988.

Appendix A
Evaluation of Some Functions Frequently Occurring in EM Scattering Studies

This appendix discusses three functions that frequently occur in radar cross section and antenna analyses: the Fresnel integral, Maliuzhinets's function, and the wedge transition function.

A.1 FRESNEL INTEGRAL

The well-known Fresnel integral has different forms. One popular form (see Section 2.5) is given by

$$F(X) = \mathrm{e}^{-\mathrm{i}\pi/4}\Big/\sqrt{\pi}\int_{x}^{\infty}\mathrm{e}^{\mathrm{i}t^2}\,\mathrm{d}t, \quad \text{for real } x \tag{A.1}$$

For a large argument x, $F(x)$ has an asymptotic expansion (eq. (2.17)):

$$F(x) = H(-x) + \hat{F}(x) + 0(x^{-3}), \quad |x| \to \infty \tag{A.2}$$

where $\hat{F}(x)$, the dominant term, is given by

$$\hat{F}(x) = \frac{1}{2x\sqrt{\pi}}\,\mathrm{e}^{\mathrm{i}(x^2+\pi/4)} \tag{A.3}$$

The modified Fresnel integral used with the UTD diffraction coefficient is given by (see (2.21))

$$F(x) = \pi^{-1/2}\mathrm{e}^{-\mathrm{i}\pi/4}\,\mathrm{e}^{-\mathrm{i}x^2}\int_{-\infty}^{x}\mathrm{e}^{\mathrm{i}t^2}\,\mathrm{d}t \tag{A.4}$$

The asymptotic expansion for a large x is

$$F(x) \sim e^{-ix^2} H(x) + \hat{F}(x), \quad x \to \pm\infty \tag{A.5}$$

as discussed in Subsection 2.5.1. A concise approximation to the Fresnel integral is given in [A.1]. The integral defined for a real argument is

$$F_{\pm}(x) = \int_x^{\infty} e^{\pm it^2}\, dt \tag{A.6}$$

For the argument $x > 3.0$, a useful approximation is

$$F_{\pm}(x) \sim \frac{1}{2x} e^{\pm i(x^2 + \pi/2)} \tag{A.7}$$

The modified Fresnel integral is defined in [A.2, A.3]. For all positive real values of x,

$$K_{\pm}(x) = \frac{1}{\sqrt{\pi}} F_{\pm}(x)\, e^{\mp i(x^2 + \pi/4)} \tag{A.8}$$

It has the properties

$$K_{\pm}(0) = 0.5$$

$$K_{\pm}(x) \sim \frac{1}{2x\sqrt{\pi}} e^{\pm i\pi/4}, \quad \text{for } x \to \infty \tag{A.9}$$

$$K_{\pm}(-x) = e^{\pm ix^2} - K_{\pm}(x)$$

An approximation for $K_{\pm}(x)$ is

$$K \pm (x) \approx \frac{e^{\pm i \arctan(x^2 + 1.5x + 1) - \pi/4}}{2\sqrt{\pi x^2 + x + 1}}, \quad x \geq 0 \tag{A.10}$$

The approximation to the Fresnel integral $F_{\pm}(x)$ can then be obtained by using (A.8). If the argument of arctan in (A.10) is greater than or equal to 1.0, the following approximation can be used:

$$\arctan(t) \approx \pi/2 - t/(0.28 + t^2), \quad |t| \geq 1.0. \tag{A.11}$$

A computer program for the Fresnel integral, in the form of (A.1), is given as Program C1 in Appendix C.

A.2 MALIUZHINETS'S FUNCTION

As discussed in Subsection 4.5.3.5, Maliuzhinets's function [A.4–A.7] occurs frequently in scattering problems associated with a half plane having two-faced impedances. Maliuzhinets's function was defined earlier by (4.61):

$$M(u) = \exp\left\{\left(\frac{-1}{8\pi}\right)\int_0^u \left[\pi \sin(t - 2\sqrt{2\pi})\sin\left(\frac{t}{2} + 2t\right)/\cos t\right] \mathrm{d}t\right\} \quad (4.61)$$

which can be rewritten as

$$M(u) = \exp\left[\left(\frac{-1}{8\pi}\right)\left(I_1 - 2\sqrt{2}\,\pi I_2 + 2I_3\right)\right] \quad (A.12)$$

with

$$I_1 = \int_0^u \pi \tan t \,\mathrm{d}t \quad (A.13)$$

$$I_2 = \int_0^u \frac{\sin(t/2)}{\cos t}\,\mathrm{d}t \quad (A.14)$$

$$I_3 = \int_0^u t \sec t \,\mathrm{d}t \quad (A.15)$$

After integrating by parts and using appropriate substitutions, these three integrals can be expressed by

$$I_1 = \pi \ln(\sec u) \quad (A.16)$$

$$I_2 = \frac{1}{2}\ln\frac{0.1715(2\cos u/2 + 1)}{2\cos u/2 - 1} \quad (A.17)$$

$$I_3 = \sum_{k=0}^{\infty} \frac{|E_{2k}|\,u^{2k+1}}{(2k+2)(2k)} \quad (A.18)$$

for $|u| < \pi/2$ and where the values of E_{2k} are Euler's numbers.

The first few terms in (A.18) are

$$I_3 = \frac{u^2}{2} + \frac{u^4}{8} + \frac{5u^6}{144} + \frac{61u^8}{5760} + \cdots \quad (A.19)$$

from which following small argument approximation is obtained:

$$I_3 \approx \frac{u^2}{2}, \quad u \to 0 \tag{A.20}$$

Now using (A.16), (A.17), and (A.20) in (A.12), the following approximation to Maliuzhinets's function is obtained:

$$M(u) \approx \exp\left[-0.125\ln(\sec u) + 0.125\sqrt{2}\,\ln\frac{0.1715(2\cos u/2 + 1)}{2\cos u/2 - 1} - \frac{u^2}{8\pi}\right],$$

for small u

A possible series expansion for Maliuzhinets's function is given by [A.5]

$$M(u) = 1.0 - 1.390038812\,E\text{-}02\,u^2 - 1.15772122\,E\text{-}05\,u^4$$
$$- 2.1129235\,E\text{-}07\,u^6 - 1.660554\,E\text{-}09\,u^8 \tag{A.21}$$

The Taylor series approximation around $u = 0$ can be expressed as

$$M(u) = 1 - 0.0139\,u^2 + 0(u^4)$$

A small argument approximation is

$$M(u) = 1 - 0.01390\,u^2 \tag{A.22}$$

If $u \gg 0$ the function has an alternative expression [A.3]:

$$M(u) = b[\cos(u - i\gamma)/4]^{1/2}\,e^{iu/2}e^{iu}$$

where

$$\gamma = \ln 2 = 0.69315 \quad \text{and} \quad b = \left[2/(\sqrt{2} + 1)^{1/2} e^{K/2\pi}\right]$$

and Catalan's constant $K = 0.91596$. Consequently,

$$M(u) = 1.05302\left[\cos\frac{(u - i\gamma)}{4}\right]^{1/2} e^{iu/2}e^{iu} \tag{A.23}$$

Equations (A.22) and (A.23) are found to give [A.7] an accurate approximation of the function.

For further details, and plots of magnitude and phase of the function, see [A.7]. It turns out [A.7] that (A.22) and (A.24) give values very close to each other for values of u up to 8.

A.3 Wedge Transition Function

The wedge transition function is given by (2.21):

$$F(x) = 2\mathrm{i}\sqrt{x}\,\mathrm{e}^{\mathrm{i}x}\int_{\sqrt{x}}^{\infty} \mathrm{e}^{-\mathrm{i}\tau^2}\,\mathrm{d}\tau$$

For small values of x,

$$F(x) \approx \left(\sqrt{\pi x} - 2x\,\mathrm{e}^{\mathrm{j}\pi/4} - 2/3x^2\,\mathrm{e}^{-\mathrm{i}\pi/4}\right)\mathrm{e}^{\mathrm{i}(\pi/4+x)} \tag{A.24}$$

For large values of x,

$$F(x) \sim \left(1 + \mathrm{i}\frac{1}{2x} - \frac{3}{4}\frac{1}{x^2}\right) \tag{A.25}$$

For intermediate values of x, $0.3 \leq x \leq 5.5$, and a linear interpolation scheme is

$$F(x) = F(x_i) + \left[F(x_{i+1}) - F(x_i)\right]\frac{x - x_i}{x_{i+1} - x_i} \tag{A.26}$$

When $x < 0$,

$$F(x) = F^*(|x|)$$

where the asterisk indicates the complex conjugate.

REFERENCES

A.1 G. L. James, "An Approximation to the Fresnel Integral," *Proc. IEEE*, Vol. 67, No. 4, April 1979, pp. 677–678.

A.2 N. A. Logan, "General Research in Diffraction Theory," Rep. LMS 288087, Lockheed Aircraft Corporation, 1959.

A.3 M. T. Abuelma'atti, "An Approximation to the Fresnel Integral," *IEEE Trans. on Antennas and Propagation*, Vol. AP-37, No. 7, July 1989, pp. 946–947.

A.4 O. M. Bucci and G. Franceschetti, "Electronmagnetic Scattering by a Half Plane with Two Face Impedance," *Radio Science*, Vol. 11, No. 1, January 1976, pp. 49–59.

A.5 O. M. Bucci, "On a Function Occuring in the Theory of Scattering From an Impedance Half-Plane," Rep. 75-1, Istituto Universitario Navale, Naples, Italy.

A.6 A. K. Bhattacharyya and S. K. Tandon, "Radar Cross Section of a Planar Structure Coated with a Lossy Dielectric," *IEEE Trans. on Antennas and Propagation*, Vol. AP-32, No. 9, September 1984, pp. 1003–1007.

A.7 J. L. Volakis and T. B. A. Senior, "Simple Expressions for a Function Occurring in Diffraction Theory," *IEEE Trans. on Antennas and Propagation*, Vol. AP-33, No. 6, June 1985, pp. 678–680.

Appendix B
Characteristics of Radar Absorbing Materials

Table B.1 lists some typical low-density, high-loss flexible foam sheet materials specially designed to absorb electromagnetic energy. The complex dielectric constant is changed by adjusting the carbon loading on the foam. K' is the dielectric constant and K'' is the product of the dielectric constant and the loss tangent. IL is the insertion loss.

Table B.2 gives the reflectivity performance of a typical flexible foam absorber.

Table B.1
Properties of ECCOSORB LS (reprinted with permission from Emerson & Cuming)

	Frequency, Hz									
	10^3	10^4	10^5	10^6	10^7	10^8	10^9	3×10^9	10^{10}	10^{11}
LS-14	11	11	11	11	11	10	2.2	1.4	1.3	1.3
K'	1900	190	19	1.9	0.19	0.029	0.091	0.17	0.11	0.01243
dB/cm	0	0	0	0	0	0	0.06	0.39	0.88	0.96
IL = dB/in.	0	0	0	0	0.01	0.25	0.52	1.1	2.2	2.5
LS-16										
K'	15	15	15	15	15	14	3.0	1.5	1.3	1.3
K''	11000	1100	110	11	1.1	0.18	0.59	0.69	0.29	0.030
dB/cm	0	0	0	0	0	0	0.31	1.5	2.3	2.4
IL = dB/in.	0.03	0.03	0.03	0.03	0.03	0.52	1.8	3.8	2.2	6.1
LS-18										
K'	19	19	19	19	19	18	4.0	1.7	1.4	1.4
K''	11000	1100	110	11	1.1	0.18	0.58	0.78	0.36	0.038
dB/cm	0	0	0	0	0	0	0.26	1.6	2.7	2.9
IL = dB/in.	0.03	0.03	0.03	0.03	0.04	0.83	2.3	4.1	7.0	7.5
LS-20										
K'	24	24	24	24	24	23	603	2.1	1.5	1.4
K''	11000	1100	110	11	1.1	0.31	1.6	1.7	0.69	0.071
dB/cm	0	0	0	0	0	0.01	0.58	3.0	5.0	5.5
IL = dB/in.	0.03	0.03	0.03	0.03	0.04	1.3	4.4	7.8	13	14
LS-22										
K'	28	28	28	28	28	27	6.5	2.2	1.6	1.5
K''	1000	1000	100	10	1.0	0.46	2.6	2.4	0.90	0.092
dB/cm	0	0	0	0	0	0.01	0.91	4.0	6.3	6.8
IL = dB/in.	0.02	0.02	0.02	0.02	0.05	1.7	5.3	10	16	17
LS-24										
K'	29	29	29	29	29	29	20	6.7	1.9	1.3
K''	7100	710	71	7.1	1.0	3.0	19	14	4.9	0.50
dB/cm	0	0	0	0	0	0.05	3.5	11	24	39
IL = dB/in.	0.02	0.02	0.02	0.02	0.05	2.3	14	32	61	100
LS-26										
K'	34	34	34	34	34	34	28	12	3.4	2.2
K''	6800	680	68	6.8	0.68	4.8	32	27	9.8	1.0
dB/cm	0		0	0	0	0.07	4.9	16	34	60
IL = dB/in.	0.02	0.02	0.02	0.02	0.05	3.0	19	46	89	150
LS-28										
K'	29	29	29	29	29	29	24	11	2.5	1.2
K''	4100	410	41	4.2	1.1	7.2	42	29	9.9	1.0
dB/cm	0	0	0	0	0	0.12	6.4	17	36	77
IL = dB/in.	0.01	0.01	0.01	0.01	0.05	2.9	23	49	93	200
LS-30										
K'	25	25	25	25	25	25	22	11	2.4	1.1
K''	3100	310	31	3.5	4.0	36	110	48	15	1.5
dB/cm	0	0	0	0	0	0.56	12	24	46	110
IL = dB/in.	0.01	0.01	0.01	0.01	0.11	6.2	41	67	120	280

Note: This information is not to be taken as a warranty or representation for which Emerson & Cuming assume legal responsibility, nor as permission or recommendation to practice any patented invention without license. It is offered solely for consideration, investigation, and verification.

Table B.2
Reflectivity Performance of ECCOSORB AN = P × P = 74: Estimated dB down from 100% Reflection Metal-Backed (reprinted with permission from Emerson & Cuming)

Polarization	Incidence Angle	50°		60°		70°		80°	
Frequency (GHz)	0°	⊥	‖	⊥	‖	⊥	‖	⊥	‖
5	17	9	7	7	4	5	2	2	1
6	22	11	10	8	6	5	2	3	1
7	25 +	13	13	10	7	6	3	3	1
8	25 +	15	16	11	9	7	4	4	1
9	25 +	17	19	12	11	8	5	4	1
10	25 +	19	21	14	13	9	6	5	1
11	25 +	21	22	15	14	10	7	5	2
12	25 +	23	21	17	16	11	8	6	2
13	25 +	25 +	21	18	16	12	9	6	2
14	25 +	25 +	21	19	16	13	10	7	3
15	25 +	25 +	21	20	15	15	11	8	3
16	25 +	25 +	22	22	15	15	11	9	3
17	25 +	25 +	24	23	15	16	11	10	4
18	25 +	25 +	25 +	24	16	17	11	11	5
19	25 +	25 +	25 +	24	17	18	10	12	5
20	25 +	25 +	25 +	25 +	18	18	10	13	6
21	25 +	24	25 +	25 +	19	18	10	14	7
22	25 +	24	25 +	24	20	19	10	15	8
23	25 +	24	25 +	24	20	19	10	17	9
24	25 +	25 +	25 +	23	20	19	11	18	9
25	25 +	25 +	25 +	22	19	19	11	19	8
26	25 +	25 +	24	22	19	19	12	20	8
27	25 +	25 +	24	22	19	19	13	21	7
28	25 +	25 +	24	22	19	19	14	21	6
29	25 +	25 +	24	22	19	19	15	20	6
30	25 +	25 +	24	22	20	19	16	19	6
31	25 +	25 +	24	22	21	19	15	18	5
32	25 +	25 +	24	23	21	18	14	17	5
33	25 +	25 +	25 +	23	21	18	14	17	5
34	25 +	25 +	25 +	24	20	18	13	6	5
35	25 +	25 +	25 +	25 +	19	18	12	15	5
36	25 +	25 +	25 +	25 +	18	18	12	15	5
37	21	25 +	24	25 +	18	18	12	14	5
38	16	25 +	23	25 +	18	18	12	14	6
39	12	25 +	23	25 +	18	18	12	13	6
40	8	25 +	23	25 +	19	18	13	13	7

Note: 25 + means that the reflectivity is down more than 25 dB.

Appendix C
Computer Programs in Fortran

C.1 FRESNEL INTEGRAL WITH REAL ARGUMENT:FRES

```
C     **  **  **  **  **  **  **  **  **  **  **  **  **
      SUBROUTINE FRNELS(X,C,S)
C     THIS IS THE FRESNEL INTEGRAL SUBROUTINE:INTEGRAL IS FROM
C     0 TO X; THE INTEGRAND IS EXP(I*T**2); THE INT. IS TO BE
C     MULTIPLIED BY EXP(-I*PI/4.)/SQRT(PI) AND OUTPUT IS C(X)-J*S(X)
      DIMENSION A(12),B(12),CC(12),D(12)
      DATA A/1.595769140,-0.000001702,-6.808568854,-0.000576361,
     1 6.920691902,-0.016898657,-3.050485660,-0.0752419,0.850663781,
     2 -0.025639041,-0.150230960,0.034404779/
      DATA B/-0.000000033,4.255387524,-0.00009281,-7.780020400
     1 -0.009520895,5.075161298,-0.138341947,-1.363729124,
       -0.403349276,
     2 0.702222016,-0.216195929,0.0195547031/
      DATA CC/0.,-0.024933975,0.000003936,0.005770956,0.000689892,
     1 -0.009497136,0.011948809,-0.006748873,0.000246420,
     2 0.002102967,-0.001217930,0.000233939/
      DATA D/0.1999471140,0.000000023,-0.009351341,0.000023006,
     1 0.004851466,0.001903218,-0.17122914,0.029064067,-0.027928955,
     2 0.016497308,-0.005598515,0.000838386/
      PI = 3.145926536
      X = X*SQRT(2./PI)
      IF (X.LE.0.0) GO TO 100
      X = PI*X*X/2.0
      FR = 0.0
      FI = 0.0
      K = 13
C     IS X < 4 ?
      IF (X-4.0) 10,40,40
   10 Y = X/4.0
```

```
C     EVALUATE FRESNEL INTEGRAL USING X < 4 APPROX.
 20   K = K - 1
      FR = (FR + A(K))*Y
      FI = (FI + B(K))*Y
      IF (K - 2) 30,30,20
 30   FR = FR + A(1)
      FI = FI + B(1)
      C = (FR*COS(X) + FI*SIN(X))*SQRT(Y)
      S = (FR*SIN(X) - FI*COS(X))*SQRT(Y)
      RETURN
C     EVALUATE INT. USING X > 4 APPROX.
 40   Y = 4./X
 50   K = K - 1
      FR = (FR + CC(K))*Y
      FI = (FI + D(K))*Y
      IF (K - 2) 60,60,50
 60   FR = FR + CC(1)
      FI = FI + D(1)
      C = 0.5 + (FR*COS(X) + FI*SIN(X))*SQRT(Y)
      S = 0.5 + (FR*SIN(X) - FI*COS(X))*SQRT(Y)
      RETURN
 100  C = -0.0
      S = -0.0
      RETURN
      END
```

C.2 COMPUTER PROGRAM WD

```
      SUBROUTINE WD (DCS, DCH, R, PHR, PHPR, SBO, FN)
C     ...
C     THIS SUBROUTINE COMPUTES THE FIRST
C     ORDER SOFT AND HARD WEDGE DIFFRACTION COEFF
C     REFER TO EQUATION (2.20) FOR THE
C     EXPRESSION FOR THE COEFFICIENT
C     THIS PROGRAM WAS DEVELOPED AND
C     ORIGINALLY REPORTED IN REF (C.1)
C     SIGNIFICANCES OF SYMBOLS USED
C     INPUT QUANTITIES
C     R—DISTANCE PARAMETER
C     PHR—ANGLE OF OBSERVATION PHI IN
```

```
C     DEGREES
C     PHPR—ANGLE OF INCIDENCE PHI PRIMED IN DEGREES
C     SBO—OBLIQUE INCIDENT ANGLE (BETA)
C     IN DEGREES
C     FN—THE N RELATED TO INTERIOR WEDGE
C     ANGLE BY EQN. (2.XX)
C     ....
C     OUTPUT QUANTITY
C     DCS AND DCH—SOFT & HARD DIFFRACTION COEFFS
C!!!
      COMPLEX D(4),CT,TERN,FT(4),FFCT
      DATA CT,UTPI/(-0.05626977,0.05626977),0.15915494/
      DATA PI,TPI,SML/3.14159265,6.28318531,0.001/
      IF (FN - 0.5.LT.SML) GO TO 10
      DKL = TPI*R
      UFN = 1./FN
      DETAR = PHR - PHPR
      TERM = CT/(FN*SBO)
      SGN = 1.
      I = O
2     I = I + 1
      DN = UFN*(0.5*SGH + UTPI*BETAR)
      N = DN + SIGN(0.5,DN)
      ANG = TPI*FN*N - BETAR
      A = 2.*(COS(0.5*ANG)**2)
      X = DKL*A
      Y = PI + SGN*BETAR
      IF (ABS(X).LT.1.E - 10) GO TO 3
      FT(I) = FFCT(C)/TAN(0.5*Y*UFN)
      GO TO 4
3     FT(I) = (0.,0.)
      IF (ABS(COS(0.5*Y*UFN)).LT.1.E - 3) GO TO 4
      FT(I) = (1.7725,1.7725)*SIGN(SQRT(DKL),Y)
      FT(I) = (FT(I) - (0.,2.)*DKL*(PI - SGN*ANG))*FN
4     SGN = SGN
      D(I) = TERM*FT(I)
      IF (SGN.LT.0) GO TO 2
      DETAR = PHR + PHPR
      IF (I.LE.3) GO TO 2
      DCS = (D(1) + D(2)) - (D(3) + D(4))
      DCH = (D(1) + D(2)) + (D(3) + D(4))
      RETURN
```

```
10    DO 5 I = 1,4
5     D(I) = (0.,0.)
C10   DI = (0.,0.)
C     DR = (0.,0.)
      RETURN
      END
```

C.3 COMPUTER PROGRAM WDP

```
      SUBROUTINE WDP (DCSS, DCHS, R, PHR, PHPR, FN, SBO)
C**   THIS SUBROUTINE COMPUTES THE COMPLEX
C**   SOFT AND HARD SLOPE DIFFRACTION
C     COEFFICIENTS
C     THIS PROGRAM WAS ORIGINALLY DEVELOPED
C     AND REPORTED IN REF. (C.2)
C     INPUT QUANTITIES
C     R—DISTANCE PARAMETER
C     PHR—ANGLE OF OBSERVATION PHI IN
C     DEGREES
C     PHPR—ANGLE OF INCIDENCE PHI PRIMED
C     IN DEGREES
C     SBO—OBLIQUE INCIDENT ANGLE (BETA)
C     IN DEGREES
C     SBO—OBLIGUE INCIDENT ANGLE (BETA)
C     IN DEGREES
C     FN—'N' THAT IS RELATED TO INTERIOR
C     WEDGE ANGLE BY EQN. (2.XX)
C     ...
C     OUTPUT QUANTITIES
C     DCSS & DCHS—ARE THE COMPLEX
C     SOFT AND HARD SLOPE DIFFRACTION
C     COEFFICIENTS
      COMPLEX D(4),CT,TERM,FT(4),FFCT,FSFCT
      DATA CT,UTPI/(-0.05626977,0.05626977),0.15915494/
      DATA PI,TPI,SML 3.14159265,6.28318531,0.001/
      IF (FN-0.5.LT.SML) GO TO 10
      DKL = TPI*R
      UFN = 1./FN
      BETAR = PHR - PHPR
      TERM = CT/(2.*FN*FN*SBO)
      SGN = 1.
      I = O
```

```
2     I = I+1
      DN = UFN*(0.5*SGN + UTPI*BETAR)
      N = DN + SIGN(0.5,DN)
      ANG = TPI*FN*N - BETAR
      A = 2,*(COS(0.5*ANG)**2)
      X = DKL*A
      Y = PI + SGN*BETAR
      IF (ABS(X).LT.1.E-10) GO TO 3
      FSFCT = (0.,2.)*X*(1. - FFCT(X))
      IF (ABS(X).GT.1000.)FSFCT = (1.,0.)
      FT(I) = FSFCT/(SIN(0.5*Y*UFN)**2)
      GO TO 4
3     FT(I) = (0.,0.)
      IF (ABS(COS(0.5*Y*UFN).LT.1.E-3) GO TO 4
      FT(I) = 0.,4.)*DKL*FN*FN
      FT(I) = FT(I)*(1. - ((0.88625,0.88625)*SQRT(DKL)*(PI - SGN*ANG)))
4     SGN = - SGN
      D(I) = TERM*FT(I)
      IF (SGN.LT.0.0) GO TO 2
      BETAR = PHR + PHPR
      IF (I.LE.3) GO TO 2
      DCSS = (D(1) - D(2)) + (D(3) - D(4))
      DCHS = (D(1) - D(2)) - (D(3) - D(4))
      RETURN
10    DO 5 I = 1,4
5     D(I) - (0.,0.)
C10   BI = (0.,0.0)
C     DR = (0.,0.)
      RETURN
      END
```

C.4 TRANSITION FUNCTION:FFCT

```
      COMPLEX FUNCTION FFCT(X)
C     WEDGE TRANSITION FUNCTION
C     THE EQUATIONS USED ARE EQNS.
C     (A.24)—(A.26)
C     THIS PROGRAM WAS ORIGINALLY REPORTED
C     AND DEVELOPED IN REF. (C.2)
      COMPLEX FXX(8),FX(8),CJ
```

```
      DIMENSION XX(8)
      DATA XX/.3,.5,.7,1.5,2.3,4.,5.5/
      DATA CJ/(0.,1.)/
      DATA FX(0.5729,0.2677),(0.6768,0.2682),(0.7439,0.2549),
      1(0.8095,0.2322),(0.873,0.1982),(0.9240,1577),(0.9658,0.1073),
      2(0.9797,0.0828)/
      DATA FXX/(0.,0.),(0.5195,0.0025),(0.3355,-0.0665),
      1(0.2187,-0.757),(0.127,-0.068),(0.0638,-0.0506),
      2(0.0246,-0.0296),(0.0093,-0.0163)/
      X = ABS(XF)
      IF (X.GT.5.5) GO TO 1
      IF (X.GT.0.3) GO TO 10
C!!!  SMALL ARGUMENT FORM
      FFCT = ((1.253,1.253)*SQRT(X)-(0.,2.)*X-0.6667*X*X)*CEXP(CJ*X)
      GO TO 20
C!!!  LINEAR INTERPOLATION REGION
10    DO 11 N=2,7
11    IF (X.LT.XX(N)) GO TO 12
12    FFCT = FXX(N)*(X-XX(N))+FX(N)
      GO TO 20
C!!!  LARGE ARGUMENT FORM
1     FFCT = 1.+COMPLX(-0.75/X,0.5)/X
20    IF (XF.GE.0.) RETURN
      FFCT = CONJG(FFCT)
      RETURN
      END
```

C.5 MALIUZHINETS'S FUNCTION WITH COMPLEX ARGUMENT: FMAL

```
C     FUNCTION SUBPROGRAM FMAL
C     Z IS THE COMPLEX ARGUMENT AND FMAL IS THE MALIUZHINETS'
C     FUNCTION
C     EQN. (A.21) IS USED IN THIS PROGRAM
      COMPLEX FUNCTION FMAL(Z)
      COMPLEX Z
      FMAL = 1.-(1.390038812E-02)*Z**2-(1.15772122E-05)*Z**4-
      1(2.1129235E-07)*Z**6-(1.660554E-09)*Z**8
C     WRITE (8,17)Z,FMAL
C     WRITE (6,17)Z,FMAL
      RETURN
   17 FORMAT (2X,'VALUES OF ARG & FMAL = '/,(4(2X,E13.6)))
      END
```

C.6 RCS OF A FINITE CYLINDER, CONE FRUSTUM, AND CONE—CYLINDER FRUSTUM: RADAR.FOR

```
C-----------------------------------------------------------------------
C>>>>PROGRAMME RADAR—TO CALCULATE THE RADAR              <<<<
C>>>>CROSS SECTION (BOTH PHASE AND MAGNITUDE)            <<<<
C>>>>OF A CYLINDER, A CONE-FRUSTRUM AND A CONE-          <<<<
C>>>>CYLINDER FRUSTRUM PHASE REFERENCE POINT             <<<<
C>>>>SITUATED AT THE CENTRE OF THE                       <<<<
C>>>>RESPECTIVE TARGETS                                  <<<<
C    THIS PROGRAM WAS DEVELOPED AT LAVAL UNIVERSITY,
C    QUEBEC, CANADA, BY V. SAMPATH AND G. Y. DELISLE
C-----------------------------------------------------------------------
C
C  INITIALIZATION OF VECTORS TO STORE THE MAGNITUDE AND THE
   PHASE
C  PER—PERPENDICULAR POLARIZATION
C  PAL—PARALLEL POLARIZATION
C
      DIMENSION SERPER(900),SERPAL(900),PHAPER(900),PHAPAL(900)
      DIMENSION TMPPER(900),TMPPAL(900)
      NMB = 900
C
C  PRINT ON THE SCREEN
C
      PRINT *,'PRINCIPAL PROGRAMME—CALCULATE THE AMPLITUDE'
      PRINT *,'AND THE PHASE OF THE FOLLOWING TARGETS'
      PRINT *,''
      PRINT *,'1-CYLINDER'
      PRINT *,'2-CONE FRUSTRUM'
      PRINT *,'3-CONE-CYLINDER FRUSTRUM'
      PRINT *,''
      PRINT *,'NO. OF THE TARGET CHOSEN ?'
      READ *,NCIB
      PRINT *,'FREQUENCY IN GHZ ?'
      READ *,FRQ
      IF (NCIB.EQ.1) THEN
        PRINT *,'RADIUS AND LENGTH OF THE CYLINDER (IN LAMBDA) ?'
        READ *,R1,CL
        PRINT *,' '
       PRINT *,'PERPENDICULAR POLARIZATION'
       PRINT *,'----------------------------------------------'
C
C  THE OP-GTD JUNCTION REFERS TO THE ANGLE ON ONE SIDE OF
C  WHICH CALCULATION IS CARRIED OUT ACCORDING TO THE PHYSI-
```

```
C  CAL OPTICS THEORY WHILE ON THE OTHER, ACCORDING TO THE
C  GEOMETRICAL THEORY OF DIFFRACTION
C
         PRINT *,'THE OP-GTD JUNCTION (IN DEGREES) ?'
         PRINT *,'IF NEGATIVE, FOUND AUTOMATICALLY'
         READ *,TJ
C
C  CALL THE SUBROUTINE TO PERFORM THE CALCULATION FOR PER-
C  PENDICULAR POLARIZATION
C
C        CALL SERCYL (SERPER,PHAPER,VMN,VMX,-1.,R1,CL,TJ)
C        PRINT *,' '
      PRINT *,'PARALLEL POLARIZATION'
         PRINT *,'----------------------------------'
         PRINT *,'THE OP-GTD JUNCTION (IN DEGREES) ?'
         READ *,TJ
C
C  CALL THE SUBROUTINE TO PERFORM THE CALCULATION FOR PARAL-
   LEL POLARIZATION
C
         CALL SERCYL (SERPAL,PHAPAL,WMN,WMX,+1.,R1,CL,TJ)
      ELSE IF (NCIB.EQ.2) THEN
         PRINT *,'RADII (R1 & R2) AND LENGTH (CL) OF CONE FRUSTRUM ?'
         READ *,R1,R2,CL
         PRINT *,' '
      PRINT *,'PERPENDICULAR POLARIZATION'
      PRINT *,'----------------------------------------------'
C
C  THERE ARE TWO JUNCTIONS IN THIS CASE: ONE WITH RESPECT TO
C  THE FRONT SURFACE AND ANOTHER WITH RESPECT TO THE REAR
C  SURFACE. THUS PHYSICAL OPTICS IS USED TWICE AS IS THE GEOMET-
C  RICAL THEORY OF DIFFRACTION
C
         PRINT *,'THE PO-GTD JUNCTIONS TJ1 & TJ2 (IN DEGREES) ?'
         PRINT *,'TJ1 = JCT OF FRONT LOBE, TJ2 = JCT OF REAR LOBE'
         PRINT *,'IF NEGATIVE, THE POINTS ARE FOUND AUTOMATICALLY'
         READ *,TJ1,TJ2
C
C  CALL THE SUBROUTINE TO PERFORM THE CALCULATION FOR PER-
C  PENDICULAR POLARIZATION
```

```
C
      CALL SERTRCO (SERPER,PHAPER,VMN,VMX,-1.,R1,R2,CL,TJ1,TJ2)
      PRINT *,' '
    PRINT *,'PARALLEL POLARIZATION'
    PRINT *,'--------------------------------------'
      PRINT *,'THE PO-GTD JUNCTIONS TJ1 & TJ2 (IN DEGREES) ?'
      PRINT *,'TJ1 = JCT OF FRONT LOBE, TJ2 = JCT OF REAR LOBE'
      PRINT *,'IF NEGATIVE, THE POINTS ARE FOUND AUTOMATI-
      CALLY'
      READ *,TJ1,TJ2
C
C  CALL THE SUBROUTINE TO PERFORM THE CALCULATION FOR PARAL-
C  LEL POLARIZATION
C
      CALL SERTRCO (SERPAL,PHAPAL,WMN,WMX,+1.,R1,R2,
      CL,TJ1,TJ2)
    ELSE
      PRINT *,'RADII (R1 & R2) & LENGTHS (H ET CL) (IN LAMBDA) ?'
      READ *,R1,R2,H,CL
      PRINT *,' '
    PRINT *,'PERPENDICULAR POLARIZATION'
    PRINT *,'--------------------------------------------'
      PRINT *,'THE PO-GTD JUNCTIONS TJ1 & TJ2 (IN DEGREES) ?'
      PRINT *,'TJ1 = JCT OF FRONT LOBE, TJ2 = JCT OF REAR LOBE'
      PRINT *,'IF NEGATIVE, THE POINTS ARE FOUND AUTOMATI-
      CALLY'
      READ *,TJ1,TJ2
C
C  CALL THE SUBROUTINE TO PERFORM THE CALCULATION FOR PER-
C  PENDICULAR POLARIZATION
C
      CALL SERTRCY (SERPER,PHAPER,VMN,VMX,-1.,R1,R2,H,CL,
      TJ1,TJ2)
      PRINT *,' '
    PRINT *,'PARALLEL POLARIZATION'
    PRINT *,'------------------------------------'
      PRINT *,'THE PO-GTD JUNCTIONS TJ1 & TJ2 (IN DEGREES) ?'
      PRINT *,'TJ1 = JCT OF FRONT LOBE, TJ2 = JCT OF REAR LOBE'
      PRINT *,'IF NEGATIVE, THE POINTS ARE FOUND AUTOMATICALLY'
      READ *,TJ1,TJ2
```

```
C
C  CALL THE SUBROUTINE TO PERFORM THE CALCULATION FOR PARAL-
C  LEL POLARIZATION
C
         CALL SERTRCY (SERPAL,PHAPAL,WMN,WMX,+1.,R1,R2,H,
         CL,TJ1,TJ2)
      END IF
C
C  CALL THE SUBROUTINE TO UNWRAP THE PHASE
C
      CALL UNW (SERPER,PHAPER,-1.,R1,R2,H,CL,FRQ,
                TMPPER,NMB,PMN,PMX)
      CALL UNW (SERPAL,PHAPAL,+1.,R1,R2,H,CL,FRQ,
                TMPPAL,NMB,QMN,QMX)
C
C  CALL THE SUBROUTINE TO PRINT THE VALUES OF THE AMPLITUDE
C  AND THE PHASE IN A FILE
C
      PRINT *,'DO YOU WANT THE VALUES OF THE AMPLITUDE & THE
           PHASE ?'
      PRINT *,'1=YES'
      READ *,IAG
      IF (IAG.EQ.1) THEN
C
C  PERPENDICULAR POLARIZATION
C
        DO J=1,900,4
          WRITE (13,20) (I,SERPER(I),I=J,J+3)
          WRITE (14,20) (I,PHAPER(I),I=J,J+3)
20        FORMAT (' ',4(I4,2X,F13.6,2X))
        END DO
C
C  PARALLEL POLARIZATION
C
        DO J=1,900,4
          WRITE (15,30) (I,SERPAL(I),I=J,J+3)
          WRITE (16,30) (I,PHAPAL(I),I=J,J+3)
30        FORMAT (' ',4(I4,2X,F13.6,2X))
        END DO
      END IF
```

```
C
C  CALL THE SUBROUTINE TO TRACE THE GRAPHS ON THE SCREEN
C
      PRINT *,'DO YOU WANT THE GRAPHS ON THE SCREEN ?'
      PRINT *,'1 = YES'
      READ *,IAG
      IF (IAG.EQ.1) THEN
        PRINT *,'DO YOU WANT THE AMPLITUDE ?'
        PRINT *,'1 = YES'
        READ *,IBG
        IF (IBG.EQ.1) THEN
          CALL AMSEN (SERPER,VMX,-1.,FRQ)
          CALL AMSEN (SERPAL,WMX,+1.,FRQ)
        END IF
        PRINT *,'DO YOU WANT THE PHASE ?'
        PRINT *,'1 = YES'
        READ *,IBG
        IF (IBG.EQ1) THEN
          CALL PHSEN (PHAPER,-1.,FRQ)
          CALL PHSEN (PHAPAL,+1.,FRQ)
      END IF
        END IF
C
C  CALL THE SUBROUTINE TO PRINT THE GRAPHS ON THE PLOTTER
C
      PRINT *,'DO YOU WANT THE GRAPHS ON THE PLOTTER ?'
      PRINT *,'1 = YES'
      READ *,IAG
      PRINT *,'PEN NUMBERS (NP1,NP2) ?'
      READ *,NP1,NP2
      IF (IAG.EQ.1) THEN
C
C  RECTANGULAR PLOTS
C
        PRINT *,'DO YOU WANT THE RECTANGULAR PLOTS ?'
        PRINT *,'1 = YES'
        READ *,IBG
        IF (IBG.EQ.1) THEN
C
C  AMPLITUDE IN RECTANGULAR CO-ORDINATES
C
          PRINT *,'DO YOU WANT THE AMPLITUDE ?'
          PRINT *,'1 = YES'
```

```
      READ *,ICG
      IF (ICG.EQ.1) THEN
        CALL AMRET (SERPER,VMN,VMX,-1.,FRQ,NP1,NP2)
        CALL AMRET (SERPAL,WMN,WMX,+1.,FRQ,NP1,NP2)
      END IF
C
C  WRAPPED PHASE IN RECTANGULAR CO-ORDINATES
C
      PRINT *,'DO YOU WANT THE WRAPPED PHASE ?'
      PRINT *,'1 = YES'
      READ *,ICG
      IF (ICG.EQ.1) THEN
        CALL PHRET (PHAPER,-1.,FRQ,NP1,NP2,R1,R2,H,CL)
        CALL PHRET (PHAPAL,+1.,FRQ,NP1,NP2,R1,R2,H,CL)
      END IF
C
C  WRAPPED PHASE IN RECTANGULAR CO-ORDINATES
C
      PRINT *,'DO YOU WANT THE UNWRAPPED PHASE ?'
      PRINT *,'1 = YES'
      READ *,ICG
      IF (ICG.EQ.1) THEN
        CALL UWRET (TMPPER,-1.,FRQ,NP1,NP2,R1,R2,H,CL,
     &                 NMB,PMN,PMX)
        CALL UWRET (TMPPAL,+1.,FRQ,NP1,NP2,R1,R2,H,CL,
     &                 NMB,QMN,QMX)
      END IF
    END IF
C
C  POLAR PLOTS
C
    PRINT *,'DO YOU WANT THE POLAR PLOTS ?'
    PRINT *,'1 = YES'
    READ *,IBG
    IF (IBG.EQ.1) THEN
C
C  AMPLITUDE IN POLAR CO-ORDINATES
C
      PRINT *,'DO YOU WANT THE AMPLITUDE ?'
      PRINT *,'1 = YES'
      READ *,ICG
```

```
          IF (ICG.EQ.1) THEN
            CALL AMPOL (SERPER,VMN,VMX,-1.,FRQ,NP1,NP2)
            CALL AMPOL (SERPAL,WMN,WMX,+1.,FRW,NP1,NP2)
          END IF
C
C  WRAPPED PHASE IN POLAR CO-ORDINATES
C
          PRINT *,'DO YOU WANT THE WRAPPED PHASE ?'
          PRINT *,'1 = YES'
          READ *,ICG
          IF (ICG.EQ.1) THEN
            CALL PHPOL (PHAPER,-1.,FRQ,NP1,NP2,R1,R2,H,CL)
            CALL PHPOL (PHAPAL,+1.,FRQ,NP1,NP2,R1,R2,H,CL)
          END IF
C
C  UNWRAPPED PHASE IN POLAR CO-ORDINATES
C
          PRINT *,'DO YOU WANT THE UNWRAPPED PHASE ?'
          PRINT *,'1 = YES'
          READ *,ICG
          IF (ICG.EQ.1) THEN
            CALL UWPOL (TMPPER,-1.,FRQ,NP1,NP2,R1,R2,H,CL,
     &                  NMB,PMN,PMX)
            CALL UWPOL (TMPPAL,+1.,FRQ,NP1,NP2,R1,R2,H,CL,
     &                  NMB,QMN,QMX)
          END IF
        END IF
      END IF
      STOP
      END
C-----------------------------------------------------------------------
C     SUBROUTINE SERCYL (SER,PHA,VMN,VMX,P,R1,CL,TJ)
C-----------------------------------------------------------------------
C  TO CALCULATE THE RCS (IN BOTH MAGNITUDE AND PHASE) OF A
   PERFECTLY CONDUCTING CYLINDER
C  METHODS USED:GTD, PO AND GO
C-----------------------------------------------------------------------
C  SER  = ARRAY CONTAINING THE VALUES OF THE MAGNITUDE
C         OF THE RCS CALCULATED FOR THETA = 0-180 IN INCRE-
C         MENTS OF 0.2 DEGREES
C  PHA  = ARRAY CONTAINING THE VALUES OF THE PHASE OF THE RCS
C         CALCULATED FOR THETA = 0-180 IN INCREMENTS OF 0.2 DE-
C         GREES
```

```
C  VMN =MINIMUM VALUE OF THE RCS MAGNITUDE (dB LAMBDA
          CARRE)
C  VMX =MAXIMUM VALUE OF THE RCS MAGNITUDE (dB LAMBDA
          CARRE)
C  P    =PERPENDICULAR (-1.) OR PARALLEL (+1.) POLARIZATION
C  R1   =RADIUS OF THE CYLINDER (IN LAMBDA)
C  R2   =ZERO FOR THE CYLINDER
C  H    =ZERO FOR THE CYLINDER
C  CL   =LENGTH OF THE CYLINDER (IN LAMBDA)
C  FRQ  =FREQUENCY (IN GHz)
C  TJ   =OP-GTD JUNCTION (DEGREES)
C------------------------------------------------------------------------
      DIMENSION SER(900),PHA(900)
      REAL* K2
      PARAMETER (PI = 3.141592654)
C
C  INITIALIZATION OF SOME OF THE REQUIRED PARAMETERS
C
      P2   =0.5*PI
      P4   =0.25*PI
      K2   =4.*PI
      VMN  =SER(1)
      VMX  =SER(1)
C
C  AUTOMATIC CALCULATION OF THE OP-GTD JUNCTION
C
      IF (TJ.LT.0.) THEN
        TJ = ASIN(0.3049/R1)
C
C  ELSE IF THE VALUE IS SPECIFIED, THEN CONVERT IT INTO RADIANS
C
      ELSE
        TJ = TJ*PI/180.
      END IF
C
C  T = ANGLE OF INCREMENT
C  CONVERSION OF THIS ANGLE INTO RADIANS
C
      DT = 0.2*PI/180.
       T = -DT
```

```
C
C  START MAGNITUDE AND PHASE CALCULATION
C
       DO I = 1,451
          T = T+DT
         T1 = K2*R1*SIN(T)
         T2 = K2*CL*COS(T)
C
C  WHEN THE ANGLE T (THETA) IS LESS THAN THE ANGLE CORRE-
C  SPONDING TO THE PO-GTD JUNCTION, WHAT THE RADAR SEES IS
C  A FLAT SURFACE
           3 ---------------------------------- 1
C          |                                  |
C          |                                  |  <<-------
C          |                                  |
C          |                                  |
C          4 ---------------------------------- 2
C   THE RCS FOUND IS THEREFORE THAT OF A DISC.
C
         IF (T.LT.TJ) THEN
            SO = SERDIS(R1,T)
             IJ = I
C
C  ONCE THE OP-GTD JUNCTION IS CROSSED, THE RADAR NO LONGER
C  SEES A DISC. THE MAJOR RETURNS APPEAR TO COME FROM THE
C  EDGES 1, 2, & 3.
C
         ELSE
C
C  CALL THE SUBROUTINE TO CALCULATE THE EM FIELD RETURNED
   (SCATTERED) BY EDGE 1
C
           PS = P4-T1-T2/2
           CALL SMC (ZR,ZI,P,P2+T,PS,T,R1,1.5,0)
C
C  CALL THE SUBROUTINE TO CALCULATE THE EM FIELD RETURNED
C  (SCATTERED) BY EDGE 2. ADD THIS FIELD TO THE PREVIOUS ONE.
C
           PS = P4+T1-T2/2
           CALL SMC (ZR,ZI,P,P2-T,PS,T,R1,1.5,1)
```

```
C
C  CALL THE SUBROUTINE TO CALCULATE THE EM FIELD RETURNED
C  (SCATTERED) BY EDGE 3. ADD THIS FIELD TO THE PREVIOUS TWO.
C
            PS = P4 - T1 + T2/2
            CALL SMC (ZR,ZI,P, T ,PS,T,R1,1.5,1)
C
C  THE RCS IN MAGNITUDE AND PHASE CALCULATED FOR THE ANGLES
C  FROM T = 0 TO 90.
C  ZR = REAL PART
C   ZI = IMAGINARY PART
C
            SO = ZR*ZR + ZI*ZI
            PHA(I) = ATAN2D (ZI,ZR)
          END IF
C
C  CONVERSION OF THE RCS INTO dB
C
          IF (I.NE.451) THEN
            SER(I) = 10.*LOG10(S0)
        ELSE
          SER(451) = SER(450)
        END IF
C
C  FINDING THE MAXIMUM AND MINIMUM VALUES OF THE MAGNITUDE
C  OF THE RCS
C
          VMN = MIN(VMN,SER(I))
          VMX = MAX(VMX,SER(I))
        END DO
C
C  CALCULATING THE VALUES OF THE RCS FROM THETA = 90 TO 180
C
        DO I = 452,900
          SER(I) = SER(902 - I)
          PHA(I) = PHA(902 - I)
        END DO
        RETURN
        END
C------------------------------------------------------------------------
        SUBROUTINE SERTRCO (SER,PHA,VMN,VMX,P,R1,R2,CL,TJ1,TJ2)
C------------------------------------------------------------------------
```

```
C  TO CALCULATE THE RCS (IN BOTH MAGNITUDE AND PHASE) OF A
C  PERFECTLY CONDUCTING CONE FRUSTRUM
C  METHODS USED:GTD, PO AND GO
C--------------------------------------------------------------------------
C  SER   = ARRAY CONTAINING THE VALUES OF THE MAGNITUDE OF
C          THE RCS CALCULATED FOR THETA = 0-180 IN INCREMENTS
C          OF 0.2 DEGREES
C  PHA   = ARRAY CONTAINING THE VALUES OF THE PHASE OF THE RCS
C          CALCULATED FOR THETA = 0-180 IN INCREMENTS OF 0.2
C          DEGREES
C  VMN   = MINIMUM VALUE OF THE RCS MAGNITUDE (dB LAMBDA
C          CARRE)
C  VMX   = MAXIMUM VALUE OF THE RCS MAGNITUDE (dB LAMBDA
C          CARRE)
C   P    = PERPENDICULAR (-1.) OR PARALLEL (+1.) POLARIZATION
C   R1   = SMALL RADIUS OF THE CONE FRUSTRUM (IN LAMBDA)
C   R2   = LARGE RADIUS OF THE CONE FRUSTRUM (IN LAMBDA)
C   H    = ZERO FOR THE CONE FRUSTRUM
C   CL   = LENGTH OF THE CONE FRUSTRUM (IN LAMBDA)
C  FRQ   = FREQUENCY (IN GHz)
C  TJ1   = OP-GTD JUNCTION (DEGREES) FRONT LOBE
C  TJ2   = OP-GTD JUNCTION (DEGREES) REAR LOBE
C--------------------------------------------------------------------------
      DIMENSION SER(900),PHA(900)
      REAL*4 K2
      PARAMETER (PI = 3.141592654)
C
C  INITIALIZATION OF SOME OF THE REQUIRED PARAMETERS
C
        P2 = 0.5*PI
        P4 = 0.25*PI
        K2 = 4.*PI
       VMN = SER(1)
       VMX = SER(1)
C
C  THE HALF-ANGLE OF THE CONE FRUSTRUM
C
      ALPHA = ATAN2(R2-R1,CL)
C
C  THE EXTERIOR ANGLES AT EDGES 1 & 3 IN RADIANS
      FN1 = 1.5-ALPHA/PI
      FN2 = 1.5+ALPHA/PI
```

```
C
C  RETURN FROM THE SPECULAR REFLECTION DUE TO THE CONE (AT
   90-ALPHA)
C
      SL = 8./9.*PI*(R2**1.5-R1**1.5)**2
      SL = SL*COS(ALPHA)/TAN(ALPHA)**2
C
C  AUTOMATIC CALCULATION OF THE OP-GTD JUNCTIONS
C
      IF (TJ1.LT.0.) THEN
        TJ1 = ASIN(0.3049/R1)
        TJ2 = ASIN(0.3049/R2)
C
C  ELSE IF THE VALUES ARE SPECIFIED, THEN CONVERT THEM INTO
C  RADIANS
C
      ELSE
        TJ1 = TJ1*PI/180.
        TJ2 = TJ2*PI/180.
      END IF
C
C  CONVERSION OF THE ANGLE OF INCREMENT INTO RADIANS
C
      DT =0.2*PI/180.
       T = -DT
C
C  START MAGNITUDE AND PHASE CALCULATION
C
      DO I = 1,900
         T = T+DT
        T1 = K2*SIN(T)
        T2 = K2*CL*COS(T)/2
C
C  FOR ANGLES LESS THAN TJ1, THE RCS IS EQUAL TO THAT OF A DISC
C
        IF (T.LT.TJ1) THEN
          S0 = SERDIS(R1,T)
          IJ = I
C
C  FOR ANGLES BETWEEN TJ1 AND ALPHA, THE RADAR SEES ALL FOUR
C  EDGES.
C  THE RCS IS CONSEQUENTLY THE SUM OF THE RETURNS FROM ALL
C  FOUR.
```

```
C
        ELSE IF (T.LT.ALPHA) THEN
C  EDGE 1
          PS = P4 - R1*T1 - T2
          CALL SMC (ZR,ZI,P,P2+T,PS,T,R1,FN1,0)
C  EDGE 2
          PS = P4 + R1*T1 - T2
          CALL SMC (ZR,ZI,P,P2 - T,PS,T,R1,FN1,1)
C  EDGE 3
          PS = P4 - R2*T1 + T2
          CALL SMC (ZR,ZI,P,ALPHA + T,PS,T,R2,FN2,1)
C  EDGE 4
          PS = P4 + R2*T1 + T2
          CALL SMC (ZR,ZI,P,ALPHA - T,PS,T,R2,FN2,1)
C
C  MAGNITUDE AND PHASE OF THE RCS BETWEEN TJ1 & ALPHA
C
          S0 = ZR*ZR + ZI*ZI
          PHA(I) = ATAN2D(ZI,ZR)
C
C  FOR ANGLES BETWEEN ALPHA & 90, THE RADAR SEES EDGES 1, 2 & 3
C
        ELSE IF (T.LT.P2) THEN
C  EDGE 1
          PS = P4 - R1*T1 - T2
          CALL SMC (ZR,ZI,P,P2 + T,PS,T,R1,FN1,0)
C  EDGE 2
          PS = P4 + R1*T1 - T2
          CALL SMC (ZR,ZI,P,P2 - T,PS,T,R1,FN1,1)
C  EDGE 3
          PS = P4 - R2*T1 + T2
          CALL SMC (ZR,ZI,P,ALPHA + T,PS,T,R2,FN2,1)
C
C  MAGNITUDE AND PHASE OF THE RCS BETWEEN ALPHA & 90
C
          S0 = ZR*ZR + ZI*ZI
          PHA(I) = ATAN2D(ZI,ZR)
C
C  AT 90 - ALPHA, IF THE SPECULAR RETURN DUE TO THE CONE IS <
C  THAN THE RCS FOUND JUST ABOVE, REPLACE THIS RETURN BY THE
C  RCS
C
          IF (S0.GT.SL) S0 = SL
```

```
C
C  FOR ANGLES BETWEEN 90 AND 180 - ALPHA, THE RADAR SEES EDGES
   1, 3 & 4
C
       ELSE IF (T.LT.PI - ALPHA) THEN
C  EDGE 1
          PS = P4 - R1*T1 - T2
          CALL SMC (ZR,ZI,P,PI - T - ALPHA,PS,T,R1,FN1,0)
C  EDGE 3
          PS = P4 - R2*T1 + T2
          CALL SMC (ZR,ZI,P,1.5*PI - T,PS,T,R2,FN2,1)
C  EDGE 4
          PS = P4 + R2*T1 + T2
          CALL SMC (ZR,ZI,P,T - P2,PS,T,R2,FN2,1)
C
C  MAGNITUDE AND PHASE OF THE RCS BETWEEN 90 AND 180 - ALPHA
C
          S0 = ZR*ZR + ZI*ZI
          PHA(I) = ATAN2D(ZI,ZR)
C
C  FOR ANGLES BETWEEN 180 - TJ2 AND 180, THE RCS IS EQUAL TO THAT
C  OF A DISC
C
       ELSE IF (T.GT.PI - TJ2) THEN
          S0 = SERDIS(R2,T)
C
C  FOR ANGLES BETWEEN 180 - ALPHA AND 180 - TJ2, THE RADAR SEES
   EDGES 3*4
C
       ELSE
C  EDGE 3
          PS = P4 - R2*T1 + T2
       CALL SMC (ZR,ZI,P,1.5*PI - T,PS,T,R2,FN2,0)
C  EDGE 4
          PS = P4 + R2*T1 + T2
          CALL SMC (ZR,ZI,P,T - P2,PS,T,R2,FN2,1)
C
C  MAGNITUDE AND PHASE OF THE RCS BETWEEN 180 - ALPHA & 180 -
   TJ2
C
          S0 = ZR*ZR + ZI*ZI
          PHA(I) = ATAN2D(ZI,ZR)
       END IF
```

```
C
C  CONVERSION OF THE MAGNITUDE INTO dB; CALCULATION OF THE
   MAX. & MIN. VALUES
C
         SER(I) = 10.*LOG10(S0)
          VMN = MIN(VMN,SER(I))
          VMX = MAX(VMX,SER(I))
      END DO
      RETURN
      END
C-----------------------------------------------------------------------
      SUBROUTINE SERTRCY (SER,PHA,VMN,VMX,P,R1,R2,H,CL,TJ1,TJ2)
C-----------------------------------------------------------------------
C  TO CALCULATE THE RCS (IN BOTH MAGNITUDE AND PHASE) OF A
C  PERFECTLY CONDUCTING CONE-CYLINDER FRUSTRUM
C  METHODS USED:GTD, PO AND GO
C-----------------------------------------------------------------------
C  SER   = ARRAY CONTAINING THE VALUES OF THE MAGNITUDE OF
C          THE RCS CALCULATED FOR THETA = 0 - 180 IN INCREMENTS
C          OF 0.2 DEGREES
C  PHA   = ARRAY CONTAINING THE VALUES OF THE PHASE OF THE RCS
C          CALCULATED FOR THETA = 0 - 180 IN INCREMENTS OF 0.2
C          DEGREES
C  VMN   = MINIMUM VALUE OF THE RCS MAGNITUDE (dB LAMBDA
C          CARRE)
C  VMX   = MAXIMUM VALUE OF THE RCS MAGNITUDE (dB LAMBDA
C          CARRE)
C  P     = PERPENDICULAR (-1.) OR PARALLEL (+1.) POLARIZATION
C  R1    = SMALL RADIUS OF THE CONE-CYLINDER FRUSTRUM (IN
C          LAMBDA)
C  R2    = LARGE RADIUS OF THE CONE-CYLINDER FRUSTRUM (IN
C          LAMBDA)
C  H     = LENGTH OF THE CONE FRUSTRUM PART (IN LAMBDA)
C  CL    = LENGTH OF THE CONE-CYLINDER FRUSTRUM (IN LAMBDA)
C  FRQ   = FREQUENCY (IN GHz)
C  TJ1   = OP-GTD JUNCTION (DEGREES) FRONT LOBE
C  TJ2   = OP-GTD JUNCTION (DEGREES) REAR LOBE
C-----------------------------------------------------------------------
      DIMENSION SER(900),PHA(900)
      REAL*4 K2
      PARAMETER (PI = 3.141592654)
C
C  INITIALIZATION OF SOME OF THE REQUIRED PARAMETERS
C
      P2 = 0.5*PI
```

```
      P4 = 0.25*PI
      K2 = 4.*PI
      VMN = SER(1)
      VMX = SER(1)
C
C  THE HALF-ANGLE OF THE CONE
C
      ALPHA = ATAN2(R2 - R1,H)
C
C  THE EXTERIOR ANGLES AT THE EDGES 1, 3 & 5 IN RADIANS
C
      FN1 = 1.5 - ALPHA/PI
      FN2 = 1. + ALPHA/PI
      FN3 = 1.5
C
C  RETURN FROM THE SPECULAR REFLECTION DUE TO THE CONE (AT
   90 - ALPHA)
C
      SL = 8./9.*PI*(R2**1.5 - R1**1.5)**2
      SL = SL*COS(ALPHA)/TAN(ALPHA)**2
C
C  DIMENSIONS OF THE SATURATION ZONE ON EITHER SIDE OF THE
   LOBE AT 90 - ALPHA
C
      BT = 3.0*PI/180.
C
C  AUTOMATIC CALCULATION OF THE OP-GTD JUNCTIONS
C
      IF (TJ1.LT.0.)THEN
        TJ1 = ASIN(0.3049/R1)
        TJ2 = ASIN(0.3049/R2)
C
C  ELSE IF THE VALUES ARE SPECIFIED, THEN CONVERT THEM INTO
   RADIANS
C
      ELSE
        TJ1 = TJ1*PI/180.
        TJ2 = TJ2*PI/180.
      END IF
C
C  CONVERSION OF THE ANGLE OF INCREMENT INTO RADIANS
C
      DT = 0.2*PI/180.
      T = - DT
```

```
C
C   START MAGNITUDE AND PHASE CALCULATION
C
      DO I = 1,900
        T = T+DT
          T1 = K2*SIN(T)
        T2 = K2*COS(T)/2
C
C   FOR ANGLES LESS THAN TJ1, THE RCS IS EQUAL TO THAT OF A DISC
C
        IF (T.LT.TJ1) THEN
          S0 = SERDIS(R1,T)
C
C   FOR ANGLES BETWEEN TJ1 & ALPHA, THE RADAR SEES THE EDGES 1,
C   2, 3, 4 & 5. THE RCS IS CONSEQUENTLY THE SUM OF THE RETURNS
C   FROM THESE EDGES
C
        ELSE IF (T.LT.ALPHA) THEN
          PS = P4-R1*T1-(H+CL)*T2
          CALL SMC(ZR,ZI,P,P2+T,PS,T,R1,FN1,0)
          PS = P4+R1*T1-(H+CL)*T2
          CALL SMC(ZR,ZI,P,P2-T,PS,T,R1,FN1,1)
          PS = P4-R2*T1+(H-CL)*T2
          CALL SMC(ZR,ZI,P,ALPHA+T,PS,T,R2,FN2,1)
          PS = P4+R2*T1+(H-CL)*T2
          CALL SMC(ZR,ZI,P,ALPHA-T,PS,T,R2,FN2,1)
          PS = P4-R2*T1+(H+CL)*T2
          CALL SMC(ZR,ZI,P,T,PS,T,R2,FN3,1)
C
C   MAGNITUDE AND PHASE OF THE RCS BETWEEN TJ1 & ALPHA
C
          S0 = ZR*ZR+ZI*ZI
          PHA(I) = ATAN2D(ZI,ZR)
C
C   FOR ANGLES BETWEEN ALPHA & 90, THE RADAR SEES THE EDGES 1,
    2, 3 & 5
C
        ELSE IF (T.LT.P2) THEN
          PS = P4-R1*T1-(H+CL)*T2
          CALL SMC (ZR,ZI,P,P2+T,PS,T,R1,FN1,0)
          PS = P4+R1*T1-(H+CL)*T2
          CALL SMC (ZR,ZI,P,P2-T,PS,T,R1,FN1,1)
          PS = P4-R2*T1+(H-CL)*T2
```

```
      CALL SMC (ZR,ZI,P,ALPHA+T,PS,T,R2,FN2,1)
      PS = P4-R2*T1+(H+CL)*T2
      CALL SMC (ZR,ZI,P,T,PS,T,R2,FN3,1)
C
C MAGNITUDE AND PHASE OF THE RCS BETWEEN ALPHA & 90
C
      S0 = ZR*ZR+ZI*ZI
      PHA(I) = ATAN2D(ZI,ZR)
C
C FOR ANGLES > 90-ALPHA-BT BUT < 90-ALPHA+BT, IF THE SPEC-
C ULAR RETURN DUE TO THE CONE IS < THAN THE RCS FOUND JUST
C ABOVE, REPLACE THIS RETURN BY THE RCS
C
        IF (T.GT.P2-ALPHA-BT.AND.T.LT.P2-ALPHA+BT) THEN
    IF (S0.GT.SL) S0 = SL
      END IF
C
C FOR ANGLES BETWEEN 90 & 180-ALPHA THE RADAR SEES THE EDGES
C 1, 3, 5 & 6. AT 90 EXACTLY, THE RCS IS THE SPECULAR RETURN FROM
C THE CYLINDER
C
    ELSE IF (T.LT.PI-ALPHA) THEN
C SPECULAR RETURN
      IF (I.EQ.451) THEN
        ZI = (-0.375+P)/(K2*R2)
        S0 = 2.*PI*R2*CL*CL*(1.+ZI*ZI)
        SP0 = 10.**(0.1*SER(450))
        IF (S0.GT.SP0) S0 = SP0
C EDGES 1, 3, 5 & 6
      ELSE
        PS = P4-R1*T1-(H+CL)*T2
        CALL SMC (ZR,ZI,P,PI-T-ALPHA,PS,T,R1,FN1,0)
        PS = P4-R2*T1+(H-CL)*T2
        CALL SMC (ZR,ZI,P,PI-T,PS,T,R2,FN2,1)
        PS = P4-R2*T1+(H+CL)*T2
        CALL SMC (ZR,ZI,P,1.5*PI-T,PS,T,R2,FN3,1)
        PS = P4+R2*T1+(H+CL)*T2
        CALL SMC (ZR,ZI,P,T-P2,PS,T,R2,FN3,1)
C
C MAGNITUDE AND PHASE BETWEEN 90 & 180-ALPHA
C
        S0 = ZR*ZR+ZI*ZI
        PHA(I) = ATAN2D(ZI,ZR)
    END IF
```

```
C
C  FOR ANGLES BETWEEN 180-TJ2, THE RCS IS EQUAL TO THAT OF A
   DISC
C
         ELSE IF (T.GT.PI-TJ2) THEN
           S0 = SERDIS(R2,T)
C
C  FOR ANGLES BETWEEN 180-ALPHA & 180-TJ2, THE RADAR SEES
   EDGES 3,5 & 6
C
         ELSE
           PS = P4-R2*T1+(H-CL)*T2
           CALL SMC (ZR,ZI,P,PI-T,PS,T,R2,FN2,0)
           PS = P4-R2*T1+(H+CL)*T2
           CALL SMC (ZR,ZI,P,1.5*PI-T,PS,T,R2,FN3,1)
           PS = P4+R2*T1+(H+CL)*T2
           CALL SMC (ZR,ZI,P,T-P2,PS,T,R2,FN3,1)
C
C  MAGNITUDE AND PHASE BETWEEN 180-ALPHA & 180-TJ2
C
           S0 = ZR*ZR+ZI*ZI
           PHA(I) = ATAN2D(ZI,ZR)
         END IF
C
C  CONVERSION OF THE RCS INTO dB; CALCULATION OF THE MAX. &
   MIN. VALUES OF THE RCS
C
         SER(I) = 10.*LOG10(S0)
         VMN = MIN(VMN,SER(I))
         VMX = MAX(VMX,SER(I))
       END DO
       RETURN
       END
```

C.7 ELECTROMAGNETIC SCATTERING FROM A LOADED DIPOLE:

```
CC   RCS ENHANCEMENT AND REDUCTION:AVRCS. FOR; THIS PROGRAM
C    WAS DEVELOPED BY JANASWAMY (C.2)
C    PROGRAM TO COMPUTE THE TUMBLE AVERAGE RCS OF A LOADED
C    STRIP DIPOLE. IT IS ASSUMED THAT THE DIPOLE IS LOADED AT TWO
C    POINTS ALONG ITS LENGTH.
```

```
C
C     INPUT PARAMETERS
C
C     F = Frequency (GHz)
C     L = Length of the dipole (cm)
C     W = Half width of the dipole (cm)
C     NSEG = Number of segments into which the dipole is divided.
C     ZLOAD1 = Impedance of the external load-1
C     NLOD1 = Position of the segment at which the load-1 is connected.
C     ZLOAD2 = Impedance of the external load-2
C     NLOD2 = Position of the segment at which the load-2 is connected.
C   Note: The segment numbering is done beginning from one end of the
C           dipole. The indexing of the segments starts from 0.
C
C     PROGRAM AVRCS
      REAL RCS,L,KO
      COMPLEX ZLOAD1,ZLOAD2
      COMMON/SEGPAR/W,H,NMODE,PI,KO
      OPEN (UNIT = 8,FILE = 'AVRCS.PRN',TYPE = 'NEW')
      READ (6,*) F,L,W,NLOD1,ZLOAD1,NLOD2,ZLOAD2,NSEG
C     F = 30.0
C     L = 1.5
C     W = 0.15
C     NLOD1 = 5
C     ZLOAD1 = CMPLX(43.2,73.4)
C     NLOD2 = 8
C     ZLOAD2 = CMPLX(43.2,73.4)
C     NSEG = 20
      PRINT *,F,L,W,NLOD1,ZLOAD1,NLOD2,ZLOAD2,NSEG
      CALL TUMBAV (F,L,W,NLOD1,ZLOAD1,NLOD2,ZLOAD2,NSEG,RCS)
      PRINT *,L*F/30.,(10.*ALOG10(RCS))
      END
C
C     PROGRAM TO COMPUTE THE NORMALIZED BACKSCATTERED ECHO
      AREA FOR A LOADED STRIP DIPOLE.
C
C
      SUBROUTINE TUMBAV (F,L,WW,NLOD1,ZLOAD1,NLOD2,
      ZLOAD2,NSEG,TUMAV)
      IMPLICIT NONE
      REAL F,L,W,THETA,PI,ETA,D,SIGMA,K0,KZH,FAC,H,WW,DEL
      REAL TUMAV,CON
```

```
      INTEGER M,NLOD1,NLOD2,NMODE,INDX(50),K,II,NSEG
      COMPLEX ZMSTR(50,50),I(50),ZLOAD1,ZLOAD2,J,SUM
      COMMON/SEGPAR/W,H,NMODE,PI,K0
      DO K = 1,50
      I(K) = (0.,0.)
      END DO
      W = WW
      J = CMPLX (0.,1.)
      CALL ZSTRDIP (F,L,ZLOAD1,NLOD1,ZLOAD2,NLOD2,NSEG,50,ZMSTR)
      CALL LUDCMP (ZMSTR,NMODE,50,INDX,D)
      ETA = 120.*PI
      FAC = (ETA*KO/(2.*PI*SIN(K0*H)))**2/PI*3/16.
      DEL = PI/90.
      TUMAV = 0.
      DO M = 0,90
      CON = 4.
      IF (JMOD(M,2).EQ.0) CON = 2.
      IF (M.EQ.0 .OR. M.EQ.90) CON = 1.
      THETA = 2*M*PI/180.
      CALL VOLT (THETA,50,I)
      CALL LUBKSB (ZMSTR,NMODE,50,INDX,I)
      KZH = K0*COS(THETA)*H
      SIGMA = 0.0
      IF (THETA.NE.0. .AND. THETA.NE.PI) THEN
      SIGMA = (COS(KZH) - COS(K0*H))/SIN(THETA)
      END IF
      SIGMA = FAC*SIGMA**2
      SUM = (0.,0.)
      DO K = 1,NMODE
      SUM = SUM + I(K)*CEXP(J*K*KZH)
      END DO
      SIGMA = SIGMA*CABS(SUM)*CABS(SUM)*SIN(THETA)*CON
      TUMAV = TUMAV + SIGMA
      END DO
      TUMAV = TUMAV*DEL/3.
      RETURN
      END
C
C     Program to compute the impedance matrix for a strip dipole
C     loaded at a single point along its length. Galerkin's method is
C     used to model current on the strip dipole resulting in a symmetric
C     Toeplitz impedance matrix.
```

```
C
      SUBROUTINE ZSTRDIP (F,L,ZLOAD1,NLOD1,ZLOAD2,
      NLOD2,NSEG,NP,ZMS)
      IMPLICIT NONE
      REAL F,L,W,WL,K0,DELX,DW,DH,PI,H
      INTEGER I,J,NSEG,NMODE,NLOD1,NLOD2,NP
      COMPLEX ZMS(NP,NP),ZP(0:49),ZLOAD1,ZLOAD2,ZT
      COMMON/SEGPAR/W,H,NMODE,PI,K0
      WL = 30./F
      PI = 4.*ATAN(1.)
      K0 = 2.*PI/WL
      NMODE = NSEG - 1
      DELX = L/FLOAT(NSEG)
      H = DELX
      DW = 0.
      DO I = 0,NMODE - 1 ! Because of symmetric and Toeplitz properties
      DH = I*DELX          ! only elements of the first row are computed.
      CALL ZSDIP (F,DELX,W,DW,DH,ZP(I))
      END DO
      DO J = 1,NMODE
      DO I = 1,J
      ZMS(I,J) = ZP(J - I)
      ZMS(J,I) = ZMS(I,J)
      IF (I.EQ.J .AND. I.EQ.NLOD1) ZMS(I,I) = ZMS(I,I) + ZLOAD1
      IF (I.EQ.J .AND. I.EQ.NLOD2) ZMS(I,I) = ZMS(I,I) + ZLOAD2
      END DO
      END DO
      RETURN
      END
```

REFERENCES

C.1 W. D. Burnside, R. J. Marhefka, and N. Wang, "Computer Programs, Subroutines and Functions for the Short Course on the Modern Geometrical Theory of Diffraction," Ohio State University, Columbus.

C.2 R. Jamaswamy and S. W. Lee, "Scattering from Dipoles Loaded with Dipoles," *IEEE Trans. on Antennas and Propagation*, Vol. AP-36, No. 11, November 1988, pp. 1649–1651.

List of Symbols

Symbols and acronyms used in the text are given in the following list. Because of the large number of quantities to be represented and the undesirability of using alphabets other than Roman and Greek, it has been necessary to use the same symbols to represent different quantities at different places. In every instance, the symbol has been defined where introduced to avoid misunderstanding of its meaning. The list also includes quantities of units where appropriate.

A	Physical area (m^2)
$\mathbf{A}$	Magnetic vector potential ($^{Wb}/m$)
A_e	Effective aperture (m^2)
a	Radius of a loop (m)
	Radius of a disk (m)
	Length (m)
$\mathbf{B}$	Magnetic flux density vector (Wb/m^2 = T)
b	Length (m)
c	Velocity of light in free space (m/s)
$\mathbf{D}$	Electric flux density vector (C/m^2)
$D^{s,h}$	Diffraction coefficients for soft (s) and hard (h) cases
d	Length (m)
dB	Decibel
dl	Scalar length element (m)
$d\mathbf{l}$	Vector length element (m)
ds	Scalar area element (m^2)
$d\mathbf{s}$	Vector area element (m^2)
dv	Volume element (m^3)
$\mathbf{E}$	Electric field intensity vector (V/m)
$\mathbf{E}^i, \mathbf{E}^r, \mathbf{E}^s$	Incident, reflected, and scattered electric field intensity vectors (V/m)

$\mathbf{E}^G(\cdot)$	Geometric optics electric field intensity vector (V/m)
$\mathbf{E}^d(\cdot)$	Diffracted electric field intensity vector (V/m)
E_x, E_y, E_z	Rectangular components of the electric field intensity vector (V/m)
E_ρ, E_φ, E_z	Cylindrical components of the electric field intensity vector (V/m)
E_r, E_θ, E_φ	Spherical components of the electric field intensity vector (V/m)
$E_{\parallel,\perp}$	Parallel ($\parallel$) and perpendicular ($\perp$) components of the electric field intensity (V/m)
$F(x)$	Fresnel integral function
$\hat{F}(x)$	Dominant term of $F(x)$
f	Frequency in Hz (c/s)
FSS	Frequency Selective Surfaces
G	Green's function
	Dyadic Green's function
	Gain of an antenna
G_R, G_T	Gains of receiving, transmitting antennas
GTD	Geometrical theory of diffraction
$\mathbf{H}$	Magnetic field intensity vector (A/m)
$\mathbf{H}^i, \mathbf{H}^r$	Incident and reflected magnetic field intensity vectors (A/m)
$\mathbf{H}^d, \mathbf{H}^s$	Diffracted and scattered magnetic field intensity vectors (A/m)
$H(x)$	Heaviside's unit step function
$H_n^{(1)}(x), H_n^{(2)}(x)$	Ordinary Hankel functions of first and second kinds and order n
$h_n^{(1)}(x), h_n^{(2)}(x)$	Spherical Hankel functions of first and second kinds and order n
$\hat{h}$	Unit vector in the direction of horizontal polarization
$\mathbf{h}_a^r, \mathbf{H}_a^t$	Receiving and transmitting mode vector effective heights of an antenna (m)
I^e	Equivalent electric current (A)
I^m	Equivalent magnetic current (V)
$\hat{\imath}$	$\sqrt{-1}$
i	Unit vector in the incident direction
$\mathbf{J}$	Volume electric current density (A/m^2)
$\mathbf{J}^i, \mathbf{J}^r$	Surface electric current densities due to incident and reflected fields (A/m)
$\mathbf{J}_{\mathrm{GTD}}, \mathbf{J}^{\mathrm{MOM}}$	Surface electric current densities obtained by GTD, MOM
$\mathbf{J}_s$	Surface electric current density (A/m)
$\mathbf{J}_s^0$	Physical optics surface electric current density (A/m)

$\mathbf{J}_s^{(1)}$	Fringe or nonuniform surface electric current density (A/m)
J_y^{GTD}	x-component of $\mathbf{J}_{\mathrm{GTD}}$
J_x^{MOM}, J_y^{MOM}	x- and y-components of $\mathbf{J}^{\mathrm{MOM}}$ (A/m)
$J_n(x)$	Ordinary Bessel function of the first kind and order n
$j_n(x)$	Spherical Bessel Function of the first kind and order n
k_0	Wave number ($= 2\pi/\lambda_0$) in free space (1/m)
k	Wave number ($= 2\pi/\lambda$) in a medium (1/m)
L	Distance (m) Length (m)
L_p	Polarization loss factor
L_{MR}	Material loss factor in the propagating medium between the target and receiver
L_{MT}	Material loss factor in the propagating medium between the target and transmitter
L_R	Loss factor in the receiver channel
L_T	Loss factor in the transmitter channel
l_1, l_2	Lengths (m)
$\hat{l}_c$	Unit vector for left-handed circular polarization
$\mathbf{M}$	Magnetic surface current density (V/m)
$M(u)$	Maliuzhinets's functions
N	Complex reproactive index for a medium
n	$(2\pi - \alpha)/\pi$ wedge angle normalized to π
n_0	Refractive index of free space
$\hat{n}$	Unit positive normal vector
P_R	Received power (W)
P_T	Transmitted power (W)
PDF	Probability density function
PTD	Physical theory of diffraction
R	Distance between the field and source points (m) Distance between the receiving antenna and the target (m) Reflection coefficient
$\mathbf{R}$	$\mathbf{r} - \mathbf{r}^1$, vector from the source to the field points (m)
$R_{\parallel, \perp}$	Complex reflection coefficients for parallel ($\parallel$) or perpendicular ($\perp$) polarizations
R_{ij}	Reflection coefficient at the interface between medium i and medium j
R_a	Antenna resistance (Ω)
R_m	Surface resistivity of fins (Ω – m)
R_S	Surface resistance of a strip (Ω)
R_T	Distance between the transmitting antenna and the target (m)

Symbol	Definition
r	Radial distance (m)
$r_{1a}, r_{1b}, r_{2a}, r_{2b}$	Radii of curvatures of surfaces (1, 2) in the two principal planes (a, b) (m)
$\hat{r}$	Unit vector in the r direction
$\mathbf{r}, \mathbf{r}'$	Position vectors of field and source points
$\hat{r}_c$	Unit vector for right-handed circular polarization
RAM	Radar absorbing material
RAS	Radar absorbing surface
RCS	Radar cross section
S, s	Distance measured along a ray (m)
S	Surface (m^2)
	Root mean square spread of RCS
S_H	Amplitude scattering function for horizontal polarization
S_V	Amplitude scattering function for vertical polarization
$\mathbf{S}(r, \theta, \varphi)$	Range dependent scattering function
$\mathbf{S}(\theta, \varphi), \mathbf{S}$	Range independent scattering function
$S^{i,r,d}(r)$	Phase paths associated with incident (i), reflected (r), and diffracted (d) rays (m)
$\hat{s}$	Unit vector in the scattering direction
TE	Transverse electric
TEM	Transverse electric and magnetic
$T_{\parallel, \perp}$	Complex transmission coefficients for parallel ($\parallel$) and perpendicular ($\perp$) polarizations
T_{ij}	Transmission coefficient for transmission from medium i to medium j
TM	Transverse magnetic
$\tan\delta$	ϵ''/ϵ' = loss tangent of a dielectric material
t	Time (s)
UAT	Uniform asymptotic theory of diffraction
UTD	Uniform theory of diffraction
W	Width of a rectangular plate (m)
X_a	Antenna reactance (Ω)
$\hat{x}$	Unit vector in the x direction
Y	Admittance
$Y_n(x)$	Ordinary Bessel function of the second kind and order n
$y_n(x)$	Spherical Bessel function of the second kind and order n
$\hat{y}$	Unit vector in the y direction
Z_0, Z_1	Impedances (Ω)
Z_a	$R_\alpha + \mathrm{i}x_\alpha$ antenna impedance (Ω)
Z_0	Intrinsic impedance of free space (Ω)
Z	Surface impedance normalized to the intrinsic impedance of free space

Z_S	Surface impedance (Ω)
$\hat{z}$	Unit vector in the Z direction
α	Electrical thickness of an absorbing layer
α_i	Phase angles associated with scattered fields
α	Interior angle of an infinite wedge
β	Bistatic angle (rad)
β	Phase propagation constant (radians/m)
β_α	Bistatic angle
β_0	Smaller angle between the direction of incidence and the tangent drawn on the edge at the point of incidence
$\Gamma(\cdot)$	Gamma function
Γ	Divergence factor
$\Gamma_\parallel, \Gamma_\perp$	Reflection coefficients for parallel ($\parallel$) or perpendicular ($\perp$) polarizations
Γ_m	Complex reflection coefficient
γ	Half angle for a cone or cylindrical structure
	Complex propagation constant
	Average RCS per square meter
δ	Angle
δ_s	Skin depth for a material (m)
ϵ	$\epsilon' - i\epsilon''$ complex permittivity of a medium (F/m)
ϵ'	Real part of ϵ
ϵ''	Imaginary part of ϵ
ϵ_0	Permittivity of free space (F/m)
ϵ_r	Relative permittivity of a medium
η	Intrinsic impedance of a medium (Ω)
η_0	Intrinsic impedance of free space (Ω)
θ	Angle in spherical system of coordinates
θ_i, θ_s	θ associated with incident and scattering directions
$\hat{\theta}$	Unit vector in the θ direction
λ	Wavelength in a medium (m)
λ_0	Wavelength in free space (m)
μ	$\mu' - i\mu''$ permeability of medium (H/m)
μ'	Real part of μ
μ''	Imaginary part of μ
μ_r	Relative permeability of a medium
ν	A design parameter
ξ_c	Measure of chirality of medium
$\xi^{i,r}(\cdot)$	Detour parameters for incident (i) and reflected (r) cases
π	($= 3.1416$)/Radius of a circle
ρ	Cylindrical coordinate
$\hat{\rho}$	Unit vector in the ρ direction

ρ_s	Surface electric charge density (C/m^2)
ρ_v	Volume electric charge density (C/m^3)
σ	Conductivity of a material (Ω/m) RCS (m^2)
$\sigma(r, \theta, \varphi)$	Range dependent bistatic RCS (m^2)
$\sigma(\theta, \varphi)$	Range independent or far field bistatic RCS (m^2)
$\sigma_{\parallel}, \sigma_{\perp}$	RCS for parallel ($\parallel$) and perpendicular ($\perp$) polarizations (m^2)
σ_V, σ_H	Backscattering RCS for vertical and horizontal polarizations (m_2)
σ_a, σ_T	Absorption and total cross section (m^2)
σ_e	Extinction cross section (m^2)
σ_α	Antenna-mode scattering cross section (m^4)
σ_S	Structural scattering cross section for an antenna (m^2)
σ_{vv}, σ_{hh}	RCS for vertical-vertical and horizontal-horizontal polarizations
σ_{vh}, σ_{hv}	RCS for vertical-horizontal and horizontal-vertical polarizations
φ	Angle in a spherical coordinate system
φ, φ'	Angle associated with the field point and source point
φ_i, φ_s	φ_s associated with the incident (i) and scattering (s) directions
φ_n	Phase angles for scattered fields
$\hat{\varphi}$	Unit vector in the φ direction
ψ_r, ψ_l	Phase angles associated with the right-handed and left-handed circular polarized scattered electric fields.
ψ^+	$\varphi + \varphi'$
ψ^-	$\varphi - \varphi'$
ω	Angular frequency ($= 2\pi f$) (rad/s).

INDEX

The Artech House Radar Library

David K. Barton, *Series Editor*

Active Radar Electronic Countermeasures by Edward J. Chrzanowski

Airborne Pulsed Doppler Radar by Guy V. Morris

AIRCOVER: Airborne Radar Vertical Coverage Calculation Software and User's Manual by William A. Skillman

Analog Automatic Control Loops in Radar and EW by Richard S. Hughes

Aspects of Modern Radar by Eli Brookner, *et al.*

Aspects of Radar Signal Processing by Bernard Lewis, Frank Kretschmer, and Wesley Shelton

Bistatic Radar by Nicholas J. Willis

Detectability of Spread-Spectrum Signals by Robin A. Dillard and George M. Dillard

Electronic Homing Systems by M.V. Maksimov and G.I. Gorgonov

Electronic Intelligence: The Analysis of Radar Signals by Richard G. Wiley

Electronic Intelligence: The Interception of Radar Signals by Richard G. Wiley

EREPS: Engineer's Refractive Effects Prediction System Software and User's Manual, developed by NOSC

Handbook of Radar Measurement by David K. Barton and Harold R. Ward

High Resolution Radar by Donald R. Wehner

High Resolution Radar Cross-Section Imaging by Dean L. Mensa

Interference Suppression Techniques for Microwave Antennas and Transmitters by Ernest R. Freeman

Introduction to Electronic Warfare by D. Curtis Schleher

Introduction to Sensor Systems by S.A. Hovanessian

Logarithmic Amplification by Richard Smith Hughes

Modern Radar System Analysis by David K. Barton

Modern Radar System Analysis Software and User's Manual by David K. Barton and William F. Barton

Monopulse Principles and Techniques by Samuel M. Sherman

Monopulse Radar by A.I. Leonov and K.I. Fomichev

MTI and Pulsed Doppler Radar by D. Curtis Schleher

Multifunction Array Radar Design by Dale R. Billetter

Multisensor Data Fusion by Edward L. Waltz and James Llinas

Multiple-Target Tracking with Radar Applications by Samuel S. Blackman

Multitarget-Multisensor Tracking: Advanced Applications, Yaakov Bar-Shalom, ed.

Over-The-Horizon Radar by A.A. Kolosov, et al.

Principles and Applications of Millimeter-Wave Radar, Charles E. Brown and Nicholas C. Currie, eds.

Principles of Modern Radar Systems by Michel H. Carpentier

Pulse Train Analysis Using Personal Computers by Richard G. Wiley and Michael B. Szymanski

Radar and the Atmosphere by Alfred J. Bogush, Jr.

Radar Anti-Jamming Techniques by M.V. Maksimov, *et al.*

Radar Cross Section by Eugene F. Knott, *et al.*

Radar Detection by J.V. DiFranco and W.L. Rubin

Radar Electronic Countermeasures System Design by Richard J. Wiegand

Radar Evaluation Handbook by David K. Barton, *et al.*

Radar Evaluation Software by David K. Barton and William F. Barton

Radar Propagation at Low Altitudes by M.L. Meeks

Radar Range-Performance Analysis by Lamont V. Blake

Radar Reflectivity Measurement: Techniques and Applications, Nicholas C. Currie, ed.

Radar Reflectivity of Land and Sea by Maurice W. Long

Radar System Design and Analysis by S.A. Hovanessian

Radar Technology, Eli Brookner, ed.

Receiving Systems Design by Stephen J. Erst

Radar Vulnerability to Jamming by Robert N. Lothes, Michael B. Szymanski, and Richard G. Wiley

RGCALC: Radar Range Detection Software and User's Manual by John E. Fielding and Gary D. Reynolds

SACALC: Signal Analysis Software and User's Guide by William T. Hardy

Secondary Surveillance Radar by Michael C. Stevens

SIGCLUT: Surface and Volumetric Clutter-to-Noise, Jammer and Target Signal-to-Noise Radar Calculation Software and User's Manual by William A. Skillman

Signal Theory and Random Processes by Harry Urkowitz

Solid-State Radar Transmitters by Edward D. Ostroff, *et al.*

Space-Based Radar Handbook, Leopold J. Cantafio, ed.

Spaceborne Weather Radar by Robert M. Meneghini and Toshiaki Kozu

Statistical Theory of Extended Radar Targets by R.V. Ostrovityanov and F.A. Basalov

The Scattering of Electromagnetic Waves from Rough Surfaces by Peter Beckmann and Andre Spizzichino

VCCALC: Vertical Coverage Plotting Software and User's Manual by John E. Fielding and Gary D. Reynolds

www.ingramcontent.com/pod-product-compliance
Lightning Source LLC
LaVergne TN
LVHW102008210926
848483LV00010B/162
* 9 7 8 0 8 9 0 0 6 3 7 1 2 *